Surface Phenomena in the Structural and Mechanical Behaviour of Solid Polymers

A. L. Volynskii and N. F. Bakeev

CRC Press
Taylor & Francis Group
Boca Raton London New York

CRC Press is an imprint of the
Taylor & Francis Group, an **informa** business

CRC Press
Taylor & Francis Group
6000 Broken Sound Parkway NW, Suite 300
Boca Raton, FL 33487-2742

First issued in paperback 2020

© 2016 by Taylor & Francis Group, LLC
CRC Press is an imprint of Taylor & Francis Group, an Informa business

No claim to original U.S. Government works

ISBN 13: 978-0-367-57504-5 (pbk)
ISBN 13: 978-1-4987-4368-6 (hbk)

Visit the Taylor & Francis Web site at
http://www.taylorandfrancis.com

and the CRC Press Web site at
http://www.crcpress.com

Contents

	Abbreviations	ix
	Preface	1
1.	**Development of the interfacial surface in deformation of polymers**	**9**
1.1.	The method for visualisation of structural rearrangements taking place during the variation of the surface area of deformed polymers	12
1.2.	Visualisation of structural rearrangements accompanying the development of the interfacial surface in deformation of rubbery polymers	19
1.3.	Visualisation of structural rearrangements taking place during annealing of amorphous polymers oriented above the glass transition temperature	29
1.4.	Rolling of glassy polycarbonate	33
1.5.	Structural rearrangement in the deformed polymer in the conditions of isometric heating	36
2.	**Healing of the interfacial surface in polymer systems**	**41**
2.1.	Healing of the interfacial surfaces in rubbery polymers	41
2.2.	Healing of the interfacial surface in glassy polymers	45
2.3.	Heterophase healing of polymer interfaces	49
2.4.	Heterochemical healing of the polymer–polymer interfaces	56
2.5.	Monolithization of powders	58
2.6.	Healing of the interfaces produced upon fracture of glassy polymers	63
2.7.	Healing of the interfacial surface in deformed polymers	68
3.	**Special features of the structure and properties of surface layers and thin (nanometric) films of glassy polymers**	**79**
3.1.	Measurement of the glass transition temperature of amorphous glassy polymers in thin films and thin surface layers	81
3.2.	Glass transition temperature of thin films of glassy polymers, deposited on solid substrates	81
3.3.	The glass transition temperature of free-standing thin films of glassy polymers	88
3.4.	Measurement of the glass transition temperature and	

	molecular mobility in surface layers of bulk glassy polymers	90
3.5.	Interaction of metal nanoparticles with polymer surfaces	91
3.6.	Possible reasons for the decrease of the glass transition temperature in thin films and surface layers of amorphous polymers	99

4.	**Role of surface phenomena in shear yielding of glassy polymers**	**108**
4.1.	Thermal ageing of polymer glasses	110
4.2.	The main features of the effect of thermal ageing on the properties of glassy polymers	111
4.4.	Physical ageing and the structure of glassy polymers	114
4.4.	Molecular mechanism of thermal ageing of glassy polymers	120
4.5.	Effect of mechanical action on the process of physical ageing of polymer glasses	124
4.6.	Properties of glassy polymers subjected to mechanical effects	126
4.7.	The spatial inhomogeneity of deformation of polymer glasses	137
4.8.	Structure of shear bands formed during deformation of glassy polymers	147
4.9.	The nature of structural-mechanical anomalies in the properties of deformed glassy polymers	152

5.	**Role of surface phenomena in the strain softening of glassy and crystalline polymers**	**176**
5.1.	Strain softening of polymer systems, taking place without the formation of porosity	177
5.2.	Strain softening of glassy polymers	180
5.3.	Factors causing force softening of glassy polymers during crazing	187
5.4.	Strain softening of crystalline polymers	190
5.5.	Mechanism of strain softening of crystalline polymers	199

6.	**Role of surface phenomena in deformation of polymers in active liquid media**	**216**
6.1.	What is the adsorption-active medium?	217
6.2.	Structural special features of deformation of polymers in adsorption-active media	227
6.3.	Crazing in liquid media – manifestation of the Rehbinder effect in polymers	228
6.4.	Mechanism of the formation of the unique structure of crazes	229
6.5.	Crazing dynamics of polymers in liquid media	232
6.6.	Main factors determining the dynamics of crazing of the polymer in the AAM	245
6.7.	The multiplicity factor of the number of areas of localised	

	plastic deformation	249
6.8.	Relationship of the crazing dynamics of polymers in liquid media with the fine structure of crazes	251
6.9.	Crazing mechanism of polymers in liquid media	254
6.10.	Delocalized crazing of the polymers in liquid media	259

7.	**The structure and properties of crazed polymers**	**276**
7.1.	Structural–mechanical aspects of deformation of crazed polymers	276
7.2.	Thermomechanical properties of crazed polymers	292
7.3.	Colloidal swelling	308
7.4.	Adsorption properties of the crazed polymers	310

8.	**Multiphase nanodispersed systems based on crazed polymers**	**320**
8.1.	Interaction of low-molecular substances with the highly developed surface of the crazed polymer	320
8.2.	Polymer–polymer nanomixtures based on crazed polymers	339
8.3.	Crazing as a method of producing nanosized porosity in polymers	341
8.4.	Special features of production of polymer–polymer nanocomposite by polymerisation *in situ* in a crazed polymer matrix	343
8.5.	Direct addition of the second polymer component to the crazed polymer matrix	355

9.	**Instability and self-organisation of polymer surfaces**	**364**
9.1.	Special features of the development of interfacial surfaces during the flow of polymer melts and solutions	364
9.2.	Loss of stability and dispersion in flow and during phase separation in polymer systems	365
9.3.	Inhomogeneous swelling of polymers	368
9.4.	Electrodynamic and thermomechanical instability of polymer surfaces	370
9.5.	Polymers with thin rigid coatings	374
9.6.	Mechanism of the formation of the regular microrelief	389
9.7.	Regular fragmentation of the coating	400
9.8.	Surface structure formation in polymers with a chemically modified surface	407
9.9.	Polymer films with nanometric coatings – 'rigid coating on a soft substratum' systems	410

| **10.** | **Evaluation of the structural and mechanical properties of nanometric surface layers** | **415** |
| 10.1. | Physical fundamentals of the method for evaluating the | |

	stress–strain properties of surface layers and nanometric coatings deposited on polymer films	416
10.2.	Modification of polymer surfaces	419
10.3.	Evaluation of the stress–strain properties of coatings deposited on polymer surfaces	428
10.4.	Evaluation of the stress–strain properties nanometric aluminium coatings	430
10.5.	Evaluation of the stress–strain properties of nanometric coatings based on noble metals	438
10.6.	The non-metallic coatings	451
11.	**Natural systems constructed on the basis of the 'rigid coating on a soft substratum' principle**	**460**
11.1.	Examples of 'rigid coating on a soft substratum' natural systems	461
11.2.	The Earth – the typical 'rigid coating on the soft substratum' system	463
11.3.	Evaluation of the thickness of the Earth's crust	470
11.4.	Evaluation of the strength and longevity of the Earth's crust	471
12.	**Perspectives for the practical application of surface phenomena in solid polymers**	**479**
12.1.	A new approach to the formation of nanocomposites with a polymer matrix	480
12.2.	Production of polymer films and fibres capable of influencing the environment	492
12.3.	Technological aspects of polymer modification by crazing	497
12.4.	Methods for increasing the efficiency of crazing	498
12.5.	Producing the transverse microrelief in polymer fibres and films	508
12.6.	Practical application of polymer films with a regular microrelief	509
Index		**520**

Abbreviations

AAM	–	adsorption-active media
APS	–	aminopolystyrene
BRF	–	birefringence
CA	–	cetyl alcohol
DEA	–	deformation of the elastic aftereffect
HDMS	–	hexamethyl disilazane
HDPE	–	high-density polyethylene
HIPS	–	high-impact polystyrene
IPN	–	interpenetrating polymer networks
ITS	–	indigo tetra sulphonate
LDPE	–	low-density polyethylene
MM	–	molecular mass
MOR	–	mid-ocean ridges
OD	–	octadecane
P4VP	–	poly-4-vinyl pyridine
PB	–	polybutadiene
PC	–	polycarbonate
PDMS	–	polydimethyl siloxane
PEEK	–	polyester ester ketone
PET	–	polyethylene terephthalate
PETH	–	polyethylene terephthalate-co-1,4-cyclohexane dimethylene terephthalate
PIB	–	polyisobutene
PMMA	–	polymethyl methacrylate
PP	–	polypropylene
PPO	–	polyphenylene oxide
PS	–	polystyrene
PSF	–	polysulfone
PTMSP	–	poly-1-trimethylsilyl-1-propane
RCSS	–	rigid coating on a soft substratum
RMR	–	regular microrelief
SAXS	–	small angle x-ray scattering
SBS	–	styrene–butadient–styrene
TDA	–	tridecanoic acid

Preface

The book generalises a wide range of scientific problems which have not as yet been adequately clarified and analysed in the scientific literature. The main fundamental physico-mechanical and physico-chemical properties of the solid polymers have been studied extensively and, therefore, have been summarised in many monographs and textbooks. The general properties of polymers, including solid polymers, are governed by the chain structure of macromolecules. In particular, the chain structure of macromolecules is responsible for a unique property of polymers – their capacity for high reversible strains. The physical mechanism of this phenomenon has been formulated in the first half of the previous century. High strain recovery is provided by the recovery of the deformed polymer chains to their equilibrium conformations via thermal motion.

At the same time, there is another aspect of this problem. Any solids, and polymers in particular, change their geometrical dimensions under the external impact (mechanical deformation, thermal expansion, electrical and magnetic fields). It is evident that the changes in geometrical dimensions of solids are accompanied by changes in their interfacial surface area. In turn, changes in the interfacial surface area proceed via mass transfer from bulk to surface when the interfacial surface area increases and, vice versa, via the mass transfer from surface to bulk when this area decreases. The rate of these processes is especially high at high strains which are typical of solid polymers. This monograph deals with the special features of mass transfer and related structural transformations in the deformed polymers, and the effect of diverse external conditions on this process is also outlined.

So far, insufficient attention has been paid to this type of mass transfer processes, even though a better knowledge on these phenomena is crucial for the development of a new approach for the description of diverse well-known phenomena. In fact, impacts on polymers leading to changes in their geometrical dimensions and especially high strains can be regarded as surface phenomena because they are provided by mass transfer from surface to bulk, and back. This book deals with many well-known properties of polymers of this type. As we are concerned with the mass transfer to the surface from the bulk and/or from the surface to the bulk in polymers, these processes can be treated as surface phenomena so that we can define the range of the investigated problems as it is described in the title of the book.

The structure of this book may be presented as follows. Chapter 1 generalises the results of studies indicating that a well-studied phenomenon of deformation of polymers is virtually accompanied by a high gain in their interfacial area. A method is presented and substantiated for the visualisation of this process, and it is shown that different deformation conditions may result in both an increase or decrease of the interfacial surface area of the polymer. In some cases, both processes can proceed simultaneously and their interplay can be detected by the method of visualisation of surface rearrangements in polymers upon deformation developed by the authors of the book.

Chapter 2 deals with the analysis of the available literature data on healing of interfacial surfaces in polymers. These processes are provided by mutual diffusion of macromolecules or their fragments through the interface and may lead to a complete polymer monolithisation. Special attention is paid to healing of interfaces upon monolithisation of powders, cracks in bulk polymers, etc.. The healing of interfaces in glassy polymers is also discussed. Large-scale molecular motion in glassy polymers is 'frozen' and there is no diffusion of macromolecules across the interface.

Chapter 3 presents the description and analysis of a relatively new scientific direction – the structure and properties of polymers in the thin nanometric films and surface layers. Of considerable interest are the studies on special features of molecular motion in thin films and surface layers of amorphous polymers which dated back to the middle of the 1990. This relates to thin films and polymer layers with a thickness ranging from tens to hundreds of nanometres, i.e., we deal with the properties of polymer surfaces. The development of various experimental methods for the assessment of molecular motion in thin layers has led to the discovery of many surprising effects. The chapter presents the experimental data on the structure and properties of polymer films and surface layers of amorphous polymers in nanoscale layers. The glass transition temperature for all polymers in thin layers is known to be depressed. This decrease can achieve tens or even hundreds of degrees. This circumstance is responsible for an unusual mechanism of interaction of polymer surfaces with, for example, metallic nanoparticles. The currently known mechanisms of this phenomenon are analysed. These studies are of obvious fundamental and applied importance and the authors believe that they should be analysed in detail and generalised.

Chapter 4 summarises experimental and theoretical data on the mechanism of inelastic deformation in glassy amorphous polymers. Abundant experimental data on special features of deformation in glassy polymers are revisited. These peculiarities are associated mostly with the analysis of the structure and properties of deformed polymer glasses. They include diverse phenomena such as stress relaxation in polymer glasses deformed at the elastic (the so-called Hookean) region of the stress–strain curve, stress rise in the temperature range below the glass transition temperature upon isometric annealing of the deformed polymer glasses, low-temperature shrinkage of the deformed polymer glasses, storage of the internal energy by the deformed glassy polymer, and many other phenomena. The experimental data on structural rearrangements accompanying deformation of glassy polymers are presented and, at the initial stage, deformation is shown to be highly non-uniform in the bulk (proceeds via nucleation and growth of shear bands). The most important result is the following: in contrast to shear bands in low-molecular mass solids, shear bands in polymers are stuffed with a highly organised fibrillised polymer with a highly developed interface. Observations with a newly advanced direct microscopic technique show that the low-temperature shrinkage of the deformed glassy polymer is also highly non-uniform and localized in shear bands. Taking into account the spatial inhomogeneity of inelastic deformation in glassy polymers and depressed glass transition temperature in thin films and surface layers (see Chapter 3), a new mechanism of deformation and shrinkage of deformed glassy polymers is proposed, and the inelastic deformation of glassy polymers can be treated as a unique surface phenomenon.

Chapter 5 is devoted to another general property of glassy and semicrystalline polymers which is the phenomenon of strain softening. Analysis of the literature data on the mechanism of this phenomenon and new information on the properties of surface layers in polymers show that specific features of crazing in semicrystalline and amorphous glassy polymers are primarily controlled by the unique properties of surface layers and provided by surface phenomena.

Chapter 6 is concerned with the analysis of the phenomena taking place upon deformation of polymers in adsorptionally active liquid media (AAM). As was shown in previous sections, inelastic deformation of rigid polymers is a surface phenomenon; hence, deformation of polymers in contact with liquid media which are able to reduce their interfacial surface energy, should have the strongest effect on the deformation mechanism. This effect of AAM on the deformation of polymers can be treated as a particular case of a more general phenomenon – the Rehbinder phenomenon. In structural aspects, the Rehbinder phenomenon is behind the development of necessary conditions providing crazing in polymers. Crazing of polymers is one of the fundamental modes of inelastic deformation of polymers. This phenomenon involves a spontaneous dispersion of polymer and formation of regular periodic structures with nanoscale dimensions . The book discusses the relationship between the parameters of the fibrillar–porous structure of the crazed polymer and deformation conditions (stress intensity, nature of AAM, temperature, etc). Chapter 6 describes fundamental conditions providing onset and development of this particular mode of structuring in amorphous polymers, and the mechanism of this phenomenon is highlighted. This phenomenon is based on the Taylor meniscus instability: there is a balance between mechanical stresses leading to the development of interfacial surface and reduction in surface tension and minimisation of interfacial surface area of the deformed polymer.

Chapter 7 deals with the study and analysis of the properties of polymers upon environmental or solvent crazing . As a result of crazing, a polymer acquires an unusual complex of mechanical, thermomechanical, physico-chemical, and sorption properties which are not characteristic of glassy polymers. A detailed analysis of these properties shows that they are provided by an increased interfacial surface which polymer acquires due to solvent crazing. This anomalous behaviour of crazed polymers can be convincingly explained by the development and/or healing of the highly developed interfacial surface of the crazed polymer.

In chapter 8, the attention is focused on interaction of the highly developed surface of crazed polymers with diverse low- and high-molecular mass components. The point is that the highly organised structure of the crazed polymer can be developed when an active liquid (an absorptionally active medium) efficiently fills up the formed nanoporous structure of crazes. In turn, this means that, firstly, the low-molecular mass component dissolved in the AAM is also dispersed within the formed porous structure of the polymer down to nanometric dimensions and, secondly, crazing is a universal method for the delivery the low-molecular mass compounds to the bulk of polymers and for the preparation of well mixed polymer – (low-molecular mass compound) blends. This chapter discusses various structural aspects of interaction between crazed polymers and diverse low-molecular-mass compounds. First, upon crazing, the polymer acquires a highly developed interfacial surface and, consequently, the polymer becomes an effective sorbent. In turn, studies on sorption of different substances by these sorbents yield important information on the structure of crazed polymers. Second, experimental data on structural organization of the low-molecular-mass components incorporated into the nanoporous crazed polymer matrices are discussed. The nanocomposites prepared by this approach allowed detailed studies on the structure and properties of the low-molecular-mass substances in a highly dispersed colloidal state. Attention is given to the experimental methods which trace the special features of phase transitions of low-molecular mass crystalline substances within the crazed polymers. The enthalpy of the related phase transitions and their positions along the temperature scale are shown to depend on the dispersion of the incorporated low-molecular mass components. The unique behaviour of the orientation of the low-molecular mass crystals in the crazed polymer matrices is described and characterised.

Chapter 8 also discusses the problems of the addition of the second high-molecular-mass component to the nanoporous matrix of the crazed polymer. This problem can be solved in two ways: first, addition of a monomer to the nanoporous crazed matrix is followed by its *in situ* polymerisation and, second, macromolecules of the second polymer can be directly delivered to the nanoporous polymer from the solution upon crazing. In both cases new types of polymer blends based on incompatible components can be prepared and mutual dispersion of both components lies within the nanometric level.

Chapter 9 is concerned with the instability and self-organisation of surface layers in polymers and diverse polymer systems. Analysis of the literature data shows that the polymer surfaces are destabilised under different external impacts (mechanical, electrical, surface, osmotic and others) and, as a result, regular periodic structures are formed. This instability has been known for many years and presents the challenge for many areas of polymer science: physics, mechanics, rheology, physical chemistry, etc. This phenomenon leads to a spontaneous formation of regular periodic structures on polymer surfaces and the corresponding wavelength ranges from the macroscopic down to nanoscopic level. The mechanism of the related structural transformation is analysed. There are, at least, two modes of instability or stability loss in polymer systems which are accompanied by the formation of regular structures: 'dynamic' and 'static'. Both types are characterised by common features: the threshold critical nature of the transition to instability and formation of regular periodic structures. At the same time, in the case of the 'dynamic' loss of stability the system is in the dynamic stationary state and self-organisation of the system requires a continuous energy and/or matter inflow. Highly organised 'dissipative' structures of this type exist in the state far from equilibrium. In the 'static' (Euler) loss of stability, the resultant highly organised ordered structure is in the equilibrium with the acting forces, regardless of internal stresses.

Special attention is paid to surface structural formation in multilayered polymer systems. These systems include polymer films and fibres with a chemically modified surface layer or with diverse rigid coatings deposited on their surface. The book describes theoretical and experimental approaches to the analysis of structural surface formation in deformed polymers with rigid nanometric coatings. It is assumed that these systems share general fundamental properties so that they are referred to as the 'a rigid coating on a soft substratum' systems (RCSS).

In chapter 10, theoretical speculations concerning structurally mechanical behaviour of the RCSS systems described in chapter 9 are presented, and the stress–strain properties of thin (nanometric) surface layers of polymers and nanometric coatings deposited on their surface are analysed. The mechanical properties of thin layers of solids are shown to be appreciably different from the properties of bulk solids. The physical nature of this difference is considered.

Chapter 11 highlights the efficacy of the approaches developed for the RCSS systems for the analysis and description of some natural phenomena. In particular, the planet Earth is shown to be a typical RCSS system and the approaches developed for these systems can be used for the analysis and evaluation of structural and mechanical parameters of macrosystems. In particular, strength and lifetime of the ocean crust are, for the first time, estimated from the parameters of the surface relief of the Earth crust.

Chapter 12 describes the applied aspects of the role of surface phenomena in structurally mechanical behaviour of polymers. Chapter 12 discusses diverse practical applications of this mode of structural self-organisation in polymers. In particular, Chapter 12 describes an earlier unknown structurally mechanical state of rigid polymers, referred to as a 'highly dispersed oriented state of polymers'. This state is shown to be characterized by a number of unusual properties. This state is spontaneously formed via environmental (or solvent) crazing of glassy polymers. Specific properties of this structural-physical state are described: high reversible strains in glassy and semicrystalline polymers, low-temperature shrinkage of crazed glassy polymers, incorporation of low-molecular-mass compounds in crazed polymers, spontaneous elongation upon annealing, etc. Fundamental studies in this direction made it possible to develop new universal methods for the preparation of porous polymer sorbents, polymeric separation membranes, new types of polymer–polymer blends, flame-retardant polymeric materials, electrically conductive polymer materials, metallopolymers, etc. Stability loss in polymer systems and, in particular, formation of regular periodic structures upon deformation of amorphous polymers with a thin rigid coating also offer evident benefits for practical applications. In particular, this approach can be used for the preparation of polymer films with a regular surface microrelief with promising optical properties. Fundamental structural– mechanical properties of these systems offer new advantages for the estimation of basic mechanical properties of solids in thin nanometric layers, and this approach can be used in many other areas of science, including geodynamics for the simulation of principal tectonic processes.

This book is the result of many years of research performed by our colleagues at the Faculty of Chemistry of the M.V. Lomonosov Moscow State University and the N.S. Enikolopov Institute of Synthetic Polymeric Materials of the Russian Academy of Sciences, especially S.L. Bazhenov, A.V. Volkov, T.E. Grokhovskaya, A.V. Efimov, N.I. Nikonorova, L.M. Yarysheva, A.S. Kechek'yan and their students and post-graduate students. We are very grateful to all of them. We have discussed many important scientific problems mentioned in the book with A.A. Berlin, V.G. Kulichikhin, A.Ya. Malkin, A.M. Muzafarov, A.N. Ozerin, E.F. Oleinik and V.S. Papkov. Fruitful discussions and comments allowed us to improve the quality of the monograph but to move forward in the development of many scientific topics and we are very grateful to them. Our special tribute is to the late Academician V.A. Kabanov, who for many years showed a genuine interest in the scientific fields developed by the authors of the monograph and gave them full support.

The authors thank the Russian Foundation for Basic Research for many years of financial support, without which many important studies involved in this monograph would be hardly possible.

A.L. Volynskii
N.F. Bakeev

Development of the interfacial surface in deformation of polymers

One of the key problems of physical chemistry of polymers is the determination of the relationship between their structure and properties. Because of the importance of this subject, fundamental and applied investigations of the mechanism of deformation of polymers have been carried out over many decades and are generalised in many monographs and textbooks [1–7]. These investigations determined many fundamental relationships between the structure and properties of polymers. The most important results have been obtained in the application of direct investigation methods, such as different types of microscopy and x-ray phase analysis. It is important to mention that the application of these methods is most efficient when investigating the mechanism of the formation of crystalline polymers, whereas the application of these methods for amorphous polymers yields only a small amount of information.

Regardless of the evident successes in the area of determination of the deformation mechanism of polymers, one of the aspects of this problem remains unclear. The point is that any deformation of the solid, and polymers in particular, is accompanied by the change of its surface area. Whilst the volume of the deformed polymer may remain constant [8], the surface area always changes. The variation of the area of the interfacial surface of the solid means that the material from the volume diffuses to the surface with the increase of the surface area and from the surface to the volume – with a reduction of the area. The set of the phenomena, taking place at the phase boundary of the polymers and different types of polymer systems,

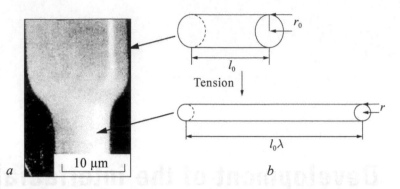

Fig. 1.1. Micrograph of a fragment of the polyester fibre in the area of transition to the neck (*a*) and the diagram of variation of the dimensions of the fibre during its orientational drawing (*b*) [11].

is of obvious fundamental and applied importance. The processes of development, increase of the area of the interfacial surface and also its reduction ('healing') [9] take place under almost any effect on the polymer system. In other words, the greatly different effects on the solid polymers are associated with the phenomena of transport of the material from the volume to the surface and vice versa, i.e., with the surface phenomena.

This situation will be demonstrated using a simple example. It is well-known that the deformation of a polymer below its glass transition temperature (T_g) is accompanied by the formation and development of a neck [10]. Figure 1.1 *a* shows a photograph of a section of a polyester fibre in the area of transition to the neck. It may easily be shown that in transition of the polymer to the neck the surface area of the polymer greatly increases. A cylinder with the initial radius r_0 and length l_0 will be investigated (Fig. 1.1 *b*). To simplify considerations, it will be assumed that the volume of the polymer remains constant during plastic deformation. Let it be that the draw ratio is equal to λ and, consequently, from the condition of the conservation of the volume we have

$$\pi r_0^2 l_0 = r^2 \lambda l_0 \tag{1.1}$$

Consequently, we obtain

$$r = r_0 / \lambda^{-1/2} \tag{1.2}$$

The surface area of the deformed fibre S is equal to

$$S = 2\pi r \lambda l_0 = S_0 \lambda^{1/2} \qquad (1.3)$$

where S_0 is the initial surface area. Or, transferring to the radius of the fibre

$$S = S_0(r_0/r) \qquad (1.4)$$

where r_0 and r are the initial and actual value of the radius of the fibre at the draw ratio λ. Consequently, this shows that as the radius of the fibre decreases, its specific surface (i.e., the surface related to its volume or mass) increases (with other conditions being equal). In a specific case, shown in Fig. 1.1, the surface area S_0 of the section of the fibre with the length $l_0 = 1$ cm and the radius $r_0 = 7.5$ μm (this radius is characteristic of the non-deformed fibre – upper part of Fig. 1.1 a) is equal to $S_0 = 47.1 \cdot 10^{-3}$ cm². If the fibre is drawn into a neck (for polyester $\lambda \approx 3$), the fibre diameter is equal to $r \sim 3.5$ μm (lower part of the fibre in Fig. 1.1 a). Equation (1.4) shows that the initial surface increases in cold drawing of the polyester fibre to the value $S = 100.8 \cdot 10^{-3}$ cm². Thus, the routine cold drawing of the fibre more than doubles its surface area. This means that extensive rearrangement takes place in the polymer associated with the migration of the material from the volume to the surface during its deformation and from the surface into the volume during strain recovery (shrinkage). However, this process is not at present considered and taken into account in the examination of the deformation mechanism of the polymers, primarily because of the absence of an efficient method of investigation.

Regardless of the large increase of the surface area of the polymer during its transition to the neck (the natural draw ratio of polyethylene terephthalate (PET) at room temperature is ~275%), the fibre subjected to such high strain retains the smooth cylindrical shape. Consequently, the local transport of the polymer, resulting in the formation of the neck, cannot be observed and characterised by direct microscopic methods. In accordance with the formulated tasks, below we describe the direct microscopic method of visualisation of the processes of variation of the interfacial surface during deformation and thermal relaxation of the oriented polymers.

1.1. The method for visualisation of structural rearrangements taking place during the variation of the surface area of deformed polymers

As mentioned previously, the interfacial surface area changes almost under any effect on the polymer. However, this change is usually evaluated by macroscopic measurements of the dimensions of the sample. These measurements do not make it possible to characterise and investigate the local processes of diffusion of the polymer from its volume to the surface and vice versa.

It is therefore important to mention that recently a direct microscopic method has been developed [12–15] for visualisation of the processes of mass transfer of the material from the surface into the volume and (or) back during mechanical deformation of the volume. This method is quite simple. Prior to deformation (shrinkage), a thin (several nanometres) solid coating is deposited on the surface of the polymer. In subsequent deformation (shrinkage) this coating cannot diffuse together with the polymer into its volume because of the incompatibility of the components. Consequently, the surface layer is characterised by a specific type of structure formation, associated in particular with the mass transfer of the polymer from the surface into the volume and (or) vice versa. The microrelief, formed in this case, can be recorded and characterised by direct microscopic methods.

We shall illustrate the efficiency of the proposed method on the example of two simplest and, at the same time, fundamental types of deformation of the polymer – planar tension and planar compression. In the context of the investigated problem, these types of deformation are relatively simple and illustrative. The point is that in planar tension of the polymer its surface only increases and in planar compression only decreases.

Planar tension

These two types of deformation can be realised using the device shown in Fig. 1.2. The polymer sample in the form of a film is strictly fixed in a circular clamp. Subsequently, the cell together with the sample is heated to the required temperature and compressed air is passed through the volume. Since the temperature of the polymer is higher than its glass transition temperature, the effect of compressed air results in the planar tension of the sample, as shown in Fig. 1.2. The process is completely identical with blowing balloons for

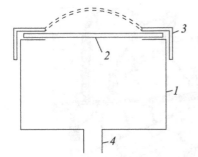

Fig. 1.2. Diagram of the cell for the planar tension of polymer films: 1) the body of the cell, 2) the polymer sample (the dotted line indicates the sample after planar tension), 3) clamping nut, 4) the nozzle for the supply of compressed air [16].

children. The method can be used to realise both types of planar deformation (planar tension and planar compression).

The first type of deformation is realised by supplying the air into the cell containing the sample with the already deposited coating. Subsequently, when the specimen is subjected to the required degree of planar tension, the cell is cooled to the temperature lower than the corresponding glass transition temperature of the polymer. As a result, the dimensions of the polymer are fixed and, therefore, the required degree of its planar tension is also fixed. To apply deformation in planar compression, the sequence of operations is slightly changed. As previously, the specimen without the coating, subjected to the required degree of planar tension in the cell, shown schematically in Fig. 1.2, is cooled to the temperature lower than the appropriate glass transition temperature of the polymer. Subsequently, the sample is withdrawn from the clamps of the cell, the coating is deposited and the sample is then heated to temperatures higher than the glass transition temperature of the polymer. Annealing results in the planar shrinkage of the polymer and, consequently, its deformation by planar compression. Thus, in the first case, structural rearrangement in the deposited coating characterises the process of planar tension of the polymer, and in the second case – planar compression.

Figure 1.3 shows the light (a) and electron (b) micrographs of a PET sample, deformed in the planar tension conditions. Figure 1.3 shows that in the planar tension conditions the surface of the polymer actually increases. The metallic coating is characterised by a considerably smaller elongation at break than the polymer and, therefore, in planar tension the coating fractures and breaks up into individual fragments. In fact, the cracks in the coating are areas of the polymer which was situated in its volume prior to deformation

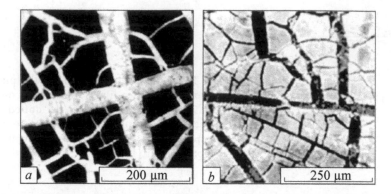

Fig. 1.3. Light (*a*) and scanning electron (*b*) micrographs of a sample of polyethylene terephthalate with a thin (10 nm) aluminium coating and subjected to 30% planar tension at 100°C [16].

and which diffused to the surface during deformation in planar tension.

Planar compression

We now examine the planar compression deformation. Figure 1.4 shows the scanning electron micrograph (*a*) and atomic force image (*b*) of a polyethylene terephthalate sample with a thin metallic coating, subjected to planar compression deformation [17].

The planar shrinkage of the polymer with the thin coating, deformed at temperatures higher than its glass transition temperature, results in the formation of a regular and extremely spectacular microrelief. The physical reasons for the formation of these microrelief is that planar compression is accompanied by the diffusion of the polymer from the surface into the volume. The coating, strongly bonded with the polymer, cannot follow the surface layer of the polymer into its volume so that it appears to be in the planar compression conditions. As a result of compression, the coating loses its stability and acquires the regular microrelief, shown in Fig. 1.4. The mechanism of formation of microreliefs, shown in Fig. 1.4, will be examined later, and here we only mention that the formation of these morphological forms is of general nature and is observed in deformation of any bilayered system, constructed on the basis of the principle of the thin solid coating on a long compliant base. Regardless of their dimensions, these systems have a complex of general structural–mechanical properties so that in [18] they are referred to as the 'rigid coating on a soft substratum' (RCSS).

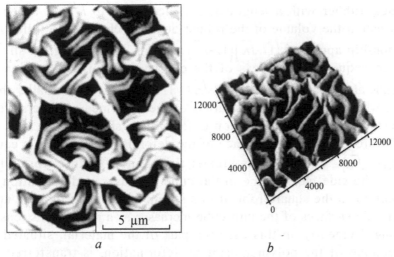

a   *b*

Fig. 1.4. Scanning electron micrograph of the polyethylene terephthalate specimen with a thin (15 nm) gold coating deformed above the glass transition temperature in the uniaxial compression conditions, after planar shrinkage in annealing by 25% (*a*) and the three-dimensional reconstruction of the atomic force microscope image of the same sample (*b*) [17].

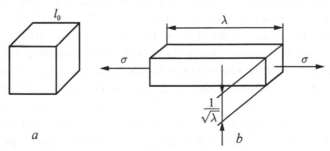

a   *b*

Fig. 1.5. Schematic representation of uniaxial tensile loading of a rubbery sample of a cubic polymer with the unit length of the face ($l_0 = 1$) prior to (*a*) and after (*b*) tensile loading with the elongation $\lambda = l/l_0$.

Uniaxial tension

We have examined the simplest and most frequent cases of deformation of polymers not aggravated by the combination of different types of deformation. However, in practice in investigating the mechanical properties of the polymers, the most frequently encountered types of deformation are uniaxial tension and uniaxial compression. From the viewpoint of the change of the surface area in deformation, these types of loading are more complicated.

We consider the case of uniaxial tension (Fig. 1.5). To simplify considerations, tensile loading will be applied to a cubic sample of an ideal rubber with a length of the face equal to $l_0 = 1$. Since in deformation the volume of the rubber does not change, the following relationship applies $\lambda\left(1/\sqrt{\lambda}\right)\left(1/\sqrt{\lambda}\right)$, where λ is the original strain, l/l_0. Consequently, the areas of the end surfaces of deformation of the cube are equal to: $\left(1/\sqrt{\lambda}\right)\left(1/\sqrt{\lambda}\right) = 1/\lambda$, i.e. they are smaller than unity, which means that these areas of the deformed polymers were subjected to planar compression deformation, shown in Fig. 1.4.

The area of each side surface of the deformed single cube is equal to: $\lambda\left(1/\sqrt{\lambda}\right) = \sqrt{\lambda}$, which is greater than unity. In other words, the area of the side faces of the initial cube increases in deformation in proportion to the square root of this strain. This means that the area of the side surfaces of the unit cube increases as a result of its tensile loading. Evidently, in this case the part of the material, situated in the volume of the polymer prior to deformation, is transferred by some mechanism to its surface.

However, this problem is not so simple as in the case of planar tension. The point is that in uniaxial tensile loading, in addition to the increase of the size in the direction of the axis of applied stress, there is also lateral contraction of the polymer, i.e., the dimensions in the direction, normal to the direction of the tensile loading axis, decrease. At the same time, the reduction of the size of the deformed polymer, as shown previously, is accompanied by migration of part of the material from the surface to its bulk. In other words, in the case of uniaxial tensile loading it should be expected that the side surfaces of the single cube (Fig. 1.5) are characterised by the co-existence of two types of mass transfer (from the surface into the bulk and from the bulk to the surface) in the deformed polymer.

We investigate how this circumstance is reflected in the processes of relief formation when deformation takes place in the polymer with the coating deposited on its surface. Figure 1.6 shows several micrographs of different polymers on which a nanometric metallic coating was deposited and which were subsequently subjected to uniaxial tension.

It may be seen that simple tensile loading of the polymer film with the nanometric coating is accompanied by the formation of a very attractive regular microrelief. The coating breaks up into a system of fragments of approximately the same size (bright bands are fragments of the fractured coating, dark bands are the cracks in the fractured coating); the coating acquires a regular microrelief. The regularity of the formed microrelief and its strict orientation in

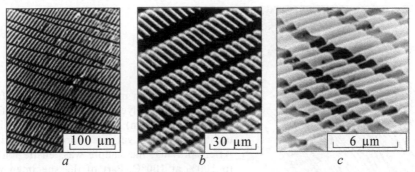

Fig. 1.6. Scanning electron micrographs of the films of natural rubber with a thin (10 nm) gold coating after 50% tensile loading at room temperature (*a*); polyvinyl chloride with a thin (10 nm) platinum coating (*b*) and polyethylene terephthalate with a thin (15 nm) gold coating after 100% tensile loading at a temperature of 90°C (c).

relation to the tensile loading axis of the polymer are surprising (its depressions and peaks are always oriented strictly along the axis of tensile loading).

It is important to note that both resultant structures (regular microrelief and the regular system of fragments of failure of the coating) are highly organised and periodic and can be easily characterised by means of direct microscopic studies.

Evidently, the resultant microrelief contains in fact the same morphological elements as the coatings subjected to planar tension (Fig. 1.3) and planar compression (Fig. 1.4). Actually, in planar tension the coating surface contains only regularly distributed folds, whereas in planar tension the folds are not found and only cracks are observed. In uniaxial tension (Fig. 1.6) both the previously mentioned morphological forms coexist on the surface of the polymer. This circumstance indicates that at apparently the simple type of deformation – uniaxial tension, a relatively complicated process of mass transfer of the polymer from the surface into the bulk and vice versa takes place.

As shown previously, deformation of the polymer is accompanied by migration of the material from the volume into the surface and vice versa from the surface into the volume. The coating, deposited on the surface of the polymer, reacts highly sensitively to the processes of this type. When the surface area decreases, the coating, strongly bonded with the polymer, can not follow the polymer into its volume together with the material from the surface. Consequently, the system appears to be in the compression conditions and, therefore, loses stability and, consequently, a regular microrelief forms. In

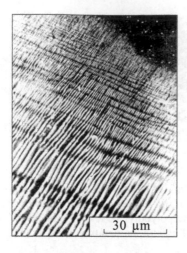

Fig. 1.7. Electron micrograph of the surface of a sample of polyethylene terephthalate film with a thin (10 nm) layer of copper deposited on its surface by vacuum thermal sputtering. After spraying, the sample was tensile loaded to 100% at 100°C. Part of the specimen was coated with a screen impermeable to the sprayed metal. The figure shows the interface between the metallic coating and the pure polymer (top right hand corner) [11].

30 μm

tensile loading of the polymer-substrate, the coating, strongly bonded with the polymer, is in the tensile loading conditions and breaks up into the fragments because its elongation at break is lower than in the polymer. The mechanism of formation of specific morphological forms on the surface of the polymer will be investigated in greater detail below and here we only mention that the procedure, developed in [11–15] can be used to visualise not only the processes of mass transfer of the polymer during its deformation but also the direction of the local stresses acting in the polymer.

The efficiency of the proposed method can be demonstrated in the following experiment. As previously, a metallic coating is deposited on the polymer surface by vacuum thermal sputtering but part of the polymer sample is coated with a screen impermeable to the sprayed metal. Subsequently, the polymer is subjected to tensile deformation. The area of the polymer at the interface where the metal was screened is shown in Fig. 1.7.

As indicated by the figure, the boundary is not sharp and the thickness of the deposited layer is gradually reduced to 0. It is clearly seen that the coating is characterised by the previously described structural formation, associated with its fragmentation and formation of a regular microrelief. In the upper right corner of the photograph (Fig. 1.7) there is the part of the polymer without the coating. It may clearly be seen that this surface has no relief and remains smooth, regardless of deformation. This means that diffusion of the polymer from the surface into the bulk and vice versa in deformation of the polymer is not accompanied by any relief formation and remains smooth and, therefore, microscopic studies of the polymer prior to and after tensile loading do not provide any information for the deformation mechanism. The data in Fig. 1.7 demonstrate the efficiency of the proposed method of visualisation of the structural rearrangement of the deformed polymer and provide new information on the internal processes of mass transfer of the polymer during its deformation. Thus, it has been established that deformation of the polymer-substrate is accompanied by the development of stresses in the coating responsible for the formation of the unique wave relief.

On the other hand, the large differences in the morphological forms and the possibility of direct evaluation of their parameters make it possible to examine separately and characterised both phenomena, described in [19–24]: regular fragmentation of the coating and formation of the regular microrelief. Analysis of this type will be carried out later and here we only mention that the deposition of the thin, nanometric coatings on the polymer films can be used to visualise the processes of mass transfer from the surface into the volume and vice versa during deformation. It is mentioned that these processes have not as yet been characterised or studied because of the fact that no efficient method is available. Below we examine several examples of such visualisation in deformation and the restoration of the dimensions of the deformed polymers in the glassy or highly elastic states.

1.2. Visualisation of structural rearrangements accompanying the development of the interfacial surface in deformation of rubbery polymers

We examine the efficiency of the method developed in [11–15] for the visualisation of structural rearrangement of the deformed polymers on the example of investigating the deformation of amorphous polymers. The method is suitable for studying structural rearrangement, accompanying deformation of the polymer in any physical states and, in particular, in the highly elastic state. Investigation of the structural special features of deformation of the amorphous polymers in the highly elastic state is of considerable fundamental and light importance. However, the problem of obtaining adequate information on the processes of this type has not as yet been solved. In particular, this is due to the fact that the rubber networks are amorphous systems and the application of the traditional structural methods of investigation, based on the phase contrast of the investigated objects (x-ray phase analysis, electron diffraction, microscopy), is not efficient. Here we present the results of investigating rubbery polymers subjected to different types of deformation using the previously described procedure for visualisation of structural rearrangements [11–15].

Uniaxial deformation
We examine structural rearrangement, taking place during deformation of rubbery polymers with a thin metallic coating. In [25] the authors carried out a comparative investigation of two rubbery polymers deformed at room temperature: plasticized polyvinyl chloride ($T_g = -15°C$) and cross-linked synthetic isoprene rubber (IR). Typical results of investigation of polyvinyl chloride are presented in Fig. 1.8.

It may be clearly seen (Fig. 1.8 a) that at low strains (10%) the metallic coating breaks up into fragments of approximately the same size (10–30 μm) and a regular wavy microrelief with a period of ~3 μm forms on the polymer surface. In this deformation stage, the resultant microrelief is not yet sufficiently regular. Many of these elements are not clear, and individual nucleated folds do not spread from one edge of the coating to another, are short and disappear.

Increase of the strain to 100% (Fig. 1.8 b) results in a large increase of the number of cracks in the coating. The size of the fracture fragments decreases. Deformation of the plasticised polyvinyl chloride with a thin metallic coating is accompanied by the same types of surface structure formation as in other systems (Fig. 1.6). The main special features of this type of structure formation are the formation of a regular microrelief, regular fragmentation of the coating and the regular variation of the parameters of the microrelief

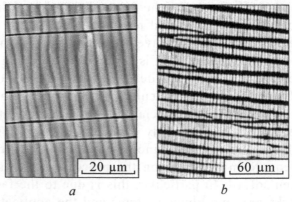

| 20 μm | | 60 μm |
| a | | b |

Fig. 1.8. Scanning electron micrographs of the specimens of the polyvinyl chloride with a thin platinum coating, tensile loaded to 10% (a) and 100% (b). The tensile loading axis is vertical [25].

and the dimensions of the fragments of the coating with the variation of the tensile strain of the polymer film-substrate. The mechanism of these phenomena (progressive disintegration of the coating and the change in the period of the microrelief) will be investigated in greater detail later.

It should be mentioned that the microrelief, formed on the surface of the rubbery polymers, can be investigated only when the dimensions of the sample after deformation are fixed. Otherwise, the polymer fully restores its dimensions and the resultant microrelief is 'smoothed out'. Nevertheless, there is a simple and efficient method of producing a stable surface microrelief on the polymer films in the rubbery state (above T_g). In this case, it is necessary initially to stretch the polymer film to the required tensile strain and then, fixing its dimensions, deposit a thin solid coating on the surface. Subsequent removal of the stress evidently results in the complete shrinkage of the polymer because the polymer is in the highly elastic state. Shrinkage is accompanied by the recovery of the geometrical dimensions and the surface area of the polymer. The coating, deposited prior to shrinkage, will be subjected to tensile and compressive stresses and this will be reflected unavoidably in the surface relief. Evidently, the resultant surface relief will be stable because the film, completely relaxed after tensile loading, will have the stable dimensions and the spontaneous change of the surface area do not occur in the film.

Such experiments have been carried out and the results are presented in Fig. 1.9. It may clearly be seen that the surface of the plasticised polyvinyl chloride shows the formation of a microrelief as a result of its shrinkage, and the microrelief is identical with that

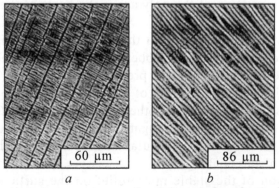

a   b

Fig. 1.9. Scanning electron micrographs of the samples of polyvinyl chloride with a thin platinum coating stretched by 10% (*a*) and 100% (*b*) with the subsequently deposited platinum coating after complete strain recovery.

formed in 'direct' tensile loading of the polymer (Fig. 1.8). Actually, deformation is accompanied by the formation of a regular microrelief and also by the relatively regular fragmentation of the coating (Fig. 1.9 a). An increase of the tensile strain of the samples of polyvinyl chloride is accompanied by an increase of the number of cracks in the coating and the wavelength of the microrelief decreases (Fig. 1.9 b). However, in the relief produced in shrinkage of the plasticised polyvinyl chloride samples there is one important difference in comparison with the reliefs produced in the 'direct' tensile loading of this polymer. The entire pattern of the microrelief, obtained in this case, is rotated by 90° in relation to the axis of tensile loading in comparison with the relief produced in 'direct' tensile loading of the polymer specimens. This is understandable because the polymer film, subjected to direct uniaxial tensile loading, is subjected to two types of deformation at the same time. The elongation in one direction is accompanied by the contraction in the perpendicular direction. In the case of shrinkage of the polymer with the coating deposited on its surface, the directions of tensile loading and subsequent shrinkage coincide and, consequently, the entire pattern of structure formation is rotated by 90° in comparison with the case of direct tensile loading.

Taking these results into account, it may be concluded that the surface structure formation in the metallic coating can be used to conclude that the process of deformation of the plasticised polyvinyl chloride is equilibrium. This is indicated by the almost complete agreement of the patterns of surface structure formation in the direct and reversed (shrinkage) deformation processes (compare Figs. 1.8 and 1.9). It should be mentioned that in particular the assumption of the uniform deformation of the polymer is the basis of the statistical theory of high elasticity [8].

Attention will now be given to the structural rearrangement of the cross-linked polymer in the rubbery state (isoprene rubber). The results of these investigations are presented in Fig. 1.10 a, b. It is clear that in 'direct' tensile loading of natural rubber the relationships are the same as in deformation of other similar systems [16–21] (Fig. 1.1 b). In the range of the investigated tensile strains deformation results in the formation of a regular microrelief and the regular fragmentation of the coating.

The formation of the stable microrelief on the surface of cross-linked rubber revealed a number of previously unknown general relationships (Fig. 1.6). As previously, a polymer sample was

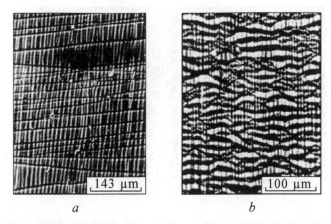

a   b

Fig. 1.10. Scanning electron micrographs of the samples of cross-linked isoprene rubber with a platinum coating after tensile drawing by 25% (a) and 400% (b).

deformed to the required tensile strain, its dimensions were recorded, a metallic coating was deposited and the sample was released from the clamps of the tensile loading device. As a result of stress relaxation the sample completely restored its dimensions and this was accompanied by the formation of a microrelief in the deposited coating.

The results show that this relief has a number of important differences in comparison with the microreliefs detected previously in the deformation of other systems and, in particular, polyvinyl chloride (Fig. 1.8). These previously unknown features of surface structure formation were especially evident at low tensile strains. Figure 1.11 *a* shows a micrograph of a rubber sample, stretched by 10%, after which a metallic coating was deposited on its surface and the mechanical stress was removed. Consequently, the sample completely restored its dimensions. It may be seen that in this case the surface of the polymer is characterised by the formation of a microrelief consisting of several structural elements. Firstly, they are parallel folds forming, as in other cases, a regular microrelief. However, these folds are assembled in relatively well-defined 'packets'. Inside these 'packets' the folds are distributed quite regularly and parallel to each other.

It should be mentioned that in direct tensile loading of the same polymer with the same coating (compare Figs. 1.10 and 1.11) both the regular cracking of the coating and formation of the regular microrelief take place by completely different mechanisms. On the basis of analysis of the results, presented in Fig. 1.11, it may

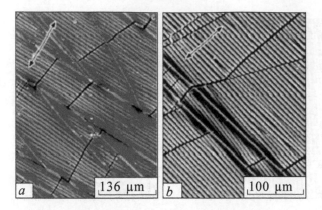

Fig. 1.11. Scanning electron micrographs of polyvinyl chloride samples tensile loaded by 10% (*a*) and 20% (*b*). Subsequently, the surface of the samples was coated with a thin (10 nm) platinum coating and the samples were released from the clamps of the tensile loading device. The arrows indicate the direction of deformation [25].

be assumed that the mechanism of deformation of the cross-linked rubber greatly differs in direct tensile loading of the polymer and its shrinkage. According to the statistical theory of high elasticity [8], deformation of the cross-linked rubber networks is affine and equilibrium and should not differ depending on whether the polymer is tensile loaded by the external force or whether it shrinks spontaneously.

Thus, the investigated experimental data indicate that in deformation and in shrinkage of the cross-linked rubber the proposed method detects a number of previously unknown special features. It should be mentioned that the surface structure formation in the rubber – metallic coating system greatly differs in direct tensile loading of the polymer with the coating and in shrinkage of the same polymer on which the coating was deposited in the deformed state (compare Fig. 1.10 and 1.11). The latter circumstance indicates that the applied method shows that in the experiment conditions used in [25] the deformation of the rubber is not equilibrium. This is indicated by the disagreement of the patterns obtained in investigating the direct (tensile loading) and reversed (shrinkage) deformation processes of the polymer. It is possible that the observed differences are associated with the orientation crystallisation of this polymer [26]. At the same time, such large differences are not detected in the deformation in shrinkage of the plasticised polyvinyl chloride, which indicates that the deformation of this polymer is far closer to equilibrium.

Planar deformation

Important information on the mechanism of deformation of the rubbery polymers is obtained by investigating their planar deformation and shrinkage. The planar tensile loading of the polymer films was realised using the following procedure [27]. A polymer film was secured in a circular frame. The film was deformed using a plunger which travelled the required distance in the orifice of the circular frame in the direction normal to the surface of the film. Consequently, the film was stretched in the plane by the required value. After tensile loading the dimensions of the specimens were fixed with a circular clamp and a thin platinum layer (10 nm) was deposited on the surface of the specimens by ion plasma sputtering. The specimens were then removed from the clamp and the surface was studied in a scanning electron microscope. It should be mentioned that after releasing the specimens from the clamps the initial dimensions of the specimens were fully restored in all cases.

Figure 1.12 shows the electron micrographs of plasticised polyvinyl chloride samples stretched in the plane to different strains followed by deposition of a thin (10 nm) metallic (platinum) layer on the surface followed by shrinkage. It may be seen that in the planar compression (shrinkage) of the film, tensile loaded in the plane of the polyvinyl chloride, the metallic coating, deposited on the polymer surface, loses its stability and acquires a regular microrelief. The nature of this relief formation corresponds completely to the morphology of the reliefs, formed in planar shrinkage of other

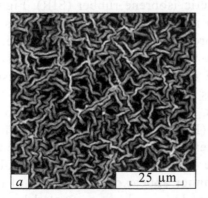 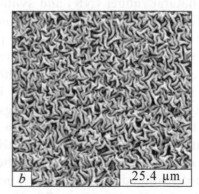

Fig. 1.12. Scanning electron micrographs of the samples of the plasticised polyvinyl chloride deformed by planar tensile loading by 14% (*a*) and 27% (*b*) and restoring their initial dimensions after deposition of a thin (10 nm) platinum layer on their surface.

polymers [27–32]. As indicated by Fig. 1.12, this microrelief consist
of close-packed randomly distributed bent folds, uniformly covering
the entire surface of the polymer. The width of these folds is highly
uniform and can be measured on the presented micrographs.

The increase of the degree of planar compression from 14 to 27%
does not result in any principal changes in the morphology of the
resultant microrelief (Fig. 1.12 a, b). Only the period of the fold
changes. As shown in [33], the number of the folds does not depend
on the degree of compression of the polymer-substrate. In the initial
moment of the loss of stability of the coating a specific number of
folds forms on the entire surface of the polymer, and with further
compression of the polymer substrate the number of these folds does
not increase. The compression of the resultant relief is similar the
compression of the bellows of an accordion.

Thus, in planar shrinkage of the plasticised polyvinyl chloride
the reduction of the surface area of the polymer results in the loss
of stability of the deposited coating which acquires the regular
microrelief. It is important to mention that this relief, covering the
entire surface of the polymer on which the coating is deposited,
is morphologically uniform. This result indicates the homogeneity
(affinity) of the stress fields, responsible for the shrinkage of the
polymer and, consequently, the relative homogeneity of the structure
of the polymer-substrate.

We now examine the results of investigation of samples of classic
elastomers prepared by the same procedure – cross-linked rubbers.
For this purpose, we examine the data obtained in the investigation
of natural rubber (NR) and synthetic isoprene rubber (SIR). Figure
1.13 shows the scanning electron micrographs of the specimens of
natural rubber prepared by the previously described procedure. It
may be seen that at relatively low values of the planar compression
a regular microrelief forms on the surface of the polymer as a result
of shrinkage. This microrelief is similar to that observed in the
polyvinyl chloride (compare Figs. 1.12 a, b and 1.13 a). However,
at high degrees of compression (Fig. 1.13 b) new morphological
features of the resultant microrelief appear. In addition to the zones
of disordered distribution of the folds, zones in which the folds
are distributed in the matched manner appear on the surface of the
polymer. The zones are 20–40 µm long and are separated by the
zones of the disordered distribution of the folds of approximately
the same size.

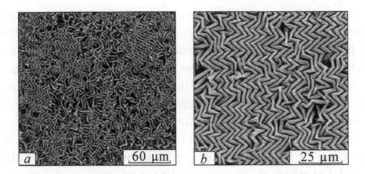

Fig. 1.13. Scanning electron micrographs of the specimens of natural rubber deformed in the conditions of planar tension by 5% (*a*) and 20% (*b*) and restoring their initial dimensions after deposition of a thin (10 nm) platinum coating on their surface.

The regular relief patterns formed in deformation (shrinkage) of the polymers with a thin coating deposited on the surface of the polymers are used for the visualisation of the stress fields responsible for these processes. It is assumed that, at least for rubbers, these fields are homogeneous (affine). This assumption is the basis of the statistical theory of high elasticity [8]. The homogeneity of the pattern of the microrelief, formed in compression of a coating on a compliant base, evidently indicates the homogeneous (affine) stress fields responsible for the planar shrinkage of the polymer. As shown previously, this pattern is detected in the planar shrinkage of the plasticised polyvinyl chloride. At the same time, when investigating the shrinkage of the cross-linked polymers, there are two types of morphological features of the regular microrelief: with a disordered and with a regular distribution of the folds (Fig. 1.13). This indicates that the stress field, responsible for the shrinkage of the polymers, is inhomogeneous and, in turn, this indicates the inhomogeneity of the spatial structure of the rubber. In addition to this, the applied procedure makes it possible to visualise and directly evaluate the dimensions of these structural inhomogeneities (20–50 μm). It should be mentioned that previously (see the preceding section) the differences in the patterns of surface structure formation between the plasticised polyvinyl chloride and the cross-linked rubbers were observed in the conditions of uniaxial tension (shrinkage) of these polymers.

It is important to examine the mechanism of the observed phenomenon or, more specifically, the reason for the differences in the patterns of the microrelief formed in the shrinkage of the plasticised polyvinyl chloride and chemically cross-linked rubbers.

In this connection, it is important to mention that the polyvinyl chloride plasticates with a high content of plasticizing agents are formed by linear unbranched macromolecules. The industrial samples of polyvinyl chloride have a low degree of crystallinity and this is explained by the small fraction of syndiotactic sequences in its chains. Nevertheless, the presence of small regular sequences of elemental links determines the possibility of local inter-chain crystallisation of the polymer and this appears to be sufficient for the formation in solutions of a spatial network of macromolecules, connected together by the crystalline nodes [34]. As a result of this circumstance, this polymer is similar to classic elastomers (reversibility of high deformation). The data presented in Fig. 1.9 indicates that this network has a spatial homogeneous structure.

The electron micrographs, shown in Fig. 1.13, confirm the literature data which can not be used to examine the classic elastomers (cross-linked natural and synthetic rubbers) in the form of topologically homogeneous networks. In particular, theoretical calculations carried out in [35] help to understand the mechanism of inhomogeneous swelling and the resultant inhomogeneous stress and strain distributions in cross-linked polymers. The microheterogeneous structure of cis-polyisoprene rubber has been confirmed by direct small-angle and wide-angle x-ray diffraction studies (SAXS and WAXS) [36]. The method of scanning electron microscopy was used to observe the heterogeneous structure of peroxide and sulphur vulcanizates of natural rubber [37]. According to [38], one of the reasons for the heterogeneity of the network may be the presence of heterogeneities in the system even prior to the vulcanisation stage. As shown in [39], the natural rubber is characterised by a non-uniform composition and contains sol and gel fractions. The gel fraction contains macromolecules with the molecular mass greater than 10^6, and also aggregates of linear macromolecules which are formed as a result of presence in them of polar oxygen-containing groups. The amount of such gel is not constant and depends on the prior history of the examined sample [39–41]. At the same time, in the studies [42, 43] it has been shown that the plasticization of natural rubber, which is carried out to improve its properties, is also accompanied by the formation of a gel fraction which evidently exists as rubber particles.

Thus, the plasticization process, which precedes chemical cross-linking (vulcanization) of rubbers results in the product with a non-uniform composition. Subsequent vulcanisation fixes this spatial inhomogeneity by chemical bonds.

The microscopic method used in this case can be used for visualisation of the structural special features of the deformation of rubbery polymers and is highly sensitive to the structural (spatial) heterogeneity of the deformed elastomer. This method not only makes it possible to determine the structural (topological) inhomogeneity of the rubber network but also evaluate the dimensions and localisation of structural inhomogeneities of this type which is very difficult or even impossible by other methods.

1.3. Visualisation of structural rearrangements taking place during annealing of amorphous polymers oriented above the glass transition temperature

It is well known that the annealing of the polymers oriented above T_g is accompanied by almost complete shrinkage. We will use the method of visualisation of structural rearrangement to examine the shrinkage of two polymers deformed above their T_g. After deformation, the polymers were cooled below their T_g, released from the clamps of the tensile loading device, thin layers of a metal were deposited on the surface, and this was followed by annealing above T_g (100°C) and by examining their surface in an electron microscope.

Figure 1.14 shows the data for polyvinyl chloride which indicate that the shrinkage of the polymer results in multiple fragmentation of the coatings and the formation of a regular microrelief. Fragmentation is associated with the lateral expansion of the samples during shrinkage of the oriented polymer. The regular microrelief is provided by uniaxial compression of the coating during its shrinkage. The results are completely identical with the pattern of surface structure formation detected in the shrinkage of the polymer heated to a

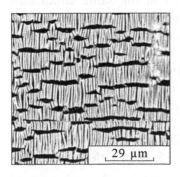

29 μm

Fig. 1.14. Scanning electron micrograph of a PVC sample tensile loaded by 100% at 90°C. After tensile loading the sample was cooled to room temperature with fixed dimensions, its surface was deposited with a thin layer of platinum and the sample was then annealed at 100°C.

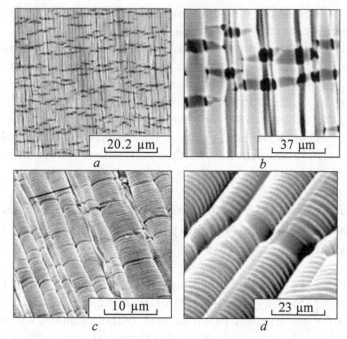

Fig. 1.15. Electron micrograph of a polyethylene terephthalate sample deformed by 100% at 90°C at a rate of 2 mm/min (*a*, *b*); *c*, *d*) the same with the sample tilted in relation to the axis of the electron beam. After tensile loading, the surface of the samples was covered with a thin (nm) platinum coating and they were annealed at 100°C. For explanation see the text [44].

temperature higher than its T_g. This is not surprising, because the shrinkage of the polymer, deformed in the rubbery state, should not differ depending on whether the specimen was cooled after deformation and then annealed or whether shrinkage takes place in the isothermal conditions. The mechanism of formation of this microrelief was discussed previously.

Completely different results were obtained in examining the polyethylene terephthalate samples, treated in the same conditions as the polyvinyl chloride samples. It is important to mention that the degree of shrinkage of the polyethylene terephthalate samples (Fig. 1.15) in annealing at 100°C was 95%, i.e., is similar to the shrinkage of the polyvinyl chloride. Regardless of this agreement, the results show that the given procedure reveals a number of morphological special features of the microrelief formed in the shrinkage of the polyethylene terephthalate samples with the coating. These features were not detected in the examination of the appropriate polyvinyl chloride samples [44]. Figure 1.15 shows clearly that the shrinkage of polyethylene terephthalate is accompanied by structural rearrangements which at first sight are qualitatively identical with those examined previously for the case of polyvinyl chloride (Fig. 1.14). All the investigated specimens are characterised by the formation of a regular microrelief and multiple fragmentation of the coating. As in the case of polyvinyl chloride (compare Figs. 1.14 and 1.15 *a*), the coatings contained a large number of short cracks oriented normal to the direction of the folds of the microrelief. Evidently, the procedure can be used to determine some structural rearrangements taking place during annealing of the polymer which are not recorded, in particular, by measuring its shrinkage.

The gradual magnification of the images of the microrelief of the polyethylene terephthalate specimens, subjected to shrinkage during annealing, made it possible to reveal the pattern of their very unusual structure formation (Fig. 1.15 *b*). Firstly, as mentioned previously, there is a large scatter in the values of the wavelength of the resultant microrelief so that the latter becomes less regular than the relief obtained in direct drawing of the polymer and in shrinkage of the polyvinyl chloride samples. Secondly, each fold of this relief is covered by a surprisingly regular microrelief with a considerably smaller period (0.27 µm). This regular microrelief covers the entire surface of irregular folds in the direction normal to their main axis. These folds are so regular that they resemble stacks of coins (Fig. 1.15 *b*).

This very attractive structure is clearly visible if the specimen is tilted in relation to the direction of the electron beam (Fig. 1.15 *c*). Under high magnification it may be seen (Fig. 1.15 *b*) that the disruptions of regularity which appear as cracks in a coating (Fig. 1.15 *a, b*) are fragments of large folds with no regular relief and have a smooth surface. It is necessary to determine the reasons for this phenomenon taking place.

To solve this problem, the dimensions of the annealed specimens during annealing were measured. The results show that regardless of the almost identical final dimensions of the specimens of both polymers obtained as a result of annealing, the paths through which they reach these dimensions greatly differ (Fig. 1.16). At the same time, whilst polyvinyl chloride shows the usual thermomechanical behaviour, characteristic of the oriented amorphous polymers (shrinkage during annealing), the shrinkage of polyethylene terephthalate is preceded by the self-elongation in the direction of the axis of preliminary tensile loading. To confirm this and

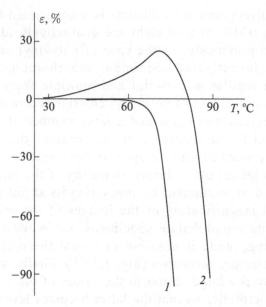

Fig. 1.16. Dependence of the relative change of the length of the specimens of polyvinyl chloride (1) and polyethylene terephthalate (2), deformed at temperatures above their T_g temperature (100°C), on annealing temperature [44].

characterise the process of the detected self-elongation, experiments were carried out with polyethylene terephthalate samples annealed at 65°C (approximately maximum self-elongation in Fig. 1.16).

The results of these studies are presented in Fig. 1.17. It is shown that the polyethylene terephthalate sample elongates spontaneously during annealing at 65°C. This is shown unambiguously by the pattern of surface structure formation. It may be clearly seen that the specimen not only elongates in the direction of the tensile loading axis, as indicated by the fragmentation of the coating, but also contracts in the normal direction, as indicated by the formation of a regular microrelief.

It is now possible to clarify the formation of the structure shown in Fig. 1.15. The point is that shrinkage takes place in the polymer whose structure is shown in Fig. 1.17. In this case, initial shrinkage takes place in the polymer not coated with the metal in the gaps between the coating fragments. Subsequently, further shrinkage of the polymer results in the compression of these fragments so that they 'swell' and the pattern shown in Fig. 1.15 is formed.

The phenomenon of self-elongation in annealing of the oriented polyethylene terephthalate has been several times detected elsewhere. In particular, it was shown that [45, 46] the amorphous polyethylene

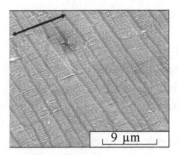

Fig. 1.17. Scanning electron micrograph of the oriented polyethylene terephthalate sample with the surface coated with a thin (11 nm) metallic coating and annealed at 65°C. The arrow indicates the direction of deformation [44].

terephthalate, stretched below T_g by the mechanism of classic crazing in the absorption-active liquid medium shows extensive self-elongation in annealing. In [47] investigations were carried out into the variation of the geometrical dimensions of the uniaxial oriented samples of polyethylene terephthalate. The samples were produced by co-extrusion through an Instron Rheometer at temperatures of 50–105°C to a draw ratio of 4.4. The deformation by extrusion made it possible to change smoothly the draw ratio of the polymer below its T_g. The results show that in annealing self-elongation takes place only in the case of the polymer deformed to 100%, whereas at other draw ratios only shrinkage takes place. There are several other studies concerned with this phenomenon [48–51].

Without discussing in detail the mechanism of this phenomenon, it is important to mention that in the previously discussed case the finite dimensions of the annealed samples of polyvinyl chloride and polyethylene terephthalate were almost identical, and the self-elongation was detected only because the given method of visualisation of structural rearrangement of the polymer during its annealing was used [44, 52].

1.4. Rolling of glassy polycarbonate

The constant volume is one of the main differences between the deformation of the rubbery polymers compared with amorphous polymers. While the deformation of rubber is homogeneous (affine) in the volume, which is one of the fundamental assumptions of the statistical theory of high elasticity [8], the deformation of glassy polymers is always inhomogeneous [33]. Macroscopically, this is manifested in the formation and growth of a neck and (or) crazes in uniaxial tension of polymers and in the development of shear bands under uniaxial compression [53]. The mechanism of deformation of amorphous glassy polymers under the rolling conditions has been studied far less extensively.

In the few studies of the structural and mechanical behaviour of glassy polymers subjected to rolling, it was established in particular that, as with other types of mechanical action, the deformed polymer completely restores its size during annealing above T_g. This allowed to use the approach described above for the visualization of structural rearrangement of the polymer deformed under these conditions. In [54], this kind of research was conducted on an example of a glassy polycarbonate (PC). PC samples were rolled at room temperature. After that, thin layers of a metal were deposited on the surface and annealed at different temperatures.

The methodology of visualization of structural rearrangements revealed that structural changes in the annealed PC start long before the polymer reaches T_g (glass transition temperature of PC is 145°C). Figure 1.18 shows a micrograph of a sample of PC subjected to rolling at room and annealed at a temperature of 110°C after depositing the metal coating on the surface. It is clearly seen that even at such a low temperature some large-scale molecular motion

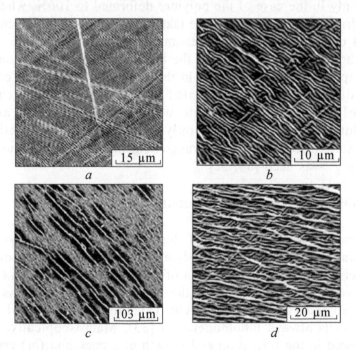

Fig. 1.18. Scanning electron micrograph images of PC samples rolled at room temperature, followed by coating their surface with a thin (10 nm) platinum coating. Annealing of the samples at 110 (*a*), 120 (*b*), 140 (*c*) and 155°C (*d*) [54].

takes place in the polymer, leading to the formation in the polymer of a large number of shear bands propagating rectilinearly across the surface of the polymer at a certain angle to the rolling axis (Fig. 1.18 *a*). In addition to shear bands, there was a regular wavy relief in the deposited coating, showing the shrinkage of the polymer in the direction of the rolling axis.

Increasing the annealing temperature to 120°C results in a further improvement of the fold relief oriented normal to the rolling axis (Fig. 1.18 *b*). The folds of this relief are relatively long and perfect, although there is some 'tortuosity'. Detailed microscopic examination can also identify new elements in the surface microrelief. It is clearly seen that at the annealing temperature of 120°C the surface of the PC shows a small number of folds, oriented normal to the axis of the 'primary' folds, i.e. along the rolling direction.

Increasing the annealing temperature to 140°C is accompanied by the 'delamination' of the general pattern of relief formation to clearly separated areas of asymmetric structures with different orientation (Fig. 1.18). Clearly visible are high folds oriented predominantly normal to the rolling direction (extended light bands in Fig. 1.18 *c*). These folds are surrounded by dark areas, the structure of which is discussed below. The rest of the polymer surface is covered with a folded structure with a smaller scale.

Finally, the annealing of the polycarbonate specimens 'rolled' at room temperature with the deposited coating at temperatures higher than its glass transition temperature (155°C) results in the formation on its surface of a uniform microrelief with two mutually perpendicular structures occupying the entire surface area of the polymer (Fig. 1.18 *d*). It should be mentioned that at this annealing temperature the initial dimensions of the polymer are completely restored.

All these differences in the morphological forms appear in the polycarbonate during annealing at temperatures below the glass transition temperature, i.e., within the 'limits' of the glassy state of the polymer, indicating the realisation of some large-scale types of molecular motion in this temperature range. The relaxation of these types of molecular motion during annealing results in the complex evolution of the internal stresses which is clearly detected using the given methods of preparation of the samples for microscopic studies.

It should be mentioned that the applied method makes it possible to detect two structural modes of the thermally stimulated relaxation of the rolled polycarbonate. Increasing annealing temperature results in the onset of strain relaxation, with the direction coinciding to that of the rolling axis. The relaxation of this deformation component starts approximately at a temperature of 100°C and appears as the formation of folds in the coating, directed normal to the rolling axis. When the annealing temperature of 130°C is reached, the second deformation component appears. The direction of this component is normal to the direction of the first component and to the rolling direction, and this part of the deformation appears in the form of folds oriented along the rolling axis. Further annealing is accompanied by the relaxation of both components, as indicated by the formation in the coating of two mutually perpendicular morphological forms of the microrelief. The results do not answer the question why rolling is accompanied only by two mutually perpendicular components of inelastic deformation, one of which coincides with the rolling direction. This result is not obvious because during the rolling the polymer film is free around its entire perimeter and there are no obstacles for its orientation in any direction.

1.5. Structural rearrangements in the deformed polymer in the conditions of isometric heating

To conclude this section, attention will be given to the investigation of structural rearrangement, taking place in the oriented polymer under isometric annealing. The isometric heating method has been used for a number of decades to obtain information on the processes taking place in the oriented polymers under temperature effects [55, 56]. The method makes it possible to record increase or reduction of the stress in the specimen of the oriented polymer with fixed dimensions. On the basis of these results it is possible to make assumptions regarding the physical reasons causing the changes of stress in the oriented polymers during annealing but so far no direct information on the processes taking place under these conditions is available. This is not surprising because the linear dimensions of the oriented polymer are fixed in isometric heating and it is difficult to imagine that some other processes, associated with internal mass transfer, will take place in these conditions.

Nevertheless, in [57] it was attempted to study these questions. In the experiments, a thin platinum coating (10 nm) was deposited on a polyethylene terephthalate sample, oriented at room temperature via necking. Subsequently, part of the neck was secured in a special frame and annealed for 1 h at 230°C.

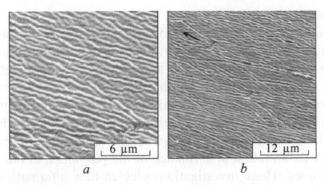

a   *b*

Fig. 1.19. Scanning electron micrographs of a specimen of oriented polyethylene terephthalate subjected to isothermal annealing at 230°C; explanation in the text [57].

Figure 1.19 *a* shows the electron micrograph of a polyethylene terephthalate sample, subjected to the previously mentioned preparation procedure. It may be seen that as a result of annealing the polymer acquires a characteristic folded relief. The majority of the folds have the uniform width (~0.5 µm) but in contrast to the patterns of surface structure formation, examined previously (Figure 1.16), not all the folds are parallel to each other. These folds are distributed along two directions positioned at a small angle (10–15°) in relation to each other.

At a low magnification (Fig. 1.19 *b*) it may be seen that the surface relief also contains other morphological elements. In addition to the previously mentioned folds, the surface of the polymer contains some straight lines extending over large distances and evidently intersecting the entire cross-section of the polymer. Both morphological elements, shown on the micrographs, cover the entire surface of the polymer. It is important to note that the microrelief, shown in Fig. 1.19, forms and is detected only if a metallic coating is deposited on the polymer surface. If isometric annealing of the oriented polymer is carried out without deposition of the metallic coating, the surface of the specimen remains smooth. It may be assumed that the resultant microrelief reflects structural rearrangement taking place in the polymer-substrate during its isometric annealing. Consequently, structural rearrangement, accompanied by internal mass transfer processes also takes place in the conditions of isometric heating of the oriented glassy polymer. The method of visualisation of structural rearrangement can be used to visualise this process.

Conclusions

Thus, the application of the microscopic method of visualisation of the structural rearrangement produces new information on the mechanism of deformation of the amorphous polymers. As shown previously, this method makes it possible to visualise the processes of mass transfer from the surface of the polymer to the volume and back, accompanying deformation of the polymers. The method was used for characterisation of the structural rearrangement, which accompanies deformation in shrinkage of the polymers, in the rubbery and glassy states. These investigations yielded new information on the self-elongation of the oriented polyethylene terephthalate during its annealing, on the structural rearrangement of the polymer, deformed in the rolling conditions, and also in annealing of the oriented glassy polymer in the isometric conditions etc. The method is universal and can be used for investigating both the amorphous polymers in the rubbery and glassy states and for studying semicrystalline polymers. On the basis of the analysis of the experimental results it may be concluded that high strains of the polymers are accompanied by the rapid development of the interfacial surface during deformation of the polymer and by interfacial healing during shrinkage. This stresses the important role of surface phenomena in these processes which previously was not mentioned.

References

1.   Lazurkin Yu.S., Dissertation, Dr. Sci. Sciences, Moscow, S.I. Vavilov Institute of Physical Problems, USSR Academy of Sciences, 1954.
2.   Tobol'skii A., Properties and structure of polymers, Ed. G.L. Slonimsky and G.M. Bartenev, Moscow, Khimiya, 1964.
3.   Gul' V.E., Kuleznev V.N., Structure and mechanical properties of polymers. Moscow, Khimiya,, 1972.
4.   Askadskii A.A., Deformation of polymers, Moscow, Khimiya,, 1973.
5.   Ferry J.D., Viscoelastic properties of polymers, 2 ed., New York, Wiley, 1970.
6.   Narisawa I., The strength of polymeric materials. Moscow, Khimiya, 1987.
7.   Bartenev G.M., Frenkel' S.Ya., Polymer physics, Leningrad, Khimiya, 1990.
8.   Treloar L., The physics of rubber elasticity, Oxford University Press, 1949.
9.   Volynskii A.L., Bakeev N.F., Vysokomolek. Soed. A, 2009, V. 51, No. 10, P. 1783.
10.   Kuleznev V.N., Shershnev V.A., Chemistry and physics of polymers, Moscow, Kolos 2007.
11.   Volynskii A.L., Yarysheva L.M., Bakeev N.F., Vysokomolek. Soed. A. 2011, V. 53, No. 10, P. 871-898.
12.   Volynskii A.L., Grokhovskaya T.E., Kulebyakina A.I., Bol'shakova A.V., Bakeev N.F., Vysokomolek. Soed. A, 2007, V. 49, No. 7, P. 1224-1238.
13.   Volynskii A.L., Grokhovskaya T.E., Kulebyakina A.I., Bol'shakova A.V., Bakeev N.F., Vysokomolek. Soed. A. 2009, V. 51, No. 4, P. 598-609.

14. Volynskii A.L., Grokhovskaya T.E., Kulebyakina A.I., Bol'shakova A.V., Bakeev N.F., Vysokomolek. Soed. A, 2007. V. 49, No. 11. P. 1946-1958.
15. Volynskii A.L., Priroda, 2011, No. 7, P. 14-21.
16. Volynskii A.L., Nechaev V.N., Kechekyan A., Bazhenov S.L., Bakeev N.F., Vysokomolek. Soed. B. 2001, V. 43, No. 12. P. 2211.
17. Volynskii A.L., Bakeev N.F., Izv. RAN, Ser. Khim., 2005, No. 1, P. 1-16.
18. Volynskii A.L., Bazhenov S.L., Bakeev N.F., Ros. Khim. Zh., 1998, V. 42, No. 3, P. 57.
19. Volynskii A.L., Chernov I.V., Lebedeva O.V., Ozerin A.N., Bakeev N.F., Vysokomolek. Soed. A, 1997, V. 39, No. 6, P. 1080.
20. Volynskii A.L., Bazhenov S.L., Lebedeva O.V., Yaminsky I.V., Ozerin A.N., Bakeev N.F., Vysokomolek. Soed. A, 1997, V. 39, No. 11, P. 1805.
21. Volynskii A.L., Bazhenov S.L., Lebedeva O.V., Ozerin A.N., Bakeev N.F., Vysokomolek. Soed. A, 1997, V. 39, No. 11, P. 1827.
22. Volynskii A.L., Chernov I.V., Bakeev N.F., Dokl. RAN, 1997, V. 355, No. 4, P. 491.
23. Bazhenov S.L., Chernov I.V., Volynskii A.L., Bakeev N.F., Dokl. RAN 1997, V. 356, No. 2, P. 199.
24. Volynskii A.L., Voronina E.E., Lebedeva O.V., Bazhenov S.L., Ozerin A.N., Bakeev N.F., Dokl. RAN, 1998, V. 360, No. 2, P. 205-208.
25. Volynskii A.L., Grokhovskaya T.E., Bol'shakova A.V., Kulebyakina A.I., Bakeev N.F., Vysokomolek. Soed. A. 2004, V. 46, No. 8, P. 1332.
26. Godowsky Yu.K., Crystallization, Encyclopedia of polymers. Moscow, Sov. entsiklopediya, 1972, V. 1, P. 1178.
27. Volynskii A.L., Grokhovskaya T.E., Bol'shakova A.V., Kulebyakina A.I., Bakeev N.F., Vysokomolek. Soed. A, 2008, V. 48, No. 12, P. 2144.
28. Bowden N., Brittain S., Evans A.G., Hutchinson J.W., Whitesides G.M., Nature, 1998. V. 393, No. 6681, P. 146.
29. Huck W.T.S., Bowden N., Onck P., Pardoen T., Hutchinson J.W., Whitesides G.M. Langmuir, 2000, V. 16, No. 7, P. 3497.
30. Volynskii A.L., Grokhovskaya T.E., Sembayeva A.D., Bazhenov S.L., Bakeev N.F., Dokl. RAN, 1998, V. 363, No. 4. P. 500.
31. Volynskii A.L., Grokhovskaya T.E., Sembayeva R.H., Yaminsky I.V., Bazhenov S.L., Bakeev N.F., Vysokomolek. Soed., 2001, V. 43, No. 2, P. 239.
32. Volynskii A.L., Grokhovskaya T.E., Sembayeva A.D., Bazhenov S.L., Bakeev N.F., Vysokomolek. Soed. A, 2001, V. 43, No. 6, P. 1008.
33. Volynskii A.L., Bakeev N.F., Structural self-organization of amorphous polymers, Moscow, Fizmatlit, 2005.
34. Guzeev V.V. Plasticates, Encyclopedia of polymers, Moscow, Sov. entsiklopediya, 1974, V. 2. P. 609.
35. Erman B. J., Polym. Sci. Polym. Phys. Ed. 1983, V. 21, No. 6, P. 893.
36. Lipatov Y.S., Mullers A.B., Privalko V., Sinder H., Angev. Makromol. Chem., 1979, V. 82, P. 79.
37. Ono K., Kato A., Int. Rubber Conf. Kyoto, Oct. 15-18, 1985, P. 1, P. 295.
38. Dusek K., Prins W. Adv. Polymer Sci., 1969, V. 6, No. 1, P. 1.
39. Dogadkin B.A., Dontsov A.A., Shershnev V.A., Chemistry of elastomers, Moscow, Khimiya, 1981.
40. Medalia A.J., J. Polym. Sci., 1951, V. 6, No. 4, P. 423.
41. Allen P.W., Bristow G.M., J. Appl. Polym. Sci., 1961, V. 5, No. 17, P. 510.
42. Baramboim N.K., Mechanochemistry of high molecular compounds. Moscow, Khimiya, 1982.

43. Reztsova E.V., Chubarova T., Vysokomolek. Soed., 1965, V. 7, No. 8, P. 1335.
44. Volynskii A.L., Grokhovskaya T.E., Bakeev N.F., Dokl. RAN 2005, Vol. 400, No. 4, P. 1.
45. Volynskii A.L., Grokhovskaya T.E., Gerasimov V.I., Bakeev N.F., Vysokomolek. Soed. A, 1976, V. 28, No. 1, P. 201.
46. Volynskii A.L., Aleskerov A.G., Grokhovskaya T.E., Bakeev N.F., Vysokomolek. Soed. A, 1976, V. 28, No. 9, P. 2114.
47. Pereira J.R.C., Porter R.S., Polymer, 1984, V. 25, No. 6, P. 877.
48. Oswald H.J., Turi E.A., Harget P.J., Khanna Y.P., J. Macromol. Sci. Phys., 1977, V. B13, No. 2, P. 231.
49. Liska E., Kolloid-Z. Z. Polym., 1973, V. 251, P. 1028.
50. Lim J.Y., Kim S.Y., J. Polym. Sci. Polym. Phys., 2001, V. 39, No. 9, P. 964.
51. Bowden P.B., Raha S., Phil. Mag., 1970, V. 22, P. 463.
52. Volynskii A.L., Grokhovskaya T.E., Bakeev N.F., Vysokomolek. Soed. A, 2005, V. 47, No. 3, P. 540.
53. Bowden P.B., Phil. Mag., 1970, V. 22, P. 455.
54. Volynskii A.L., Grokhovskaya T.E., Kulebyakina A.I., Bol'shakova A.V., Bakeev N.F., Vysokomolek. Soed. A, 2007, V. 49, No. 11, P. 1946.
55. Laius L.A., Kuvshinskii E.V., Vysokomolek. Soed., 1964, V. 6, No. 1, P. 52.
56. Shoshina V.I., Nikanovich G.V., Tashpulatov Yu.T., Isometric method studies of polymer materials. Tashkent, FAN. Uzbek. SSR, 1989.
57. Volynskii A.L., Grokhovskaya T.E., Bolshakova A.V., Kechekyan A.S., Bakeev N.F., Physical chemistry of polymers. Synthesis, structure and properties, Issue 12. Collection of scientific works, Tver', 2006, P. 66

Healing of the interfacial surface in polymer systems

Thus, any effect on the polymer results in a change of its interfacial surface area. This change takes place by the diffusion of the material from the bulk to the surface (the development of the interfacial surface) or vice versa, from the surface into the bulk (healing of the interfacial surface). Although the process of the development of the interfacial surface upon polymer deformation has not been studied, the reverse process – healing of the interfacial surface in the polymers – has been the subject of various investigations. We will investigate in greater detail the most general relationships governing the process of healing the surface in the polymer systems.

2.1. Healing of the interfacial surfaces in rubbery polymers

When two flat surfaces of the same polymer come into contact above the glass transition temperature of the polymer (T_g), the interface between them usually gradually disappears and the material becomes monolithic. This phenomenon was described for the first time in the fundamental monograph by S.S. Voyutskii, who suggested the following definition: 'The autohesion or self-adhesion is the ability of two surfaces of the same substance, brought into contact, to form a strong bond preventing their separation in the contact area' [1]. The terms 'autohesion' and 'self-adhesion' have a somewhat restricting nature because they relate to the interaction of two flat surfaces of the same polymer. In this review, attention will be given to the heterophase polymer contacts and also changes in the area of the interfacial surface of the polymer as a result of the effect of mechanical stress and hydrostatic pressure. Similar processes of the spontaneous reduction of the area of the interfacial surface are sometimes referred to as 'coagulation' – adhesion and 'coalescence' –merger. When examining the processes of formation of block glassy polymers from their powders, it is rational to use the term 'monolithization'. To denote all these processes, we shall use mostly the more general term 'healing' which is usually employed for describing any processes, associated with the change (reduction) of the free interfacial surface area.

In the general form, the variation of the interfacial surface area of the polymer is associated with the diffusion of macromolecules through the area of contact of the identical polymer surfaces. To support this assumption, S.S. Voyutskii stated that '... All factors, increasing the rate of thermal motion, support 'autohesion' and vice versa, the factors reducing the diffusion rate, reduce the 'autohesion'. In fact, the process is identical with the displacement of two volumes of the same fluid. The uniqueness of the polymer system in this case is represented by the special features of the molecular motion of the polymer chains in this displacement.

These factors have been observed most extensively and adequately in the so-called reptation model of molecular motion [2]. The concept of the diffusing separate chain in the gel ensures the simple solution of complex multidimensional problems and, primarily, the dynamics of polymer melts. Assuming that each chain exists in a tube, formed by restrictions of the bonds, the complex multi-chain dynamic problem was efficiently solved for the separate chain as the one-dimensional diffusion problem. The solution is described by the formula $L^2 = 2D_1 t$, where L is the curvilinear diffusion length of the chain along the axis of its tube, $D_1 \sim 1/M$ is the one-dimensional diffusion coefficient. The reptation time depends on the molecular mass as $\tau_d \sim M^3$. During this period, the mass centre travels the distance R_g (the radius of gyration of the macromolecule). Since $R_g^2 = 2D\tau_d$, the diffusion coefficient D behaves as $D \sim M^{-2}$. The resultant scaling law for the dynamics of the melts can be efficiently used in the terms of the developed model [2] and is used widely for a large set of experimental data, characterising the behaviour and properties of the polymer fluids and solids. On the basis of the good agreement between the theory and experiment it may be assumed that the approach, proposed by DeGennes [2] for describing the diffusion processes in the solutions and melts of polymers adequately describes the actual physical processes.

The molecular interpretation of the process of healing is associated with the phenomenon of self-diffusion in the block polymers, but these phenomena are not identical. In self-diffusion, polymer coils move over distances many times greater than their average diameter. The healing process is completed in the time required for the displacement of the coils by the value equalling half their size on the path through the interface. This process results in the complete monolithization of the interface and in restoration of the integrity of the polymer. The healing time may be compared with the relaxation time of the polymer chain configuration. In the theory of thermal motion, described by the reptation model [2], the reptation time is the time required for a macromolecule to travel the distance whilst the macromolecule still 'remembers' its tube.

It is well known [3] that the polymers with a temperature higher than T_g are characterised by different types of molecular motion, determined by the irreversible displacement of the macromolecules. It is therefore necessary to determine which kinetic units are responsible for the process of healing the interfacial surface of the polymer. Important results on this subject were obtained in [4] in which two dispersions of latex particles based on copolymers PMMA and PIB were obtained by an independent method. One of them contained 0.37 moles of cross-linked naphthalene links (donor) producing the N-marked (dispersion 1) latex, and the other one – 0.37 moles of cross-linked pyrene links (acceptor) producing the P-marked (dispersion 2) latex of the same polymer. Thin films (2–3 µm) were produced by mixing the dispersions 1 and 2 of these latexes in heptane, followed by drying on the surface of a silicon sheet. Subsequently, the fluorescence method was used to examine the interaction of 1 and 2 in the produced film, taking place during annealing in the temperature range 110–220°C. The process of monolithization (healing of the interfaces) in the produced film was studied in parallel on the basis of the transparency of the film. These experiments determined the healing time at different temperatures. At the same time, the fluorescence method was used to determine the energy transfer between the naphthalene and pyrene labels in the molecules of the PMMA and PIB copolymers in the same temperature and time ranges. The experimental results were used to determine the activation energy of two processes: healing of the interphase boundaries and self-diffusion. The results show that the activation energy of healing is 41.1 kJ/mol, whereas that of the self-diffusion is 125.4 kJ/mol. Previously [5], the fluorescence method was used for the process of monolithization of the particles of copolymers based on PMMA and PIB to obtain a slightly higher value of activation energy (154.7 kJ/mol) in annealing PMMA above T_g. The activation energy for the coagulation of the latex particles based on PBMA was 158.8 kJ/mol [6]. As assumed a long time ago by Ferry [7], these values of the activation energy characterised the movement of the main chain of the polymer. Since the activation energy of healing is three times smaller than that of self-diffusion, this result shows that the process of healing of the interfacial boundary in the polymer with a temperature higher than T_g is sustained by the molecular motion on a considerably smaller scale.

In [8] the photon transmission method was used for examining the evolution of transparency (healing) of films of latexes of high- and low-molecular mass samples of PMMA. The films were produced in annealing of the latexes at temperatures higher than T_g for different periods of time. The experimental results show that the intensity of the penetrating photons from these films increases with increase of the annealing time. The curves of the time dependence of the intensity of the penetrating photos show two different stages of healing. Firstly, it is the stage associated with closing of the walls of the microcavities and, second, the process of self-diffusion through the interfacial boundary. The activation energies of the viscous flow (ΔH) and movement of the main chain (ΔE_b) were determined: ΔH = 150 and 134 kJ/mol, and ΔE_b = 142–199 and 59–98 kJ/mol in annealing of the low- and high-molecular mass films, respectively. The determination of the conditions in which the healing process was completed enabled the principle of temperature–time superposition to be used for the evaluation of the activation energy of healing as 188 and 117 kJ/mol for the low- and high-molecular mass specimens.

In [9] the process of healing of the surfaces was investigated by measuring the variation of the strength between two polystyrene (PS) films brought into contact at a temperature 20°C higher than the T_g temperature of the films. This method was used for the first time for examining the processes of autohesion in [10]. The results show that the time of development of strength follows the law $\sim t^{1/4}$ which is in agreement with the molecular dynamics of the DeGennes reptation model [2]. The activation energy E = 401 kJ/mol is close to the value E = 389.5 kJ/mol, predicted by the Williams–Landel–Ferry equation using C_1 = 13.7 and C_2 = 50 at T_g = 100°C. The duration of complete healing at 118°C was 256 min, which corresponds by the order of magnitude to the viscoelastic relaxation time required for the diffusion of the polymer chain over the distance equal to the RMS distance between the ends of the chain.

The phenomena of the same type were detected when examining the healing of the interfacial boundaries of two thermodynamically compatible polymers. For example, in [11] investigations were carried out into the processes taking place in contact of plates of two compatible polymers cis-1,4-PI and 1,2-PB. The results show that the adhesion between them increases with time and after 130 min reaches the values corresponding to the autohesion of pure PI and then remains constant. It has been assumed that the increase of adhesion and compatibility in the mixtures of PB and PI are not determined by the interaction parameter which is very small for the two polymer hydrocarbons. The authors assume that the driving force of the mutual diffusion of the components in this case is the gain in the combinatorial entropy [11].

As indicated by the experimental results, the process of healing of the interfacial surfaces is determined by the high intensity of molecular motion characteristic of the rubbery state of the polymer at the phase boundary. The change of the intensity of molecular motion has the strongest effect on the process of healing the interphase boundary. For example, in [12] the samples of the styrene copolymer with acrylonitrile were subjected to γ-radiation with a dose of 0–800 mrad. This resulted in the molecular cross-linking of the volume which depended on the radiation dose. The experimental results show that the presence of cross-links greatly increases the healing time in comparison with the initial polymer to values above the T_g of the initial styrene–acrylonitrile copolymer. It is assumed that these effects are associated with the suppression of molecular mobility in the cross-linked specimens in comparison with the linear polymer.

On the basis of these results it may be concluded that the healing of the interfacial surfaces in rubbery polymers is determined by the diffusion transfer through the interface which is directly associated with their molecular mobility. These processes follow the well-known laws of diffusion and self-diffusion of macromolecules and (or) their fragments

2.2. Healing of the interfacial surface in glassy polymers

Thus, the healing of the interfacial surfaces in the rubbery polymers is controlled by diffusion processes, characteristic of the polymer chains and (or) their fragments. In fact, this mechanism of mass transfer does not differ at all from the appropriate processes taking place in the melts and concentrated solutions of the polymers.

The question is: can the processes of healing of the interfacial surfaces take place in the polymers which are in the glassy state where, it would appear, the large-scale molecular motion is frozen?

The answer to this question was obtained in a series of investigations carried out recently by Yu. Boiko et al [13–19]. In these studies, the processes of healing of the interfacial surfaces were investigated by the experimental approach used in the above book for healing the interfaces in the rubbery polymers [19]. The process of healing was evaluated on the basis of the increase of the strength of the bond formed in superposition of two fragments of the polymer film on each other, followed by holding under a low compressive stress. Subsequently, the samples were extracted from the compression device and the shear strength of the contact was measured.

For example, in [13] attention was given to the bonding of PS/ PS (M_w = 2.25·10^5) surfaces, brought into contact, and the fracture energy G was measured as a function of healing time t_h and healing temperature T_h. The results show that G increases in proportion to $t_h^{1/2}$ at $T_h = T_g - 33°$ and log G is proportional to $1/T_h$ at $T_g - 43° > T_h < T_g - 23°$. The lowest measured value $G = 1.4$ J/m^2 was at least an order of magnitude greater than the work of adhesion of the PS surfaces. Similar values $G = 8–9$ J/m^2 were obtained for the surfaces of polydisperse PS and monodisperse PS with a different molecular mass. This fact indicates that the process of healing of the interfacial surfaces in the PS does not depend on its polydispersion. These observations indicate that the increase of G on the PS–PS surfaces in the investigated temperature range (below T_g) is controlled by the diffusion of segments of chains, although the large-scale molecular mobility should be suppressed in these conditions.

The effect of molecular mass on the processes of healing of the surface of the PS was studied in detail in [14]. The strength of the resultant contact below T_g in the time range from 10 min to 24 hours was measured for the conventional and ultrahigh molecular mass PS (1.1 · 10^6). The strength of the contact for the ultrahigh molecular mass PS was always higher than for conventional PS. However, the difference at healing temperatures lower than $T_h = T_g - 30°C$ was very small. The investigations of the kinetics of increase of strength

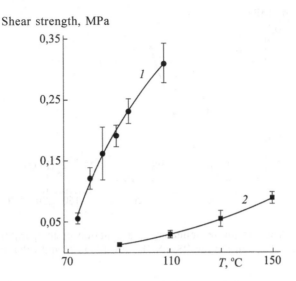

Fig. 2.1. Dependence of the shear strength on the healing temperature for amorphous – amorphous (1) and crystalline–crystalline (2) contacts of the PET films. Healing time 30 min [15].

of two PS phases, brought into contact at T_h 33 and 23°C lower than T_g, led the authors to the conclusion [14] that the healing of the interfacial boundary takes place by diffusion of the segments of the chains, as is the case at temperatures higher than T_g.

The healing process is characteristic of the interfacial surfaces of any glassy polymer. For example, in [16] the healing of the interfacial surface was investigated for two samples of amorphous PET and two samples of crystalline PET of different molecular mass at temperatures higher and lower than the T_g of the block polymer. The results show that the strength at the interface of amorphous PET increases without any sudden jump at T_g and the development of strength at the amorphous/amorphous and amorphous/crystalline PET interfaces takes place in accordance with a power law with the parameter equal to 1/4. This indicates that the diffusion-controlled process takes place in both cases. At the same time, the rate of increase of the strength at the boundary of the crystalline surfaces of the PET was at least one order of magnitude lower in the same conditions than the rate of increase of the strength of the amorphous surfaces (Fig. 2.1). The results show that the process of healing is determined by the mutual diffusion of the macromolecules through the interphase boundary with the formation of a network of load-bearing bonds. It is natural that this process is less pronounced for

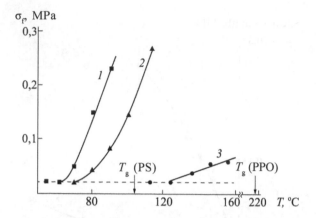

Fig. 2.2. Strength at the phase boundary as a function of temperature of contact at a contact time of 2 min for the PS–PS (1), PS–PPO (2) and PPO–PPO (3) [17].

the semicrystalline PET in which part of the chains at the interface is included in the crystal lattice and cannot take part in the diffusion process.

Similar processes were also detected if compatible polymer pairs were brought into contact. For example, in [16–20] investigations were carried out into the increase of strength when two thermo-dynamically compatible polymers were brought into contact – polystyrene (PS) and polyphenylene oxide (PPO).

For this purpose, the surfaces of the films of these polymers was joined and held for different periods of time in contact at different temperatures, but always lower than T_g. The results show that as a result of this contact the strength increases in proportion to $t^{1/4}$ both between the contacting surfaces of the same nature (PS–PS and PPO–PPO) and in the formation of the heterophase contact (PS–PPO), i.e., the process is controlled by the diffusion molecular motion of the reptation type (Fig. 2.2). The results show unambiguously the high segmental mobility on the polymer surfaces in the temperature range considerably lower than bulk T_g.

Thus, it has been shown convincingly that the surfaces of the glassy polymers brought into contact are characterised by the processes of healing of the interfacial surface. The relationships governing these processes are controlled by the processes of diffusion and do not differ at all from those observed in the polymers with a temperature higher than T_g. At first sight, this is unusual because it is assumed that below T_g the large-scale molecular motion in the polymers is 'frozen'.

However, recent studies solve this contradiction quite easily. Of obvious interest are the investigations directed at examining the special features of molecular movement in the thin films and the surface layers of the amorphous polymers which were started in the middle of the 1990s. This concerns the films and polymers layers with a thickness of tens to hundreds of nanometres. The development of a number of new investigation methods, which determine reliably the nature of molecular movement in such thin layers, has resulted in discovering a large number of surprising effects. The special features of molecular movement in the surface layers of the glassy polymers will be discussed in greater detail in the following chapter and here we only mention that these layers actually have the properties greatly different from the properties of the bulk polymers.

2.3. Heterophase healing of polymer interfaces

The previously discussed data indicate that regardless of the physical state of the polymer diffusion processes take place at interfacial surfaces of the polymer leading to efficient healing of these surfaces. It is therefore important to examine the question whether similar phenomena can also take place at heterophase contacts between polymers, when polymers are incompatible. The formation of molecular contact between incompatible polymers, if this process takes place, consists of a variety of the processes of healing of the free interfacial surfaces of the polymers discussed in the previous sections.

The experimental approaches, developed in the above-mentioned studies, make it possible to characterise the processes of this type of healing. In [21] the films of high-molecular mass PS ($T_g = 97°C$) and PMMA ($T_g = 109°C$) were brought into contact in the temperature range between 44 and 114°C for 1 to 24 hours and subsequently the specimens were tensile tested at room temperature. The development of strength at the boundary of the incompatible polymers PS–PMMA took place by the mechanism identical with the mechanism of the appropriate process at the PS–PS and PMMA –PMMA compatible surfaces. The strength of the PS–PMMA and PMMA–PMMA contacts was 2–3 times lower than for the PS–PS pair under the same conditions. Analysis of the values of strength on all three surfaces, obtained within 24 hours at 44°C, i.e., below bulk transition temperature T_g of both polymers, was used to conclude that, firstly, the formation of contact at the interface leads the formation of some interfacial layer in which the macromolecules of the incompatible polymers (intermediate phase) are entangled, and, secondly, the incompatibility at the level of the segments of PS and PMMA has only a slight effect on the formation of this intermediate phase.

The problem of the formation and properties of the intermediate phase of the incompatible polymers is of obvious applied importance because it includes the interfacial tension, wetting, dewetting, adhesion, etc. Naturally, over many years, the phenomena, taking place on the polymer surfaces, have been the subject of special attention of experts in different areas of the chemistry and physics of polymers.

The key concept, forming the basis of the interaction of the incompatible polymers is the formation of the interfacial layer, which has been discussed previously. This process was investigated in the greatest detail in the monograph by V.N. Kuleznev [22]. The author presented convincing proofs regarding the existence of the interfacial layer at the boundary between the incompatible polymers and assumed that this interfacial layer forms at the level of the compatibility of the individual segments and not macromolecules as a whole. 'The thicknesses (of the interfacial layer), greater than the root-mean-square distance between the ends of the macromolecules, are doubtful because to obtain these thicknesses there should be diffusion of the whole macromolecules and not of the parts of the segments forming them. This is an indication of the extensive mutual solubility of the components' [22]. However, the length (thickness) of the interfacial layers is difficult to determine so that the properties of such an intermediate phase have not been studied sufficiently up to now. The most interesting results have been obtained in the examination of the properties of the intermediate phase in block copolymers, as a result of the mutual high dispersion of the components forming them. In [23], the authors presented data for the temperature dependence of the mechanical losses of a butadiene–styrene block copolymer. The results show that in addition to the T_g of both components, the intermediate temperature range is characterised by the formation of a small peak attributed to the interfacial layer.

It is therefore important to analyse the results of investigations carried out in recent years which would make it possible to explain the structure and properties of the interfacial layer in the incompatible polymer pairs. In particular, it is important to examine the data for the composition and length of the interfacial layer. Interfacial mixing is caused by the entropy gain for the chains diffusing through the interphase boundary which is compensated by the repulsion between the incompatible segments of the chains. The results also show that when two polymers are brought into contact, the boundary between them is not strictly sharp, because some intermediate layer forms at this boundary. In the literature, importance is attached to the intermediate phase for polymer adhesion and compatibility of the polymer mixes and melts [24]. The theory of the self-consistent mean field, improved by Helfand et al, is the basis of the quantitative relationships between the composition of the intermediate phase and the thermodynamic interaction parameter χ [25–28]. In these studies, the following equation was also derived for the thickness d_1 of the interfacial layer:

$$d_1 = 2b/(6\chi)^{1/2} \tag{2.1}$$

where b is the length of the statistical chain segment.

It is rational to regard the intermediate phase of the incompatible polymer mixtures as a third phase with its own characteristic properties [29]. Experimental studies of the properties of the intermediate phase are associated with considerable difficulties because the length of the phase is very small so that the volume fraction of the intermediate phase in the mixture of the incompatible polymers is also very small. Even the determination of the dimensions of the phase causes problems. The currently available estimates indicate that the size of the phase may reach several hundreds of nanometres, depending on the interaction parameter [26–28]. Other physical properties, such as density, free volume and permeability of the intermediate phase, have not as yet been determined.

A breakthrough in the investigation of the properties of the interfacial layers of the incompatible polymers came recently with the detected phenomenon of the so-called force assembling of the multilayer polymer systems. The process of layer multiplication is based on the utilisation of the viscoelastic properties of the polymer melts during coextrusion of two polymer flows leading to a gradual increase of the number of the layers of the incompatible polymers in the produced film. Equipment used for this assembling is shown schematically in Fig. 2.3. Equipment, consisting of N consecutive dies, produces the number of layers in the film equal to $2^{(N+1)}$. The typical number of layers is expressed is several hundreds and the thickness of the layer is of the order of 1 μm.

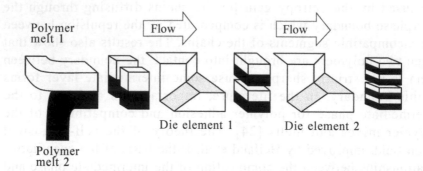

Fig. 2.3. Diagram of the device for coextrusion of the melts for force assembling of the nanolayered films on the basis of incompatible polymer pairs. The figure shows the initial stage of the process of formation of the films, containing from 2 to 8 layers [30].

The produced multilayer systems based on the incompatible polymer pairs represent a rich source of experimental studies of the polymer mixtures [31–36]. The large interfacial surface area of the multilayer films is used for studying interphase phenomena, associated with the compatibility of the polymers [37–39]. Strict conditions of the microlayer coextrusion ensure the rare possibility of joining incompatible polymers on the nanolevel with a very small degree of molecular mixing or completely without it.

Recent successes in multilayer coextrusion made it possible to reduce the thickness of the layers by two orders of magnitude – from micro- to nanolevel [40]. Nanolayered films, consisting of thousands of continuous layers of two incompatible polymers with a thickness of each layer less than 10 nm have been produced. This means that the thickness of the individual layer becomes comparable with the thickness of the intermediate phase so that it is possible to use the conventional methods of analysis of the polymers for studying the size-dependent properties of the nanolayers.

In the initial stage, the effect of the thickness of the layers on the general properties of the systems, produced from two incompatible polymers PC and PMMA, was described [40]. The conventional DSC thermograms shows thats as the layer thickness decreases, two glass transition temperature of the polymers come closer to each other; when the thickness of the layers reached 10 nm or less there is only a single T_g. This result indicates that as soon as the thickness of the layer becomes comparable with the dimensions of the intermediate phases, the layers loose their individuality and the film becomes

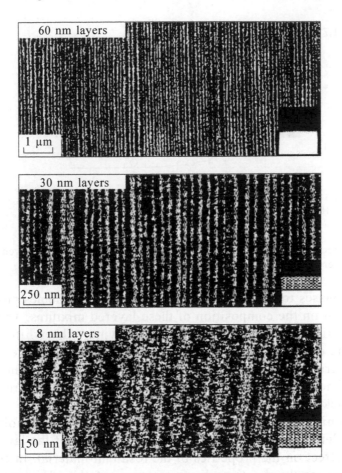

Fig. 2.4. Atomic force microscope images of the cross-section of multilayer films based on PETH–PS, with different numbers of layers [30].

in fact a homogeneous intermediate phase i.e., a thin single-phase nanomixture of incompatible polymers.

In [30], the properties of the layer systems based on the amorphous copolyester polyethylene terephthalate-co-1,4-cyclohexane dimethylene terephthalate (PETH) with the styrene–acrylonitrile copolymers of different composition were evaluated by measuring the permeability of oxygen. The layered nature of the resultant co-extruded products was determined directly by atomic force microscopy (Fig. 2.4). It can clearly be seen that as soon as the thickness of the layers becomes equal to 10 nm, the initially sharp boundaries between them become blurred and the system transforms to a nanoscale mixture of two incompatible polymers with a homogeneous structure. The bright phases belong to PETH,

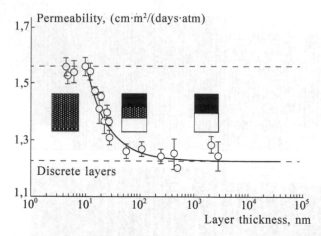

Fig. 2.5. Effect of the thickness of the layers on the permeability of oxygen through the multilayer film based on PS–styrene–acrylonitrile copolymer [41].

the dark phases to PS. The same figure shows the schemes providing information on the composition of these layered mixtures. The black colour indicates the phase of the PS polymer, the white colour PETH, the crosshatched region provides information on the amount of the intermediate phase in the multilayer film. The amount of the intermediate phases in the system rapidly increases with a reduction of the thickness of the layers.

This unique variation of the composition of the nanolayered systems, determined by the change of the thickness of the layers, has a strong effect on the properties of the resultant material. In addition to the previously mentioned convergence of the glass transition temperature of the components, the permeability of the gases through such films also greatly changes. The gas transport, like the glass transition temperature, is a probe which can be interpreted at the molecular level by the concept of the free volume [41, 42]. In contrast to the measurements of T_g which require heating the glassy film to transition to the rubbery state, gas transport is measured at a constant temperature and the only condition is that the compound parts differ in gas permeability. If the layers are relatively thick, the permeability is determined primarily by the component with the lowest permeability. The situation changes if the thickness of the layers decreases and reaches the size of the intermediate phase. For example, in the case of the layer system based on PC–PMMA, the oxygen permeability greatly increased when the thickness of the layer was decreased from 100 to 10 nm [41].

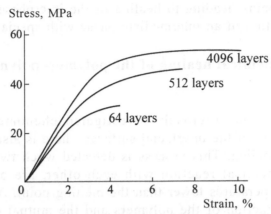

Fig. 2.6. Strain–stress curves of multilayer specimens based on PS–PMMA polymer system. Tests were carried out at room temperature with a tensile loading rate of 2 mm/min [43].

Figure 2.5 shows the dependence of the oxygen permeability through the multilayer films based on PETH and the styrene–acrylonitrile polymer on the thickness of the layers. Starting at a layer thickness of ~100 nm the permeability of oxygen rapidly increases, and at a thickness of ~10 nm and less the material becomes fully the intermediate phase so that its permeability no longer depends on the thickness of the layers. The appropriate structural special features of the multilayer films are presented in the form of reccoils in which the black colour denotes the phase of the styrene–acrylonitrile polymer, the white colour PETH and the intermediate colour indicates the intermediate phase.

The application of the multilayer films made of the incompatible polymers makes it possible in particular to evaluate the mechanical properties of the intermediate phase. For example, in [43] investigations were carried out into the mechanical properties of the PMMA–PS multilayer films. Typical results of this investigation are presented in Fig. 2.6. It may be seen that with increasing number of the layers or, which is the same, with the reduction of the thickness of each layer the elastic modulus, strength and elongation at break of the multilayer compositions increase. It is important to note that both polymers are brittle at room temperature because their elongation at break does not exceed 2–3%. In may be seen that the reduction of the thickness of the coexisting phases is accompanied by some synergism in the mechanical properties of the relatively brittle polymers.

Thus, the contact of even incompatible polymer pairs, as in many previously examined cases, is characterised by the mutual penetration of polymer chains, leading to healing of the interphase boundaries and the formation of an intermediate phase with special properties.

2.4. Heterochemical healing of the polymer–polymer interfaces

In addition to the previously investigated phenomenon of the physical healing of the interfacial surfaces, there is also some kind of chemical healing. This process is detected when two polymers, capable of chemical reaction with each other, are brought into contact at temperatures lower than the melting point. Although the chemical interaction of the polymers and the mutual diffusion of their macromolecules are completely different processes, the final result of both types of healing is the same – disappearance of the interface between two polymer films and the increase of mechanical strength at the polymer–polymer interface.

However, the following question arises in each case: which of these processes is dominant for a given polymer under the given conditions? If we try to separate the two types of healing, then in this case it is necessary to minimise the mobility of the macromolecules, i.e., suppress the diffusion processes. In this case, it may be expected that healing will take place by the chemical mechanism. For this purpose, it is possible to cross-link, for example polyamides (PA-11, PA-12, PA-6,6) by the chemical process in their amorphous regions. Subsequent healing should take place exclusively by the chemical process because the mutual diffusion, which is possible only in the amorphous regions of the polymer, is suppressed as a result of cross-linking.

A process of this type was demonstrated for the first time in [44] using the example of healing both single-phase surfaces (PET–PET, PA-66–PA-66) and heterophase contacts (PET–PA-6 and PET–PA-66). Depending on the conditions, it was shown that the healing, determined by the occurrence of the chemical reactions at the phase boundary, takes place in all cases. This also means that at the temperatures lower than the melting point the chemical composition of the macromolecules at the phase boundary does not remain constant as a result of the chemical reaction of the exchange type. It was found unexpectedly that the rate of the chemical reactions on the single-phase surfaces was even lower than in the case of heterochemical contact. Heterochemical healing of the interfaces in the polymers indicates the occurrence of chemical diffusion and mass transfer as a result of the solid phase chemical reaction at the phase boundary.

The chemical mechanism of healing the interphase surface was applied in [45] in which rigid-chain polymers were studied (Kapton-H and Upilex-R). These polymers are characterised by restricted molecular motion and by a low susceptibility to diffusion processes. In the low-temperature range (below the glass transition temperature) the process is determined mainly by the diffusion of macromolecules through the phase boundary resulting in a relatively low strength of the healed surfaces. The abrupt change in the strength on the temperature dependence of shear fracture stress of the contacts was found at high temperatures. It is associated with the increase of the activation energy of the healing process and is explained by the chemical reaction on the contacting planes.

In [46] experiments were carried out to verify the contributions to the interfacial healing by the conventional diffusion mutual penetration of the macromolecules and the heterophase chemical reaction. Films of linear and cross-linked polymers PA-11, PA-12 and PA-6,6 and of the same polymers with limited mobility of the chains in the surface layer determined by crystallisation and chemical cross-linking, were brought into contact. The contacting films were annealed at different temperatures and this was followed by determining the shear strength of the resultant bonds. The results show that in the same annealing conditions, more efficient healing is found in the cross-linked samples so that in this aspect the phenomenon greatly differs from the processes of healing of the interfacial surfaces, which take place by mutual diffusion of the polymer chains and were considered in the previous sections. The results also show that the main contribution to the increase of the strength of the interfacial contact comes from the solid phase chemical reaction at the interface.

The chemical reactions at the phase boundary form the basis of a technological method – the so-called reaction matching, directed at reducing the interfacial tension and improving the adhesion between the domains in the polymer mixtures [47]. For this purpose, the polymers with reactive groups, are brought into contact. The reaction of this type at the phase boundary between the PS with the OH end groups, and the statistical copolymer PMMA with methacryloyl chloride groups in the main chain, was studied in [48] with the results indicating the high efficiency of the reaction.

2.5. Monolithization of powders

A special case of healing the interfacial surfaces is observed in the monolithization of polymer powders. In this case it is necessary to apply hydrostatic pressure to produce tight contact between the polymer particles. In other words, here we cannot discuss the spontaneous healing process because a certain amount of work must be carried out on the system prior to the start of self-diffusion of the polymer chains. Main special features of this process will be discussed.

Amorphous polymers

The processes of monolithization of the latexes are used widely in processing to produce components and, consequently, they have been clarified in detail in the literature (see, for example, [49]) and in this book this question will not be discussed in detail.

In this section, attention will be given to the set of the physical-chemical phenomena taking place during monolithization of the powders based on glassy polymers. To produce a true monolith from the powder of the glassy polymer, it is necessary, as in all other cases, to create suitable conditions for the self-diffusion of the macromolecules through the interface. In the case of powders, this can be achieved only if the following conditions are satisfied: it is necessary to produce an efficient contact between the particles and ensure high molecular mobility of the macromolecules at the boundary between the particles. The first condition is satisfied by applying hydrostatic pressure to the initial powder. This is accompanied not only by the simple compression of polymer particles but also by their plastic deformation. It is well-known [50–56] that plastic deformation of the glassy polymers is a relatively complicated phenomenon, and one of the features of this phenomenon is the existence of two contributions and its thermally stimulated relaxation (shrinkage). If the self-diffusion at the interface between the powder particles has not taken place to a sufficiently large depth, subsequent thermally stimulated relaxation of each particle will be accompanied by the shape recovery, i.e., the powder-like product. In turn, this relaxation may take place during annealing in different temperature ranges [50–56] and they should be accompanied by the dispersion of the monolith, produced by the external pressure, into individual particles.

Another factor, influencing the processes of monolithization of the powders is the well-known dependence of T_g of the polymer on hydrostatic pressure [57]. Since the T_g temperature of the polymer is directly connected with its large-scale molecular mobility, this has the strongest effect on the process of self-diffusion of the macromolecules and, consequently, the processes of healing of the interphase boundaries as a whole.

The first systematic investigation in this area was carried out by S.A. Arzhakov [58, 59]. In the studies, the process of monolithization of the glassy polymers (PMMA, PVC, PAN, PS, etc) was investigated in a wide temperature range and a wide pressure range during deformation in a closed pressing mould. The healing of the interphase boundaries was evaluated visually and, consequently, the specimens became transparent under the effect of temperature and force.

The experimental results of these studies showed that the process of monolithization of the polymer powders is a complicated phenomenon. Figure 2.7 shows the diagram in the pressure–temperature coordinates which can be used to characterise the range of phenomena taking place during the monolithization of PMMA powders. The region III of the p–T diagram is characterised by complete true monolithization of the PMMA powder, and this product does not differ at all from the polymer, produced by bulk polymerisation. In the region IV there is also the formation of monolithic transparent samples but only when they are heated to T_g of the polymer (~400 K). The latter lose their transparency as a result of the recovery of the interphase boundaries between the powder particles. The formation of the transparent specimens of PMMA

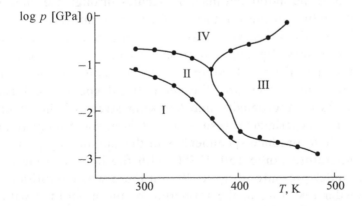

Fig. 2.7. Diagram of production of transparent samples of PMMA, explanation in the text [59].

can also take place in region II. In cooling under the pressure in the pressing moulds, the samples remain transparent but heating to room temperature results in the loss of transparency. Finally, in the range I the monolithization of the PMMA powder does not take place and the powder regains the shape and dimensions of the particles without any change.

The authors assume that the region I is characterised by the elastic Hookean deformation of the polymer particles which relaxes quite rapidly after stress relaxation. The region III corresponds to the conditions in which the thermal–force effect results in the diffusion of macromolecules on the whole through the interface leading to the formation of a true monolith. The region II is characterised by the increase of the density of the polymer powder as a result of the plastic deformation of the individual particles of the powder. This process is accompanied by a large increase of the internal stresses which relax in the range of low temperatures (below T_g temperature of the polymer). The interfacial boundaries are restored in this case. The region IV does not differ from the region III but since in this case deformation of the polymer takes place at a higher temperature, the relaxation of the polymer also starts only in the glass transition range in which the large scale molecular motion is observed. According to the authors of [58, 59] in the regions II and IV the mutual penetration (healing of the interphase boundaries) takes place at the level of the segments of macromolecules and, consequently, the stored internal stresses restore the dispersion of the produced pseudo-monoliths, whereas the region III corresponds to the conditions in which the effect of temperature and force results in the diffusion of the individual macromolecules through the interfacial boundaries leading to the formation of a true monolith.

In [60] investigations were carried out into the monolithization of the PS powders. Special attention was given to the effect of the rate and magnitude of compression, the compaction time of the specimen under pressure, the geometrical special features of the specimens on the density and breaking stress of the resultant monoliths. The experimental results show that at room temperature it is not possible to produce specimens with the satisfactory properties. In the temperature range 150–173°C with the holding time of the specimen under pressure of up to 30 min there was monolithization of the powder resulting in the formation of the monoliths with the optimum properties. The authors noted the effect of the dispersion of the initial powders on the properties of the monoliths and also the effect of the thickness of the samples on the strength properties of the final product. The latter effect is associated with the effect of the gas in the interparticle space of the powder which after stress relaxation may lead to the irreversible restoration of the porous structure.

In [61] it was shown that the monolithization of the polycarbonate (PC) powders cannot take place at temperatures below the T_g temperature of the polymer, but the increase of the pressure to 50 kPa greatly improves the properties of compacts and suppresses efficiently the relaxation of the polymer after removing the load.

Another factor, influencing the quality of the produced monoliths in the monolithization of the powders of the glassy polymers, is the process of their physical ageing. For example, in [62] it was shown that the difference in the degree of physical ageing in the processes of drying dispersions and long-term storage of the PC powders has the strongest effect on their capacity for monolithization under pressure at room temperature.

The process of monolithization on the polymer powders is very important in the formation of components of polymers with a relatively low temperature of chemical dissociation, for example polyester ester ketone (PEEK) – semicrystalline thermoplastic with $T_g = 143°C$ and $T_m = 334°C$. Upon melt extrusion the required temperature is of the order of 370–400°C at which the PEEK starts to show the processes of branching and cross-linking of the macromolecules. In this connection, it is sufficient to attempt to produce monolithic components from PEEK by monolithization of its powders. In [63] the PEEK powders were consolidated by compression at room temperature followed by annealing of the produced compact. The authors noted the effect of the particle size and temperature on the properties of the produced monoliths.

Monolithization of reactor powders of ultra-high-molecular-weight polyethylene (PE)

Recently, the attention of the researchers has been concentrated on the ultra-high-molecular-weight polyethylene. Processing of this compound by the gel technology resulted in the extremely high strain–stress parameters close to the theoretical values for the individual macromolecules [64]. However, the obvious shortcoming of gel technology is the application of a large number of toxic solvents creating serious ecological problems. The alternative method is the direct processing of the reactor powders of the ultra-high-molecular-weight polyethylene without using the melts and solutions [65, 66]. At the present time, the values of the elastic modulus and tensile strength of the fibres, formed by this method, reach 220 and 6 GPa, respectively [67], which is fully comparable with the values of the fibres of the ultra-high-molecular-weight polyethylene, produced by gel technology.

Naturally, the controlling effect on the formation of the fibres by the direct method from the reactor powders is exerted by their monolithization before orientational drawing [68–72]. In the general form, the process of production of the high-strong fibres from the reactor powder of the ultra-high-molecular-weight polyethylene consists of the following stages: the powder is placed in a closed pressing mould and compressive stress is applied. In the first stage of compaction, the powder particles come into contact with each other and, consequently, the bulk density of the powder increases. When the number of contacts becomes sufficiently high, plastic deformation of the particles starts to take place. This is the only stage of the process in which monolithization of the reactor powder takes place. A further increase of pressure is accompanied by the elastic deformation of the compact and does not provide any contribution to the properties of the product. Naturally, the strongest effect on the process of monolithization of the reactor powder is exerted by temperature. In subsequent stages, the monolith, produced by this method, is stretched in order to produce oriented fibres with the optimum stress–strain properties.

The experimental results show that the density and strength of the produced monoliths are influenced strongly by the applied pressure and the compaction time of the reactor powder under load, even at room temperature. The process of monolithization can be divided into two stages. The initial stage consists of the formation of adhesion bonds between the individual particles of the reactor powder and this is followed by the inter-diffusion of the macromolecules through the grain boundaries (process of healing of the interfacial surfaces). Naturally, both these processes are localized in the amorphous regions of the ultra-high-molecular-weight polyethylene.

The properties of the high-strength filaments, produced from the reactor powders, are strongly influenced by the specific surface of the initial product. As the dispersion of the initial powder increases, it becomes easier to process the powder into a high strength, high-modulus product. This effect is associated with the possible formation of a large number entanglements in the amorphous regions of the fine particles of the ultra-high-molecular-weight polyethylene. The entanglements already form in the stage of synthesis of ultra-high-molecular-weight polyethylene and prevent its subsequent orientational drawing.

2.6. Healing of interfaces produced upon fracture of glassy polymers

The application of mechanical load to glassy and crystalline polymers, with the magnitude of the load exceeding the tensile strength of the polymers, results in the failure of the polymers which is accompanied by the formation of interphase boundaries on the newly formed interfacial surfaces. In the case of polymers, such interphase boundaries formed by cracks are susceptible to healing during thermal treatment, and also under the effect of solvents.

Attention will now be given to the special features of healing of cracks in glassy polymers. In [73] the molecular diffusion of chains, taking place during heat treatment of the cracks above T_g, was detected for the first time. This process was described in greater detail in [74] in which the fractured parts of a PMMA sample were bonded under a press at a low pressure and annealed at a temperature higher than T_g. The experimental results show that in the cyclic mechanical testing of the specimens at room temperature the strength of the joint increases with time at a rate which depends on the annealing temperature. After 1–20 hours (depending on the annealing temperature), the interphase boundaries, visible in the microscope, disappear. The surfaces of repeated failure of the healed specimens contain the plastically deformed material. Analysis of the experimental data [57] has made it possible to characterise the process of mutual penetration of the macromolecules through the interphase boundaries, responsible for the formation of a strong bond. It was also shown for the PMMA and the styrene–acrylonitrile copolymer SAN that the mutual penetration of the macromolecules takes place at a distance of ~110 Å and the diffusion coefficient of $D = 32.5 \cdot 10^{-20}$ m²/s. It has been assumed that the healing mechanism is associated with the mutual diffusion of the chains and the formation of the entanglement network.

In [76] extensive investigations were carried out into the processes of healing the cracks in the PMMA and the styrene–acrylonitrile polymer. The fracture surfaces were bonded under a pressure of 1 bar and annealed at a temperature 1–15° higher than the glass transition temperature. This was followed by the determination of the bonding strength in these samples. The results were analysed in terms of fracture mechanics. Analysis of the experimental data shows that after annealing above T_g physical entanglements of macromolecules form on the fracture surfaces as a result of their mutual diffusion in the polymer. The entanglements are capable of stress transferthrough the interfaces and can induce marked plastic deformation of the bonded interface layer during repeated deformation.

Since the processes of healing of cracks are determined by the processes of self-diffusion at the interfaces, the strongest effect on this process is exerted by the molecular mobility of the polymer chains and the neighbouring surfaces. In particular, this is confirmed by the fact that, as mentioned previously, the healing of the cracks is most efficient at temperatures higher than T_g.

In [77], the molecular mobility of the styrene–acrylonitrile copolymer was changed by γ-irradiation of the copolymer with the dose of 0–800 mrad on notched specimens. The presence of cross-links greatly increases the time required for complete healing. Examination by fracture mechanics shows that in the case of the non-irradiated samples, the factor of the energy release rate G_1 increases in proportion to the square root of the healing time t_h until the initial value G_0 of the initial material is reached. For the cross-linked specimens, this value increases to some limit and then ceases to change (reaches a plateau). Further restoration of the healing process is detected after the first plateau in the period from several minutes to several hours. The two-stage mechanism of healing of the styrene–acrylonitrile copolymer of this type [78] is characterised by the fact that in both stages of the process at a temperature 10°C higher than the T_g, the healing time is proportional to $t^{1/4}$, although the healing rate was considerably lower than for the pure PC.

It is well-known [79] that the molecular mobility of the polymer may be strongly affected by the low-molecular solvent. Therefore, a number of investigations were carried out to investigate the effect of the low-molecular liquids on the process of healing of cracks formed in glassy polymers. In contrast to thermal healing, examined previously, healing under the effect of the solvent may take place at temperatures below T_g. The solvent penetrates into the polymer at relatively low temperatures, causes healing and is then removed from the final product. This circumstance is of the obvious applied importance.

In [80] investigations were carried out into the healing of cracks in PMMA in treatment with methanol at temperatures lower than the glass transition temperature (40–60°C). This treatment was followed by repeated mechanical tests of the specimens subjected to the healing treatment. According to the results of the mechanical tests, there are two healing stages. The first stage corresponds to the healing caused by wetting which ensures the constant rate of closure of the crack walls at a given temperature. The second stage corresponds to easy diffusion of the macromolecules through the interface. The data obtained in the examination of the surface morphology of the fracture surfaces of the healed regions confirm the presence of these two stages. The experimental results also show that the tensile strength of PMMA after treatment with methanol may be restored to the level of the initial material. It is assumed that healing can take place only at temperatures higher than T_g which is reached in the investigated temperature range as a result of plastification of PMMA by methanol. It is important to note that when using a fluid with a low activity (ethanol) [81] there is no complete healing of the cracks in the PMMA in the given temperature range.

The special features of healing of the cracks in the glassy polymers under the effect of solvents are determine to a large extent by the nature of their penetration into the polymer. There are three types of such transport. Case 1 – Ficks's [82]. Case 2 – the transport, characterised by the co-existence of the swollen layer and the dry core. The formation of stresses at the polymer–swollen layer is of great importance in this case [83]. Finally, the anomalous case representing the combination of the cases 1 and 2. The anomalous behaviour was also extensively studied and equations were derived to characterise this type of process of the solvent transfer in the polymers. The equations are in good agreement with the experimental data [84, 85].

In [86] investigations were carried out into the healing of cracks in PC at 40–60°C in CCl_4 which is able to depress the T_g of the polymer. Healing takes place due to the fact that the effective T_g temperature of the polymer decreases below the test temperature. According to the results of repeated mechanical tests and fractographic studies, as in [80], two stages of healing were detected. The first stage corresponds to the healing as a result of diffusion of the solvent and swelling of the polymer. Consequently, the rate of closure of the crack walls is constant. The first stage is immediately followed by the second stage corresponding to self-diffusion of the polymer chains through the interface. The transport of CCl_4 takes place by diffusion as in the cases 1 (controlled by the concentration gradient) and 2 (controlled by relaxation of the polymer as a result of the buildup of internal stresses at the boundary between the dry polymer and the swollen polymer shell), and these processes are directed in the opposite directions. The point is that with increasing temperature the solubility of CCl_4 in the polymer decreases (the system has the upper critical solution temperature) and characteristics such as the rate of mass transfer, the rate of closure of the crack walls, the speed of movement of the diffusion front – increase. The transport of CCl_4 in the polymer changes the mechanical behaviour of PC from ductile to brittle.

In investigating the healing of cracks in the PMMA, swollen by the mechanism 2 in methanol, the stresses, induced in the compression of the internal glassy core, were made visible [87]. The elastic stresses were analysed by the finite element method. Initially, attention was given to the growth of the crack from the tip of an artificial notch in methanol and the effect of constant load at 20°C with the evaluation of the internal compressive stresses in the vicinity of the interface between the swollen layer and the glassy core. The relatively long cracks, induced in a thinner swollen surface layer, completely disappeared in methanol at 40°C for 3 min in the absence of external load. However, complete healing of the crack was not detected in methanol at temperatures below 20°C nor in the completely swollen sample even at 40°C. These results can be used to assume that healing of the cracks requires both the thermal effect at temperatures above T_g and high internal compressive stresses.

In [88] investigations were carried out into the healing of cracks in the PMMA by treatment with a mixture of more active (methanol) and less active (ethanol) solvent. The transport of this mixture in the PMMA was anomalous and this corresponded to the mixed mechanism of the cases 1 and 2. The healing, including the stages of wetting and interdiffusion, was investigated by analysis of the strength data in repeated loading of the samples, and also by investigating the morphology of the resultant fracture surfaces. The results show that the rate of closure of the crack walls is constant with time in the wetting stage. The mechanical strength of the healed specimen increases, with other conditions being equal, with the reduction of the volume fraction of the ethanol in the mixture with methanol. The fractographic data confirm the reversibility of the mechanical strength in treatment with the mixture of these solvents.

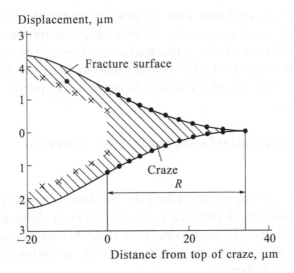

Fig. 2.8. Schematic representation of the tip of a growing crack in a glassy polymer [90].

One of the most important factors, which determine the mechanism of healing of the cracks, formed in the glassy polymers, is their surface structure. It is well-known (see, for example, [89]) that in fracture of the glassy polymers the growing crack has important structural special features.

Figure 2.8 shows the pattern of structural rearrangement taking place during the growth of a fracture crack in a glassy polymer (PMMA), theoretically determined in [90] on the basis of direct microscopic data [91]. It can be clearly seen that in the process of the growth of the true crack the area ahead of the tip of the crack is characterised by the development of the zone of the plastically deformed, loosened polymer (crazing); healing takes place in the surfaces coated with the layer of the fractured craze material. Finally, this loosened material is far more suitable for healing by mutual diffusion through the interface.

It is important to note that such an interfacial surface contains a large number of the so-called weak polymer chains which form as a result of the fracture of the macromolecules on the fracture surface of the polymer. As shown in [92], the resultant weak chains play a significant role in the process of crack healing. The molecular special features of the chains of the fracture surface greatly differ from those for the case in which the two non-fractured surfaces are brought into contact for interdiffusion.

Thus, it may be concluded that the healing of the fracture cracks in the glassy polymers, as in many other cases, is determined by the processes of diffusion of the macromolecules through the interface followed by the formation of an entanglement network. However, the structure of the fracture surfaces brings in its special features into this process.

2.7. Healing of the interfacial surface in deformed polymers

In the previous section, we described the method of visualisation of structural rearrangement (surface mass transfer) in deformation of the polymers [93–97]. In particular, this stage makes it possible to visualise the process of healing of the surface in shrinkage of the deformed rubbery polymers.

Figure 2.9 shows the atomic force microscopy data, which can be used for the visualisation of the process of planar shrinkage of the polymer (PET) which takes place in the rubbery polymer. It is important to note that when the size of the interfacial surface is reduced, the deposited metallic coating acquires the regular and effective microrelief. The mechanism of formation of such a microrelief will be examined in greater detail below and here we only mention that this relief uniformly covers the entire surface of the polymer, which again confirms the assumption of the homogeneity (affinity) of high elastic deformation [98].

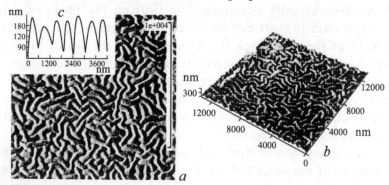

Fig. 2.9. Atomic force image (*a*), its three-dimensional reconstruction (*b*) and the appropriate profile pattern (*c*) of the surface of a PET sample, deformed in plane by 19% at 100°C (at temperatures higher than the glass transition temperature of the polymer). After tensile loading the sample was cooled to room temperature with fixed dimensions and a thin (10 nm) platinum coating was deposited on its surface. Subsequently, the sample was annealed in the free state at 105°C so that the initial dimensions were completely restored [96].

Important information on the molecular structural rearrangement, taking place in the surface layers of the deformed polymer, is provided by the method of x-ray photoelectronic spectroscopy which makes it possible to investigate quantitatively the surface chemical composition and its evolution in the process of deformation of the polymer so that the processes of transport of the polymer from its surface into the volume and back can be recorded. This investigation should be carried out on polymers with small additions of a second polymer component, with a different chemical composition.

For example, in [99, 100] investigations were carried out into the mixtures of the polydimethyl siloxane (PDMS) (molecular mass MM = 2500 and 62500) and non-cross-linked polychloroprene (PCP) (MM = 200 000 and the degree of crystallinity 30%). The films were stretched to ~700–800%. After stress relaxation these samples showed a low strain recovery which disappeared after annealing at 35–40°C. The experimental results show that mechanical deformation (uniaxial tension) may have the strongest effect on the surface composition of the multicomponent polymer. In the initial film the surface contains mainly PDMS with a lower surface energy. The tensile loading of the film, accompanied by the increase of the area of the interfacial surface, resulted in the enrichment of its surface with the PCP component. The mechanism of surface heating of the siloxane component depends on the MM and the initial surface concentration of the siloxane. In the case of the oligomeric siloxane (MM = 2500) surface splitting (PCP component) starts at a certain draw ratio, which depends on the initial surface concentration of the siloxane. When the concentration increases, the surface splitting of the siloxane starts at higher values of the draw ratio. A special feature of these mixtures is the irreversibility of their surface composition with the surface concentration of the siloxane, stretched to the given draw ratio. For the higher molecular PDMS (MM = 62500) its splitting on the surface is far less distinctive and is fully restored (healed) in shrinkage.

It is important to mention that the complete restoration of the dimensions of the specimen in annealing completely restores its surface chemical composition only in the case of high molecular PDMS, whereas complete restoration of the dimensions of the specimen, containing low molecular PDMS, does not result in the complete restoration of its chemical composition.

In the mixtures of PCP with PDMS-polysulphone bulk copolymer, the effect of stretching on the surface composition is regulated by the length of the copolymer blocks. In the case of the 2500/3500 block-polymer the surface layer follows stretching of the elastic substrate, whereas in the mixtures of the more rigid copolymer (2500/44000) the deformation of the surface layer, consisting of the block copolymer, greatly lags behind the PCP substrate. The surface regions (2500/44000) of the copolymer show extensive surface depletion in the siloxane component with the draw ratio, whereas for the mixture with 2500/3500 block polymer tensile loading has almost no effect on the distribution of the copolymer component in the subsurface region.

Investigation of the mixture of the PCP with the surface active addition PDMS/polysulfone (PSF) block polymer shows [100] that the initial specimens was characterised by a high content of the addition in the surface. The concentration of the additive at a distance of 50 Å from the surface was considerably lower than at 25 Å. The experimental results also show that with increase of the draw ratio (up to 700%) the content of the PSF groups decreases all the time whilst that of PCP increases. The reverse process starts during shrinkage (healing). However, this was observed only for the copolymer 2500/44000 (with a higher content of the rigid block), whereas for the 2500/3500 copolymer (with a higher content of the flexible block) this relationship did not change at tall. This takes place in addition to the fact that after unloading the specimen completely restores its dimensions. Thus, the reversibility of the process of tensile loading–shrinkage from the viewpoint of the surface composition is regulated by the ratio of the rigid and flexible blocks. As the amount of the rigid block increases, the reversibility of the restoration of the initial composition becomes less marked.

Thus, the deformation of the rubbery polymer is accompanied by the change of its surface area, and the reverse process (shrinkage) results in a decrease of the size of its surface area. In turn, these processes are determined by the diffusion of polymer molecules from the volume to the surface in deformation and spontaneous process of diffusion from the surface into the volume (healing) in shrinkage.

Further, we examine in detail the processes of healing of the interfacial surfaces in the deformed glassy polymers, and in this section we pay attention briefly to the appropriate processes in the crazed polymers. Crazing of polymers is one of the fundamental types of inelastic deformation of the polymers. Its special feature is the development of a high level of the interfacial surface. Naturally, these systems are characterised by the spontaneous processes of healing of the interfacial surface. There are two types of crazing: 'dry' crazing and solvent crazing (environmental crazing). The latter type will be studied in a separate section of this book (chapter 9), and here we examine briefly some of the special features of the processes of healing of the interfacial surface, developed during 'dry' crazing.

The crazing phenomenon of the polymers was detected a relatively long time ago [101] because it often accompanies the shear yielding and failure of glassy and crystalline polymers. As a result of the distinctive nanoporous structure of the crazes they are quite easy to record and examine in microscopy experiments of different type. The high susceptibility of the crazed polymer to the spontaneous healing of its nanoporous structure in the temperature range below the glass transition temperature was also observed a relatively long time ago. For example, in [102] it was reported that if the crazed polymer is stored for a long period of time, it gradually becomes monolithic, i.e. not only closure (convergence of the walls) of the crazes but also complete healing of its porous structure takes place. Here we again face the presence of large-scale molecular motion responsible for the healing of the interfaces in the temperature range below the T_g of the bulk polymer.

The possibility of the large-scale molecular motion in the crazed polymers below the glass transition temperature of the bulk PC has also been confirmed by the data in [103]. In this study, the method of measuring the thermally stimulated depolarization currents was used to examine the molecular mobility in the crazed and bulk specimens of PC and brominated PC. The experimental results show that in the highly crazed specimens the dielectric signal is much stronger at temperatures almost 100° below the glass transition temperature. In low-temperature annealing (10–20°) below the glass transition temperature, the low-temperature relaxation becomes far more difficult, indicating the healing of the high dispersion material of the crazes (coalescence of the interfaces) and the restoration of the block structure of the polymer. Evidently, this effect is associated with the higher molecular mobility in the highly developed surface of the crazes in the temperature range of the glassy state of the block PC. Although no direct experimental studies have been carried out as yet, which would be concerned with the special features of molecular mobility in the thin surface layers of the polymers, the authors of [103] explained the result showing that the molecular mobility of the polymer molecules in the crazes is higher by the fact that they are close to the free surface.

Finally, it is important to mention the data obtained in [104]. In this study, crazes were initiated in thin PS films, the dimensions of the crazes were recorded and the distance between fibrils and the diameter of the fibrils were measured by electron diffraction directly in the video chamber of a transmission electron microscope. The results show that the distance between the fibrils increases from 25 to 350 nm over a period of 250 hours at room temperature. The diameter of the fibrils increases in the same fashion. These changes are accompanied by the coalescence of the fibrils, and after 750 hours fibrils were no longer detected. It is clear that this phenomenon requires the high mobility of the molecular chains. The estimates show that at a fibril diameter of 10 nm each fibril contains no more than 10 chains. From this it follows that all the molecules in the fibrils of the crazes are distributed in the surface layer.

Important data for the process of healing of the crazes were obtained in [105]. In this study, the formation of crazes in the PS was initiated at different temperatures, and measurements were taken of the time to the formation of the first craze and its growth rate using the light microscopy data. This was followed by recording the stress and by relaxation of the polymer, containing the crazes, for different periods of time at different temperatures. The same stress was then again applied and the initiation time and the rate of growth of the same crazes were measured. With increasing relaxation time at the given temperature the duration of secondary initiation increased and the rate of repeated growth decreased and new crazes could form in the sample. The processes of healing was expressed in particular by the fact that the craze, selected for the investigation, disappeared visually during the healing cycle. The experimental results were used to construct the temperature–time diagram indicating (determining) the time required at the given temperature for complete healing. For example, according to these data at a temperature of 70°C (T_g = 100°C) the total closure time (healing time) is longer than 10^5 s. The experimental results indicate the high molecular mobility in the fibrils of the crazes in the temperature range not exceeding the glass transition temperature of the polymer. Important results were obtained for the secondary initiation of crazing after, it would appear, complete healing in an optical microscope.

Figure 2.10 shows the dependence obtained in [106] of the initiation strain of the crazes in the PS on the deformation temperature

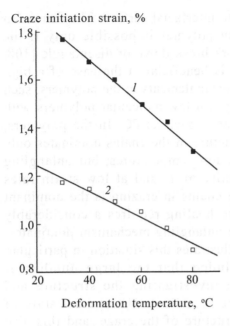

Craze initiation strain, %

Deformation temperature, °C

Fig. 2.10. Dependence of the initiation strain of the crazes in the PS on temperature: 1) prior to annealing; 2) after annealing for 10 minutes at 130°C [106].

in the initial (1) polymer and in the same polymer after annealing for 10 min at 130°C resulting in the complete disappearance of the images of crazing in the light microscope (2). It may be seen clearly that the repeated initiation of the crazing takes place at considerably lower values of strain, regardless of the temperature of tensile loading of the PS. The results show unambiguously that the restoration of the optical transparency of the polymer does not yet indicate its complete healing which should be accompanied by the complete restoration of its properties. This phenomenon is associated with 'geometrical necessity' of the loss of entanglements in crazing. The glassy polymer is a set of entangled macromolecules bonded by the system of physical nodes. Deformation is accompanied by the change of the conformation chains between the sides of this physical lattice and by the displacement of its sites in relation to each other without any excessive failure of the sites. The existence of these stable entanglements explains, in particular, the phenomenon of the natural draw ratio of the polymer which the polymer acquires in the process of formation and development of a neck [107]. Upon annealing of the oriented glassy polymer above T_g, the entropy forces result in shrinkage of the chain and the entanglements regain their equilibrium positions.

In addition to the orientation of the macromolecules, the crazing is accompanied by the development of a nanoporous structure, i.e., the development of a high level of the interfacial surface. The formation of the free surface in the glassy polymer is possible only if the chains in the entanglement network break down or disentangle [108, 109]. The fracture of the chains is beneficial in the case of a high molecular weight, low density of entanglements in the polymers, such as PS, and disentanglement [109] – in low molecular polymers with a high density of the entanglements, such as PC. In the polymers, such as low molecular PS, the fracture of the chains dominates only at high strain rate and relatively low temperatures, but untangling may also take place in the vicinity of T_g and at low strain rates [108, 109]. If the fracture of the chains in crazing is the dominant process, the process of complete healing requires a considerably longer annealing time than if the untangling mechanism dominates. The data presented in Fig. 2.10 illustrates this situation in particular.

Therefore, it may be concluded that the large number of experimental data, obtained in investigating the structure and properties of the crazed polymers, indicate the special state of the polymer, dispersed in the structure of the craze, and this also determines the susceptibility of the crazes to the processes of healing the interfacial surface, in particular in the temperature range below the T_g temperature of the bulk polymer.

Conclusion

The literature data on the processes of healing of the interfacial surface in the polymers have been analysed. Special features of the healing of the interfaces of the surfaces of the rubbery and polymer films brought into contact, and also healing in heterophase polymer contacts have been discussed. The phenomena of heterochemical healing of the polymers, capable of chemical reaction with each other, have been examined. The processes of monolithization of the polymer powders have been studied. Special attention has been given to the analysis of the phenomena taking place during the healing of the new interfaces, formed during the deformation of the polymers in different physical states. Special attention has been given to the processes of healing the shear bands and crazes in annealing of the deformed polymer glasses. The experimental results also show that the interfaces in the glassy polymers are capable of healing. This contradicts the assumption on the absence of large-scale molecular motion in the amorphous polymers below their glass transition temperature. The healing of the interfaces in the polymers is a unique surface phenomenon determined by their large-scale molecular mobility. Possibilities of the practical application of the processes of healing of the interfaces in the development of new types of nanocomposites with a polymer matrix have been indicated.

References

1. Voyutsky S.S., Autohesion and adhesion of high polymers, Moscow, Khimiya, 1960.
2. DeGennes P.G., J. Chem. Phys., 1971, V. 55, P. 572.
3. Ferry J.D., Viscoelastic properties of polymers, 2 ed. New York, Wiley, 1970.
4. Canpolat V., Pekcan O., Polymer, 1995, V. 36, No. 10, P. 2025.
5. Pekcan O., Winnik M. F., Croucher M.D., Macromolecules, 1990, V. 23, P. 2673.
6. Zhao C.L., Wang Y., Hruska Z., Winnik M.F., Macromolecules, 1990, V. 23, P. 4082.
7. Ferry J.D., Strella S., J. Colloid Sci., 1958, V. 13, P. 459.
8. Arda E., Pekcan O., Polymer, 2001, V. 42, No. 17. P. 7419.
9. Kline D.B., Wool R.P., Polym. Eng. Sci., 2004, V. 28, No. 1, P. 52.
10. Voyutsky S.S., Margolina Yu.L., Usp. Khimii, 1949, V. 18, P. 449.
11. Nguen T.Q., Kauch H.H., Jud K., Dettenmaier M., Polymer, 1982, V. 23, No. 5, P. 1305.
12. Roland C.M., Macromolecules, 1987, V. 20, No. 10, P. 2557.
13. Boiko Y.M., Bach A., Lyngaae-Jorgensen J., J. Polym. Sci., Polym. Phys., 2004, V. 42, No. 10, P. 1861.
14. Boiko Y.M., Lyngaae-Jørgensen J., Polymer, 2004, V. 45, No. 25, P. 8541.
15. Boiko Y.M., Guerin G., Marikhin V.A., Prud'homme R.E., Polymer, 2001. V. 42, No. 21, P. 8695.
16. Boiko Y.M., Lyngaae-Jørgensen J., Polymer, 2005, V. 46, P. 6016.
17. Boyko Y.M., Vysokomolek. Soed. B, 2000, V. 42, No. 3, P. 542.
18. Boyko Y.M., Vysokomolek. Soed. A, 2002, V. 44, No. 7, P. 1160.
19. Boiko M.Y., Prud'homme R.E., J. Appl. Polym. Sci., 1999, V. 74, P. 825.
20. Boyko Y.M., Adhesive interaction between the glassy amorphous polymers, Abstract of dissertation, Dr. Sci. Sciences, St. Petersburg, 2011.
21. Boiko Y.M., Lyngaae J., Irgensen J., J. Macromol. Sci., Phys., 2004, V. 43, No. 3, P. 695.
22. Kuleznev V.N., Mixtures of the polymers, Moscow, Khimiya, 1980.
23. Os'kin V.N., Yanovsky Yu.G., Malkin A.Ya, Kuleznev V.N., Al'tzitser V.S., Tutorskii I.A., Vysokomolek. Soed. A, 1972, V. 14, No. 10, P. 2120.
24. Polymer Blends, V. 1, 2, Ed. by Paul D.R., Bucknall C.B., New York: Wiley, 2000.
25. Helfand E., Tagami Y., J. Polym. Sci., Polym. Lett., 1971, V. 9, P. 741.
26. Helfand E., Tagami Y., J. Chem. Phys., 1972, V. 56, P. 3592.
27. Helfand E., Sapse A.M., J.Chem. Phys., 1975, V. 62, P. 1327.
28. Helfand E., J. Chem. Phys., 1975, V. 63, P. 2192.
29. Utraki L.A., Polymer Alloys and Blends: Thermodynamics and Rheology. Munich, Hanser Publ., 1990, P. 118.
30. Liu R., Bernal-Lara T., Hiltner A., Baer E., Macromolecules, 2004, V. 37, No. 18, P. 6972.
31. Baer E., Hiltner A., Keith H.D., Science, 1987, V. 235, P. 1015.
32. Kerns J., Hsieh F., Hiltner A., Baer E., J. Appl. Polym. Sci., 2000, V. 77, P. 1545.

76   **Chapter 2**

33.   Kerns J., Hsieh F., Hiltner A., Baer E., Macromol. Symp., 1999, V. 147, P. 15.
34.   Ronesi V., Cheung Y.W., Hiltner A., Baer E., J. Appl. Polym. Sci., 2003, V. 89, P. 153.
35.   Poon B.C., Chum S.P., Hiltner A., Baer E., J. Appl. Polym. Sci., 2004, V. 92, P. 109.
36.   Poon B.C., Chum S.P., Hiltner A., Baer E., Polymer, 2004, V. 45, P. 893
37.   Pollock G., Nazarenko S., Hiltner A., Baer E., J. Appl. Polym. Sci., 1994, V. 52, P. 163.
38.   Mueller C.D., Nazarenko S., Ebeling T., Shuman T.L., Hiltner A., Baer E., Polym. Eng. Sci., 1997, V. 37, P. 355.
39.   Stepanov E.V., Shuman T.L., Nazarenko S., Capassio G., Hiltner A., Baer E., Macromolecules, 1998, V. 31, P. 4551.
40.   Liu R., Jin Y., Hiltner A., Baer E., Macromol. Rapid. Commun., 2003, V. 24, P. 943.
41.   Hu Y.S., Liu R.Y.S., Zhang L.O., Rogunova M., Schiraldt D.A., Nazarenko S., Hiltner A., Baer E., Macromolecules, 2002, V. 35, P. 7326.
42.   Liu R.Y.S., Hiltner A., Baer E., J. Polym. Sci., Polym. Phys., 2004, V. 42, P. 493.
43.   Ivankova E.V., Krumova M., Michler G.H., Koets P.P., Colloid. Polym. Sci., 2004, V. 282, P. 203.
44.   Fakirov S., Polym. Commun., 1985, V. 26, No. 5, P. 137.
45.   Avramova N., Polymer, 1993, V. 34, No. 9, P. 1904.
46.   Fakirov S., Avramova N., J. Polym. Sci., Polym. Phys., 1987, V. 25, No. 6, P. 1331.
47.   Pernot H., Baumert M., Court F., Leiber L., Nat. Mater., 2002, V. 1, P. 54.
48.   Harton S.E., Stevie F.A., Ade H., Macromolecules, 2005, V. 38, No. 9, P. 3543.
49.   Trofimovich D.P., Chernaya V.V., Shepelev M.I., Latex products, Encyclopedia of polymers, Moscow, Sov. entsyklopediya, 1974, V. 2, P. 40.
50.   Oleynik E.F., High performance polymers, Ed by E. Baer, S.Moet, Berlin, Hauser Verlag, 1991, P. 79.
51.   Oleinik E.E., Salamatina O.B., Rudnev S.N., Shenogin S.V., Vysokomolek. Soed. A, 1993, V. 35, No. 11, S. 1819.
52.   Shenogin S.V., Hohne G.W.H., Salamatina O.B., Rudnev S.N., Oleinik E.F., Vysokomolek. Soed. A, 2004, V. 46, No. 1, P. 30.
53.   Arzhakov S.A., Kabanov V.A., Vysokomolek. Soed. B, 1971, V. 13, No. 5, P. 318.
54.   Arzhakov S.A., Bakeev N.F., Kabanov V.A., Vysokomolek. Soed. A, 1973, V. 15, No. 10, P. 1154.
55.   Arzhakov M.S., Lukovkin G.M., Arzhakov S.A., Dokl. RAN, 1999, V. 369, No. 5, P. 629.
56.   Arzhakov M.S., Lukovkin G.M., Arzhakov S.A., Dokl. RAN, 2000, V. 371, No. 4, P. 484.
57.   Bartenev G.M., Nikol'skii V.G., Encyclopedia of polymers, Moscow, Sov. entsyklopediya, 1977. V. 3. P. 489.
58.   Structural and mechanical behavior of glassy polymers, Arzhakov M.S., Arzhakov S.A., Zaikov G.E. (editors) Nova Science Publishers, Inc. Commack, New York, 1997, P. 62-83.
59.   Arzhakov S.A., Dissertation, Dr. Chem. Sciences, Moscow, Karpov Institute, 1975.
60.   Jayaraman G.S., Wallace J.F., Geil P.H., Baer E., Polym.Eng. Sci., 2004, V. 16, No. 8, P. 529.
61.   Linda W.V., Kander R.G., Polym. Eng. Sci., 2004, V. 38, No. 11, P. 1824.
62.   Linda W.V., Kander R.G., Polym. Eng. Sci., 2004, V. 37, No. 1, P. 120.
63.   Brink A.E., Jordens K.J., Riffle J.S., Polym. Eng. Sci., 2004, V. 35, No. 24, P. 1923
64.   Penings A.J., Zwijenburgf O., J. Polym. Sci., Polym. Phys., Ed. 1979, V. 17, No. 6, P. 1011.

65.   Aulov V.A., Chvalun S.N., Ozerin L.A., Bakeev N.F., Dokl. RAN, 1997, V. 34, No. 2, P. 198.
66.   Aulov V.A., Bakeev N.F., Dokl. RAN, 1998, V. 360, No. 2, P. 202.
67.   Porter R.S., Kanamoto T., Zachariades A.E., Polymer, 1994, V. 35, No. 23, P. 4979.
68.   Aulov V.A., Makarov S.V., Kuchkina I.O., Ozerin A.N., Bakeev N.F., Vysokomolek. Soed., A, 2000, V. 42, No. 11, P. 1843.
69.   Aulov V.A., Makarov S.V., Kuchkina I.O., Pantyukhin A.A., Ozerin A.N., Bakeev N.F., Vysokomolek. Soed. A, 2002, V. 44, No. 8, P. 1367.
70.   Aulov V.A., Makarov S.V., Kuchkina I.O., Pantyukhin A.A., Akopian E.L. Ozerin A.N., Bakeev N.F., Vysokomolek. Soed. A, 2001, V. 43, No. 10, P. 1766.
71.   Aulov V.A., Shcherbina M.A., Chvalun S.N., Makarov S.V., Kuchkina I.O., Pantyukhin A.A., Bakeev N.F., Pavlov Yu.S., Vysokomolek. Soed. A, 2004, V. 46, No. 6. P. 1005.
72.   Sinevich E.A., Aulov V.A., Bakeev N.F., Doklady RAN, 2006, Vol. 408, No. 5, P. 1–4.
73.   Jud K., Kaush H.H., Polym. Bull., 1979, V. 1, P. 697.
74.   Kaush H.H., Jud K., Prepr. Short Commun. IUPAC Makro Mainz 26 Int. Symp. Macromol., Mainz, 1979, V. 2, P. 1426.
75.   Kaush H.H., Jud K., Plastic Rubber Proc. Applic., 1982, V. 2, No. 3, P. 265.
76.   Jud K., Kaush H.H., Williams J.G., J. Mater. Sci., 1981, V. 16, No. 1, P. 204.
77.   Kausch H.H., Petrovska D., Landel R.F., Monnerie L., Polym. Eng. Sci., 2004, V. 27, No. 2, P. 149.
78.   Kausch H.H., Petrovska D., Landel R.F., Monnerie L., Polym. Eng. Sci., Oxford, Oxford University Press, 2004.
79.   Kozlov P.V., Papkov S.P. Physico-chemical basis of plasticizing polymers, Moscow, Khimiya, 1982.
80.   Lin C.B., Lee S., Liu K.S., Polym. Eng Sci., 2004, V. 30, No. 21, P. 1399.
81.   Wang P.P., Lee S., Harmon J.P., J. Polym. Sci., Polym. Phys., 1994, V. 32, P. 1217.
82.   Crank J., The mathematics of diffusion. Oxford, Oxford University Press, 1975.
83.   Govinjee S., Simo J.C., J. Mech. Phys. Solids, 1993, V. 41, P. 863.
84.   Harmon J.P., Lee S., Li J.S.M., J. Polym. Sci., Polym. Chem., 1987, V. 25, P. 3215.
85.   Harmon J.P., Lee S., Li J.S.M., Polymer, 1988, V. 29, P. 1221.
86.   Wu T., Lee S., J. Polym. Sci., Polym. Chem., 2003, V. 32, No. 12, P. 2055.
87.   Kawagoe M., Nakanishi M., Qiu J., Morita M., Polymer, 1997, V. 38, No. 24, P. 5969.
88.   Hsieh H.C., Yang T.-J., Lee S., Polymer, 2001, V. 42, No. 3, P. 1227.
89.   Passaglia E., J. Phys. Chem. Solids, 1987, V. 48, No. 11, P. 1075.
90.   Weidmann G., Dull W., Int. Fract. J., 1978, V. 13, P. R189.
91.   Kambour R. P., J. Polym. Sci., Macromol. Rev., 1973, V. 7, P. 1.
92.   Kim H. J., Lee K.-J., Lee H. H., Polymer, 1996, V. 37, No. 20, P. 4593.
93.   Volynskii A.L., Grokhovskaya T.E., Kulebyakina A.I., Bolshakova A.V., Bakeev N.F., Vysokomolek. Soed. A, 2007, V. 49, No. 7, P. 1224-1238.
94.   Volynskii A.L., Grokhovskaya T.E., Kulebyakina A.I., Bolshakova A.V., Bakeev N.F., Vysokomolek. Soed. A, 2007, V. 49, No. 11, P. 1946-1958.
95.   Volynskii A.L., Grokhovskaya T.E., Kulebyakina A.I., Bolshakova A.V., Bakeev N.F., Vysokomolek. Soed., A, 2009, V. 51, No. 4, P. 598-609.
96.   Volynskii A.L., Yarysheva L.M., Bakeev N.F., Vysokomolek. Soed. A, 2011, V. 53, No. 10. P. 871-898.
97.   Volynskii A.L., Priroda, 2011, No. 7, P. 14-21.
98.   Treloar L. Physics of rubber elasticity, Moscow, Mir, 1975.

99.   Gorelova M.I., Pertsin A.J., Volkov I.O., Filimonova L.V., Obolonkova E.S., J. of
        Appl. Polymer Sci., 1996. V. 60, P. 363-370.
100.   Gorelova M.I., Pertsin A.J., Volkov I.O., Sanches N.B., Gomes A.S., J. of Appl.
        Polymer Sci., 1998, V. 69, P. 2349.
101.   Sauer I.A., Marin I., Hsiao C.C., J. Appl. Phys., 1949, V. 20, P. 507.
102.   Kambour R.P., Kopp R.W., J. Polymer Sci., 1969, pt. A-2, V. 7, P. 183.
103.   Berger L.L., Sauer B.B., Macromolecules, 1991, V. 24, No. 8, P. 2096.
104.   Yang A.S.M., Kramer E.J., J. Polym. Sci. A, 1985, V. 23, P. 1353.
105.   Wool R.P., O'Connor K.M., Polym. Eng. Sci., 1981, V. 21, P. 970.
106.   Plummer C.J.G., Donald A.M., J. Mater. Sci., 1989, V. 24, P. 1399.
107.   Donald A.M., Kramer E.J., Polymer, 1982, V. 23, P. 1183.
108.   Donald A.M., Kramer E.J., J. Polym. Sci., Polym. Phys. Edn., 1982, V. 20, P. 899.
109.   Kramer E.J., Fdv. Polym. Sci., 1983, V. 52/53. Ch. 1.

Special features of the structure and properties of surface layers and thin (nanometric) films of glassy polymers

Prior to discussing the fundamental properties of the surface layers of solid polymers, it is necessary to mention briefly the most general properties of the surface layers of liquid and solid bodies. It is well-known that these bodies have a free surface and the structure and properties of the surface greatly differ from the structure and properties of these phases in the volume. The molecules in the surface of the liquid and the solid body, in contrast to the molecules in the bulk, are not surrounded by other molecules from all sides.

In the volume, the forces of intermolecular interaction, acting on one of the molecules from the side of the adjacent molecules, are on average mutually compensated. In the surface there is no such compensation and, consequently, a certain equally acting force, directed into the liquid or the solid body, appears. If the molecule travels from the surface into the volume, the forces of intermolecular interaction carry out a positive work. On the other hand, in order to transfer a certain number of the molecules from the volume to the surface (i.e., to increase the surface area), it is necessary to carry out a positive work of external forces, proportional to the change of the surface area. For this reason, the free drop of the liquid becomes spherical. The liquid behaves as if forces, which appear to shorten (contract) the surface, act along the tangent to the surface. In other words, as a result of the surface tension forces the molecules on the surface layer appear to be 'attracted' to each other to a greater extent than in the bulk.

These forces are referred to as the surface tension forces. The presence of the surface tension forces makes the surface of the liquid similar to an elastically stretched film, with the only difference being that the elastic forces in the film depend on the surface area of the film (i.e., on the mode of deformation of the film), and the surface tension forces are independent of the surface area of the liquid or solid body. From the viewpoint of the surface phenomena there is no significant difference between the properties of solid and liquid bodies. The qualitative difference is that in the liquids, the surface tension forces can change the dimensions and shape of the liquid and this makes it not only possible to determine their presence but also carry out a quantitative evaluation of some surface characteristics. In the case of the solid bodies, the surface forces also exist but they can not overcome the forces of intermolecular interaction and, consequently, they can not have any significant effect on the volume and shape of the solid body. Nevertheless, the actual existence of the surface forces in the solids is the basis of an entire set of surface phenomena which have been for many years and still are the subject of extensive investigations [1, 2].

Thus, the surface of the liquid or solid body appears to be some container which, like the previously mentioned film, contracts the underlying volume and tries to minimise the surface of this volume.

The above described fundamental considerations regarding the structure and properties of the interfaces are general. Consequently, the discovery in the 90s of highly unusual surface properties of the amorphous glassy polymers is really surprising. Here, we are talking about the thickness of the films and polymer layers of tens or hundreds of nanometres. The development of a number of investigation methods, which make it possible to determine reliably the nature of molecular motion in such tin layers, has resulted in the discovery of a large number of unusual effects.

The results were so unexpected and important that, to consider them, it is necessary to review many considerations regarding the structure and properties of the polymers which now appear to be complete and accepted. Some of the aspects of these problems are also the subject of this book.

3.1. Measurement of the glass transition temperature of amorphous glassy polymers in thin films and thin surface layers

The first systematic investigations, concerned with the precision measurement of the glass transition temperature (T_g) in thin films of a glassy polymer, were published in 1994 [3]. In this study, ellipsometric measurements were carried out to determine T_g of thin polystyrene (PS) films, deposited on a hydrogen-passivated silicon substrate. The thickness of the films was varied from 3000 to 100 Å. Evidently, the authors showed for the first time that, starting at a thickness of 100 Å, the values of T_g of polystyrene starts to decrease rapidly in comparison with the T_g of the bulk polymer. At a film thickness of 100 Å this reduction was 25 K which was far outside the range of the error of the determination of the glass transition temperature by the ellipsometric method. In this work, M_w of the polystyrene was $(120–2900)\cdot10^3$, whereas the corresponding dimensions of the coil changed from 200 to 1000 Å. In all cases, there was no dependence of T_g on M_w and the experimental results were satisfactorily described by the relationship $T_g = T_g^{block} [1 - (a/h)^\delta]]$, where T_g^{block} is the glass transition temperature of the polystyrene in the bulk, h is the thickness of the film, $a = 32$ Å, $\delta = 1.8$. Later, a large number of investigations, concerned with this problem, were carried out and at the present time, they can be divided into several separate groups. In particular, work was carried out to determine the method of production and examination of the properties of thin films of the polymers, deposited on hard substrates, free thin films, and also the properties of thin surface layers of the bulk polymers.

3.2. Glass transition temperature of thin films of glassy polymers, deposited on solid substrates

Further development of this direction was accompanied by the expansion of both the range of experimental methods and the number of studied objects. In most cases, experiments were carried out on thin films of classic amorphous polymers, PS, PMMA, and others. One surface of these films is in contact with the solid substrate, and the other surface faces the air. The authors of these studies determine T_g of the polymer at the polymer–air interface.

In [4], surface T_g was determined by measuring the relaxation of birefringence in ultrathin polystyrene films on glassy substrates. The experimental results show that when the thickness of the film is reduced from 10 μm to 5.8 nm, T_g decreases by 15–20°C. In thick films (10 μm), the molecules, close to the polymer–air interface, relax more easily (at a higher rate) than the molecules, facing the interface. The combination of photon correlation spectroscopy with the method of the quartz balance enabled the authors of [5] to investigate the dynamics of relaxation of ultrathin ($h < 400$ Å) polystyrene films. The results show that in these films the T_g temperature is reduced by more than 70°C in comparison with the bulk polymer. Both the shape of the relaxation function and the temperature dependence of the relaxation time above T_g are identical with those observed for the bulk polymer although for the films these functions are shifted by ~70°C on the temperature scale

Similar results were also obtained in [6] in which a quartz resonator was used to measure the high frequency shear compliance in very thin (10 and 93 nm) polystyrene films on the quartz substrates. The experimental results show that with increase of the thickness of the films above 10 nm the films show a reduction of the shear compliance, whereas at a thickness of ≤10 nm the increase of the compliance is greater than 50%. It is assumed that the softening is caused by geometrical restrictions forcing the polymer chains to parachute-like conformations.

Very important and effective was the application of the annihilation of positrons for studies of the free volume in the thin films. In [7, 8] measurements were taken of the annihilation time of the positrons and Doppler shift of annihilation radiation for the polystyrene film using a monoenergetic weak positron probe. Large changes in the signal of positron annihilation were detected at a small distance from the surface of the sample (200 Å). The ortho-positron lifetime in the polymer increases in the vicinity of the surface, whereas its intensity decreases. The results obtained for the lifetime were interpreted as the local expansion of the holes of the volume in the vicinity of the surface. The distribution of the free volume and the holes in the vicinity of the surface was found to be wider than in the bulk. The authors of [9] obtained, for an epoxy polymer, the range of the lifetime of positron annihilation with the energy of 0.2–3.0 keV using a weak positron electrical engineering device. The experimental results show that the ortho-positron lifetime in the volume increases, and in the vicinity of the surface decreases.

The method makes it possible to determine the fraction of the free volume f_v of the polymer using the principle of temperature–time superposition [10]:

$$f_v = f_T + (T-T_g)\Delta\alpha \qquad (3.1)$$

where f_T is the fraction of the free volume of the glass transition temperature; α is the difference in the thermal expansion coefficients above and below the glass transition temperature, T and T_g are the experiment temperature and the glass transition temperature, respectively. The experimental results show that the subsurface layer is characterised by the increase of the size of the holes of the free volume and of the fraction of the holes in the volume (up to 4%). In this case, the reduction of T_g was 40°C (from 50 to 10°C).

A similar behaviour was also found for PET [11] which, in contrast to PS, is capable of crystallisation. In [11], ellipsometric measurements were carried out to determine the glass transition temperature in thin PET layers. The experimental results show that the glass transition temperature of PET also depends strongly on the distance of the polymer from the polymer–air interface. For example, if T_g in the bulk is 71°C, then at a depth of 25 nm it is equal to 65°C, and at a depth of 15 nm it is 56°C.

Important information on the structure and properties of the surface layers of the polymers is provided by many varieties of probe microscopy. For example, in [12] direct measurements were taken of the molecular surface motion of the monodispersed polystyrene by a lateral force microscope, a scanning viscoelastic microscope and differential x-ray spectroscopy. The experimental results show that the intensity of the molecular motion in the surface layer is higher than in the volume. In [13], force modulation scanning microscope and lateral force microscopy were used to examine the surface molecular motion in the monodispersed polystyrene. Force modulation scanning microscopy can be used to measure the dynamic modulus E' and tg φ in the surface layers. It is shown that E' at the surface is lower and tg φ higher than the appropriate values for the bulk for $M < 2.6 \cdot 10^4$. The glass transition temperature of the polystyrene at the polymer–air surface is lower than room temperature. A large reduction of T_g in the surface layer of the polystyrene was also reported in [4] by the method of temperature-dependent scanning viscoelastic microscopy. In [15] the monodispersed polystyrene was studied by scanning viscoelastic microscopy and lateral force microscopy. Measurements were taken of the dynamic modulus E' and the tangent of the losses

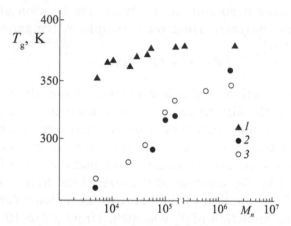

Fig. 3.1. Dependence of T_g of the bulk PS, determined by the DSC methods (1), surface T_g of the PS, determined by lateral force microscopy (2) and scanning viscoelastic microscopy (3) on its molecular mass M_n [5].

tg φ of the surface of the polystyrene for the molecular mass fractions, equal to 10^3–10^7. For the bulk samples (Fig. 3.1), according to DSC data, there is a week dependence of T_g on molecular mass (MM), described by the Fox–Flory equation [16] $T_g = T_{g\infty} - K/M_n$, where T_g is the glass transition temperature for the infinite MM, K is the constant of the polymer. Figure 3.1 shows that for the surface T_g this dependence is far stronger

The large decrease of the glass transition temperature was also found in the mixtures of a number of fractions of PS, with greatly differing MMs. In [17], the molecular surface movement of the monodispersed PS films and of their binary and ternary mixtures was investigated by scanning force microscopy. The results show that in the surface layer T_g is reduced below room temperature. This reduction becomes greater with increase of the fraction of the low-molecular component.

In a number of studies [18–21] investigations were carried out into the special features of molecular motion in symmetric diblock-copolymers PS–PMMA. Within the limits of each copolymer, the molecular masses of PS and PMMA were identical but they were changed in wider ranges for different copolymers. A large number of experimental methods were used in these studies: lateral force microscopy, scanning viscoelastic microscopy, differential x-ray photoelectronic spectroscopy, temperature-dependent x-ray photoelectronic spectroscopy, DSC. The results show that in the surface layers of the PS–PMMA block-copolymer, like in the homopolymers, T_g of both blocks decreases. This reduction is very large (many tens of degrees) and depends strongly on the distance of the blocks from the surface.

It is important to note that the T_g of both blocks also depends on the general MM of the block-copolymer. For the specimens with $M \leq 3.0 \cdot 10^4$ the entire surface at room temperature is in the rubber-like state. The effect of the MM on the surface T_g is not so strong as the effect of the distance from the interface, but is clearly visible.

The reduction of the glass transition temperature in the thin films was also detected for the mixtures of compatible polymers. In [22] investigations were carried out into the dependence of T_g on the thickness of the films and the composition of the compatible mixtures of poly-2,6 dimethyl-1,4-phenylene oxide (PPO) and PS. Thin films were produced by evaporation from the solution in toluene on a silicon substrate with ellipsometric measurements of T_g. The results show that in both the thin films and in the bulk polymer, the composition of the mixture influences T_g. At the same time, T_g decreases with a reduction of the thickness for all compositions (Fig. 3.2).

In the case of thin films, situated on the substrates, the intensity of the molecular motion is strongly influenced by the nature of

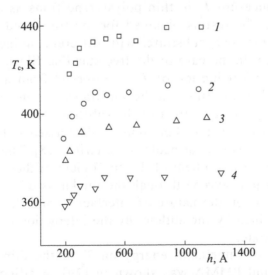

Fig. 3.2. Dependence of the glass transition temperature T_g of the mixtures of poly-2,6-dimethyl oxide with the polystyrene on the thickness of the film. The composition of the PPF/PS mixtures: 7/3 (1), 5/5 (2), 3/7 (3) and pure polystyrene (4) [21].

interfacial contact [23]. This effect was clearly detected in, for example, [24] in which the method of reflective x-ray spectroscopy was used to determine the electronic density in the surface layers of the thin films of PS, PMMA and poly-4-vinyl pyridine (P4VP) on SiO_2 substrates. The thickness of the films was equal to approximately $4R_g$, where R_g is the radius of rotation of the macromolecule. It was shown that both the PS and PMMA show a reduction of the electronic density in the vicinity of the free surface of the film, whereas in the case of P4VP the electronic density increases. The authors assume that the detected effect is determined by different intensities of the interaction of hydrophobic (PS and PMMA) and hydrophilic (P4VP) polymers with a polar substrate.

.A strong effect of the nature of the substrate on the molecular mobility of the polymer in the adjacent thin layer was detected in [25]. In this study, investigations were carried out into the mobility of the PS chains, marked at the end with fluorescent markers, on quartz substrates. The fluorescence method was used to measure the diffusion coefficient of the macromolecules (D). The results show that at a film thickness smaller than 1500 Å the value of D decreases. As in [26], this reduction is explained by the strong interaction with the substrate, restricting the molecular mobility of the chains.

The same conclusion was reached by the authors of [27], in which the methods of Brillouin light scattering and ellipsometry was used to measure T_g in thin polystyrene films as a function of the thickness. The results obtained for the free-standing films and the films on the SiO_x substrate, with different molecular masses, were compared. In the case of the free-standing films, the large (by more than 40°C) reduction of T_g is observed from a thickness of $h \leq R_{ee}$. The T_g of the films on the substrates also decreased, but by only 4°C. The strongest effect of the substrate on the coefficient of thermal expansion in the thin films of deuterised polystyrene was detected by the methods of neutron scattering [28]. The results show that there is a steep gradient of the coefficient of thermal expansion for deuterised polystyrene through the thickness of the film which, in turn, depends on the nature of interfacial contact. The observed effect is explained by the authors by the interaction of the polymer with the substrate.

The effect of interfacial energy on T_g of the thin (18–89 nm) films of PS and PMMA was shown in [29]. A silicone substrate for the thin surface films was produced by x-ray irradiation of octadecyl trichlorosilane $[(CH_2)_{17}SiCl_3]$. Under irradiation, the surface layer contained oxygen-containing groups so that it was possible to regulate the interfacial energy (calculated on the basis of the measurement of the contact angles) in the range from 0.5 to 6.48 mJ/m^2. T_g was measured by three methods: local thermal analysis, ellipsometry and reflective x-ray spectroscopy, which provided the same results. It was shown that the magnitude of reduction of T_g depends a linear manner on the interfacial energy for the films of different thickness.

The results of investigations of the relaxation of birefringence were used to determine T_g in ultrathin polystyrene films on glass substrates [30]. The results show that when the thickness of the film is reduced from 10 μm to 5.8 nm T_g decreases by 15–20°. In the thick films (10 μm) the molecules, close to the polymer–air interface, relax at a higher rate than the molecule situated in contact with the substrate. Large restrictions of the molecular mobility in the thin polystyrene films on the silicon substrate were also detected in [31], regardless of the fact that the thickness of the films was considerably greater than the radius of rotation of the molecule. It was assumed that the detected restrictions in the molecular mobility formed as a result of the strong interaction with the substrate. In [32], x-ray diffraction analysis was used to measure the gradient of the electronic density in the direction normal to the plane of the thin (20–80 nm) films of stereoregular PMMA on a silicon substrate. The method can be used to evaluate weak fluctuations of density. The results show that in the polymer, which was in contact with the substrate, the density increases the respective of the tacticity of the polymer and film thickness. The detected effect is explained by the better packing of the macromolecules at the interface as a result of the interfacial interaction and selective adsorption of stereoregular sequences in chains of macromolecules.

In [33], dielectric spectroscopy was used to examine the dynamics of isotactic PMMA at temperatures of 273–392 K. The dynamics shows two peaks α and β, where β is independent of the film thickness, and α depends and decreases (by two orders of magnitude) with increasing thickness. The glass transition temperature of the polymer in contact with the silicon substrate gradually increases with the decrease of thickness according to the ellipsometric results which, according to the authors, indicates the strong interaction with the substrate.

The adsorption effects at the phase boundary, leading to the restriction of the mobility of the polymer chains, can be so strong that they may even increase the glass transition temperature of the polymer. In [34], modulating scanning differential calorimetry was used to examine the behaviour of the PMMA, absorbed on a silicon substrate from the solution. The results show that T_g of the absorbed PMMA increases from 108°C (T_g of the bulk polymer) to 136°C.

3.3. The glass transition temperature of free-standing thin films of glassy polymers

As shown previously, the interaction of the polymer with the substrate may have a strong effect on the nature of the molecular motion in thin films. Therefore, procedures have been developed for producing thin films (hundreds of angströms) with no substrate, and also methods for measuring their T_g temperature [35]. These films can be produced by two methods. Firstly, the film can be produced on the substrate and floated on the water surface. Subsequently, the film is trapped on a support mesh. Secondly, such a film can be produced by evaporation of the thin layer of the solution of the polymer deposited on the surface of a liquid with a higher density, usually water. In fact, this is the method of producing support films-substrate for transmission electron microscopy. After formation of the film on the water surface, the film is placed in a special holder for investigating its properties.

Since the majority of investigations of the 'free-standing' films have been carried out in studies of polystyrene, the experimental data, obtained by different investigators, can be generalised on a single figure [36].

It has been shown (Fig. 3.3) that the reduction of T_g in this case is considerably larger than for the films on the hard substrates. In addition to this, in the case of polystyrene the dependence of T_g on the film thickness is not only stronger but also qualitatively different. These dependences can be divided into two groups. At a relatively small values of MM (up to $3.5 \cdot 10^5$), the dependence of T_g on the thickness of the film is qualitatively similar to the dependence obtained for the polystyrene films on the substrates. In this case, T_g decreases gradually, starting at a film thickness of 500–700 Å, and is independent of the molecular mass of polystyrene. However, at a higher value of the molecular mass of the polystyrene, this effect is very clear. For each molecular mass, the dependence of T_g on the film thickness becomes linear and there is some thickness of the film starting at which T_g starts to rapidly decrease in accordance with

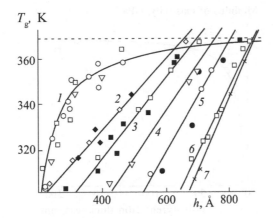

Fig. 3.3. Dependence of the glass transition temperature of free-standing polystyrene films on their thickness. The molecular mass of polystyrene, $M \cdot 10^{-3}$:116–347 (1), 541 (2), 691 (3), 1250 (4), 2077 (5), 6700 (6), and 9000 (7) [36]. The dotted line shows the glass transition temperature of block polystyrene.

a linear law. As the molecular mass of the polymer increases, the reduction of T_g becomes larger with decreasing film thickness, and the start of the reduction of T_g is recorded at a smaller film thickness for higher molecular masses.

Thus, for the free-standing films, like for the films on the substrates, the glass transition temperature is considerably lower than the appropriate temperature for the bulk. This reduction is far greater the in the case of the films located on the substrate, and depends mostly on the molecular mass of the polymer.

To conclude this section, it should be mentioned that the large reduction of the mechanical characteristics in the surface layer and the thin films of the glassy polymers is directly confirmed by measurements of their elasticity modulus. In a series of studies [37–40] the approach used for evaluation of these characteristics was based on the analysis of the microrelief which forms in deformation of the two-layer polymer systems [41, 42]. This method of evaluating the mechanical characteristics of solids of the nanometric dimensions will be examined in greater detail later (chapters 9–11), and here we present only one figure demonstrating the dependence of the elasticity modulus of a glassy polymer (polystyrene) on the thickness of the film (Fig. 3.4).

Figure 3.4 shows that as soon as the thickness of the polystyrene film drops below 50–70 nm, the elasticity modulus of the film rapidly decreases, irrespective of the molecular mass of the polymer. This

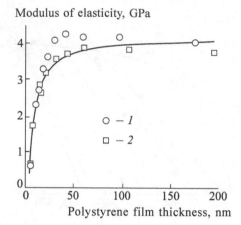

Fig. 3.4. Dependence of the elasticity modulus on the thickness of the polystyrene film. 1) M_w = 1800 kg/mol; 2) M_w = 140 kg/mol [40].

reduction becomes very large (the modulus decreases at least 4–5 times). The results are in complete agreement with the previously described results for the measurement of T_g in the thin films of glassy polymers (compare Figs. 3.3 and 3.4).

3.4. Measurement of the glass transition temperature and molecular mobility in surface layers of bulk glassy polymers

The experimental results, discussed previously, indicate that the thin films of the amorphous polymers (i.e., the polymer in the thin surface layer) have a number of unusual properties, greatly different from the appropriate properties of the bulk polymers. It is therefore necessary to answer the question whether the properties of the thin surface layers of the bulk polymers are the same? To answer this question, a large number of investigations were carried out to investigate the properties of the surface layers of amorphous polymers of the nanometre thickness. These investigations slightly differed in the procedure from the appropriate investigations of the thin films. They are based mainly on the mechanical probing of the surface of bulk polymers. The most efficient methods used for this purpose were those based on the principle of the atomic force microscope [43]. The point is that the mechanical response of the polymer surface depends more strongly on its physical state.

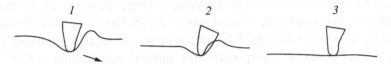

1   *2*   *3*

Fig. 3.5. Schematic representation of the deformation of the surface layer under the effect of a cantilever probe microscope. The polymer is in the rubbery state (1), in the glass transition range (2) and in the glassy state (3) [43].

Figure 3.5 shows schematically the situation formed in the interaction of the indentor (cantilever) with the polymer surface. This difference makes it possible to determine not only the local glass transition temperature but also estimate the dynamic mechanical characteristics of the surface layer of the polymer, such as E' and tg φ. It is evident that, knowing these characteristics, we can easily determine the physical state of the surface layer of the polymer. In particular, on the basis of these data it was established in [18] that the glass transition temperature in the thin surface layers of polystyrene and the PS–PMMA copolymer is reduced by many tens of degrees. Recently, a new method has been developed for examining the properties of the polymer surface which actually combines the advantages of probe microscopy and differential thermal analysis [44]. In this study, the authors used an atomic force microscope with a thermal probe in order to carry out delocalized thermal analysis and determine T_g of the thin surface layer of the bulk polymer. The analysis of the resultant atomic force images made the authors to conclude that the surface of the bulk polymer has a layer with a thickness of several nanometres in which the T_g is greatly reduced.

3.5. Interaction of metal nanoparticles with polymer surfaces

As shown previously, the surface layer of the glassy polymers have a very unusual structure with the main feature being the high content of the free volume. This circumstance creates suitable conditions for the formation of mutually penetrating interphase layers of the 'polymer–the component of a different chemical nature incompatible with the polymer' type, because the penetration and inclusion of metals and other compounds in the polymers is controlled mostly by the fraction of the free volume [45].

Evidently, the metals are incompatible to a high degree with organic polymers. Nevertheless, it is necessary to develop different metal–polymer systems for solving a large number of practical problems. For example, considerable effort has been developed to produce polymer films of different type with metallic coatings [46] or polymer–metallic composites for optical applications [47]. The metallised films are used for reflectors, CDs, for electrical protection of computers, and also for the production of packing materials [48]. Therefore, it is very important to investigate the mechanism of diffusion of metals into polymers and study the structure and properties of the metal–polymer interfacial surfaces [49].

The capacity of metals to penetrate into the volume of the polymer is very low; for example, the diffusion coefficient of gold and silver in industrial polycarbonates in the temperature range 147–235°C is in the range from $1 \cdot 10^{-16}$ to $5 \cdot 10^{-14}$ cm^2 [43]. In [50], attention was given to the penetration of metallic Ga into polymers (polyethylene, polystyrene, polyimide, polyethylene terephthalate and polytetrafluoroethylene). The results show that Ga penetrates into the polymers only from the liquid phase (T_m = 30°C). The amount of Ga in the polymers increases with temperature and time, indicating the diffusion (molecular) nature of its penetration into the volume of the polymer. The amount of Ga in the surface layer of the polymer is inversely proportional to the density of the polymer. The Ga concentration rapidly decreases from the surface of the polymer into the bulk to a depth of 200 nm and then remains constant to a depth of 2000 nm. The diffusion processes of this type of penetration of the metals into the bulk of the polymer are mostly of scientific importance. Practical value is attached to the processes of relatively rapid interaction between a polymer and a metal.

It is far simpler and more efficient to produce metal–polymer systems by introducing the already formed nanoparticles of the metals into the polymer surface. In this case, the nanoparticles should be capable of penetrating into the surface layer of the polymer (Fig. 3.6).

For the process, shown in Fig. 3.6, to take place, the macro-molecules of the polymer must have a sufficiently high large-scale mobility. Otherwise, the metal nanoparticles cannot be efficiently introduced into the surface layer of the polymer. In the case of the polymers in the rubbery state the mobility of this type does exist and, therefore, there are no serious difficulties in introducing the metallic particles into the polymer.

It is rational to expect that the polymer in the rubbery state is uniform throughout the volume and its properties on the surface

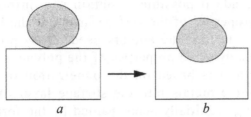

a   *b*

Fig. 3.6. Schematic representation of the interaction of a metallic nanoparticle with the hard surface (below T_g of the polymer-substrate) (a) and with the rubbery surface (above T_g of the polymer-substrate) (b).

and in the bulk are identical. This is confirmed by recent studies of the x-ray photon correlation radiation [51] where it was shown that the surface viscosity of the rubbery polymer is the same as in the volume. For example, in the polystyrene–gold nanoparticles system, the metallic nanospheres should penetrate into the polymer until the angle between the flat surface and the sphere becomes equal to the contact wetting angle of the polystyrene–Au system [52–54]. Experimentally, this fact was demonstrated by investigating the penetration of gold nanospheres with a diameter of 10 and 20 nm into polystyrene. The results show that the ratio of the apparent height of the particles immersed in the surface layer of the polymer to the initial value (the size of the sphere) as a function of time (i.e. $h(t)/t_0$) does not depend on the size of the nanoparticles.

On the scale of the gold nanoparticles, the main driving force of the indentation forms as a result of surface tension which acts on the contact line between the particles and the polystyrene meniscus. Assuming that the polystyrene wets the gold nanoparticles, with the contact angle, equal to the equilibrium contact angle of PS–Au (θ) at temperatures higher than T_g, it is possible to calculate in the functional form the force acting on the particle.

When the polymer is in the glassy state, the situation is much more complicated. If in the glassy polymer the large-scale molecular motion was 'frozen', the indentation of the particles into the surface of the polymer would not be possible (case *a* in Fig. 3.5). However, the penetration of the nanoparticles of this type is not found. In other words, investigation of the penetration of the metallic nanoparticles into the surface not only provides information on the interaction between the polymer and the metal but is also used as an efficient probe for evaluating the state of the polymer in the surface layers. Further improvement of the methods of surface probing of the polymers has made it possible to obtain new information on the structure and properties of the surface layers of the bulk polymers.

For example, in [55] the authors used a new probe method of investigating the dynamic properties of the polymers in the vicinity of T_g. The method is based on the penetration of nanoparticles (1–4 nm) of noble metals into the surface layer of the polymer. Vacuum spraying is initially accompanied by the formation, on the surface, of disperse spherical particles with the size dependent on the polymer–metal pair and the spraying conditions [56, 57]. Using the method of x-ray photoelectronic spectroscopy it is possible to estimate the depth of penetration of these particles into the surface layer of the polymer [58]. For polystyrene and polycarbonate it has been shown that the surface layer of these polymers has a lower T_g temperature than the bulk. It is assumed that the molecular motion at the surface is facilitated by additional degrees of freedom.

Very important results were obtained in [59] where the process of penetration of the gold nanoparticles with the size of 20 and 10 nm into the polystyrene surface was studied. Typical results of these investigations are presented in Fig. 3.7. The figure shows

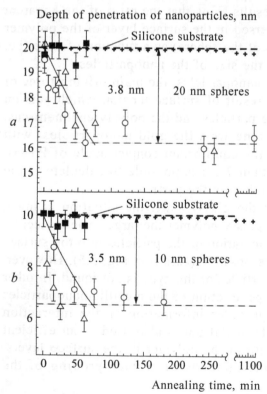

Fig. 3.7. Kinetics of penetration of gold nanoparticles with the size of 20 (a) and 10 nm (b) to the polystyrene surface in annealing below T_g [59].

the kinetics of penetration of gold nanoparticles of different sizes into a polystyrene substrate at a temperature of 363 K, i.e., at a temperature lower than T_g of the bulk polymer. As shown by Fig. 3.7, at temperatures below T_g both types of spheres (20 and 10 nm) penetrate into the polystyrene to a depth of the 3–4 mm and are then arrested. This means that the liquid-like surface layer on the polystyrene does actually exist and its thickness is 3–4 nm. In the same study, the authors developed a method for deposition of metallic coatings on the polymer without thermal spraying of the metal in vacuum. In this procedure, the metal was initially sprayed on the surface of a NaCl crystal and then the NaCl crystal with the metallic coating was placed on the polymer and the crystalline substrate was dissolved. This produced a polystyrene film with a metallic coating deposited without vacuum spraying of the metal. This procedure eliminated the need for heating the polymer surface during vacuum thermal spraying of the metal. The results show that this procedure prevents the reduction of T_g of the polymer in the nanometric films. However, subsequent removal of the metal from the polymer surface completely restores the effect of the reduction of T_g in the surface layer of the polymer. On the basis of these experiments, the authors of [59] concluded that to reduce T_g of the polymer in the surface and in the thin films it is necessary to ensure that the free surface of the polymer exists.

Various methods can be used to deposit the metal particles on the polymers. One of the most widely used methods is the thermal or ion-plasma spraying of metals in vacuum on polymer films. In [60] the authors presented a detailed review of the phenomena accompanying the vacuum spraying of metals on polymers of this type. The main types of interaction of the metal atoms with the polymer surface are shown schematically in Fig. 3.8.

Spraying is accompanied by the adsorption of metal atoms on the polymer surface, their surface diffusion, nucleation and growth of clusters, bulk diffusion, bulk aggregation, and the penetration of the metallic clusters into the polymer surface. Discrete metal nanoparticles form in this period and their size and concentration depend on several factors. The formation of critical clusters is important for describing the nucleation and growth processes. If the clusters themselves travel on the polymer surface, they can also grow by means of agglomeration [61]. The main problem here is the aggregation of the atoms in the clusters which takes place even at very low metal concentrations on the surface [62]. The results

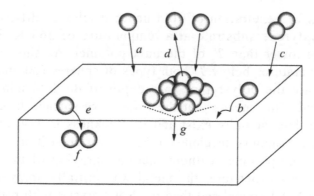

Fig. 3.8. Diagram of the main processes taking place in the initial stages of deposition of metal vapours on the polymer surface: a) adsorption; b) surface diffusion; c) nucleation and growth of clusters; e) bulk diffusion; f) bulk aggregation; g) penetration of metallic clusters into the polymer [60].

of this diffusion to bulk polymers have been recently studied in a detailed review [63]. The aggregation of the atoms in the clusters prevents their diffusion from the surface into the bulk. Nevertheless, the diffusion in the volume can be greatly intensified if the coating is deposited at higher temperatures (to increase the diffusion coefficient) or by using low spraying rates (to slow down aggregation of the atoms in the clusters). And yet, the intensity of the diffusion of the metals in the polymer system is 10 orders of magnitude smaller than that of the gases with similar dimensions and mass. The driving force of the penetration of the metallic nanoparticles from their dispersions into the bulk polymer is the decrease of their high surface energy [64]. However, this penetration of the nanoparticles into the polymer can take place when the mobility of the polymer chains is sufficiently high. In this case, it should be mentioned that, as shown previously, the T_g temperature greatly decreases in the vicinity of the surface.

When examining the surface phenomena, taking place during the deposition of the metallic nanoparticles on the polymers, it is important to note that, regulating the parameters of spraying, it is possible to deposit on the polymer a family of isolated nanoparticles, and the dimensions and number of these particles can be regulated in a wide range. For example, in [65] the mobility of the chains on the polystyrene surface was characterised when investigating the penetration of the metal nanoparticles into the surface. The process was studied *in situ* by x-ray photoelectronic spectroscopy. The penetration of the nanosized clusters requires the large-scale mobility of the chains in the surface of the polymer and, therefore, the penetration process is in fact a probe for evaluating T_g on the surface. The threshold temperature at which the penetration of the nanoparticles into the polymer surface starts T_p was determined. This temperature increases in heating of the polymer-substrate and is comparable with the T_g of the bulk polymer, determined by the DSC method; T_p also increases with increase of the average cluster size. The value T_p is in qualitative agreement with the Fox–Flory equation, but with the saturation temperature 8 K lower than the block value. This difference $\Delta T = T_g$ (bulk) $-T_p$ increases with the molecular weight of the polymer.

In [47], the authors investigated the kinetics of the process of penetration of the nanoparticles into the surface layer of the polymer to determine its viscosity which was considerably lower than the appropriate characteristics for the same polymer in the bulk. The structure of the polymer with the nanoparticles on the surface layer is influenced by: substrate temperature, the deposition rate of the metal and the number of defective areas on the polymer surface. For example, in [45] a 1.5 nm gold layer was deposited on the surface of a polycarbonate film by thermal evaporation in high vacuum. The film with the coating was then held in acetone vapours for 303 K for different periods of time. Acetone greatly accelerated the crystallisation of polycarbonate [66]. Consequently, the surface of the samples showed distinctive spherulites with a diameter of 40 μm. The size of the spherulites depended only slightly on the holding time in acetone vapours in the time range 10–30 h. The sections showed the population of gold nanoparticles with the size of 8–10 nm in the crystallised film of the polycarbonate. Indirect evaluation showed that the treatment of the polymer with acetone helps to insert the gold nanoparticles into the polymer to a depth of 150 nm±20 nm. The authors assume that the layer of the gold clusters in the surface of the amorphous polymer can penetrate into the polymer block during the crystallisation induced by the vapours.

The process of penetration of the nanoclusters into the surface layer of the polymer influences most markedly the adhesion of the metallic coatings on the polymer substrate. In [67] the combination of transmission and atomic force microscopy showed that long-term annealing above T_g results in the aggregation of the clusters in the surface and their penetration into the surface layer of the polymer. For example, in [60] it was shown that a copper coating, deposited on polycarbonate, greatly increased the peel strength of the metal from the substrate in annealing above T_g (250°C). The detected effect is explained by the increase of the mobility of the polymer chains leading to the healing of some microcavities and a chemical reaction between the metal and the polymer.

To conclude this section, it is important to mention the studies in which the metallic nanoparticles were deposited on the polymer surface by their direct absorption from dispersions. In [68] it was reported that the nanoparticles of a gold hydrosol, sorbed on the surface of polystyrene, form two-dimensional ordered ensembles. A suspension of the gold nanoparticles with the size of 80–20 nm was sorbed on the polystyrene surface and the film was then annealed at 80°C (i.e., lower than T_g), and examination in the atomic force microscope showed that these nanoparticles form on the polystyrene surface a two-dimensional ordered lattice of the nanoparticles, almost completely penetrated (to a depth of 15– 17 nm) into the polystyrene surface. It is assumed that this can take place only if the surface layer of the polymer is devitrified. Identical conclusions were published by the authors of [69, 70] where the atomic force microscope was used to study the dynamic of interaction of the nanoparticles of the gold hydrosol with the polystyrene surface. The authors used the softened state of the surface layer of the polymer for 'embedding' in the layer the colloidal particles of the metals in order to produce the two-dimensional nanocomposites. An example of such 'embedding' of the nanoparticles in the polymer surface is shown in Fig. 3.9.

The dimensions of the sorbed nanoparticles on the polymer surface can be controlled. In [72] the authors reported the additional growth during annealing of the gold nanoparticles, adsorbed on the surface of the polymer film and 'immersed' in the surface layer. Sorption of

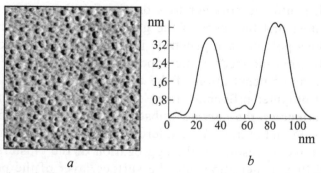

a b

Fig. 3.9. Atomic force microscope image (scanning area 3×3 μm²) (a) and the appropriate cross-section of the polystyrene surface with the deposited platinum nanoparticles (b) [71].

the particles takes place in such a manner that a percolation network, detected by electrical conductivity measurements, forms.

Precision scanning of the properties in the depth for the bulk polymer can be carried out by neutron scattering [73]. The neutron scattering of deuterised polystyrene shows the strong dependence of the thermal expansion coefficients on the distance from the surface in the thin films and in the surface layers of the bulk polymers. A large change of the thermal expansion coefficient for polystyrene was observed in the thin (35–60 nm) layers.

Important results were obtained in a series of studies [74–76) in which attention was given to the processes of diffusion of macromolecules through the phase boundary below T_g. In these studies, the polymer surfaces were joined (PS and PPO) and held for different periods of time in contact at different temperatures, but always below the glass transition temperature. The results show that holding results in adhesion between the contacting surfaces of the same type (PS–PS and PPO) and in the formation of a heterophase contact (PS–PPO). The strength of the adhesion contact is proportional to time $t^{1/4}$ i.e., it is controlled by the diffusion of polymer chains or their fragments through the interface. The results show unambiguously the high segmental mobility in the polymer surfaces in the temperature range well below the bulk T_g.

Summing up the data, obtained in the study of the molecular mobility in the properties of the thin (units or tens of nanometres) surface layers of the bulk polymers, it may be concluded that, as in the case of the thin films, these layers are also characterised by the high-speed molecular motion as a result of which the surface layers of the bulk polymers have a considerably lower glass transition temperature than the bulk polymer. In addition, some authors assume [15, 74] that at room temperature all glassy polymers are coated with the thinnest layer of the 'devitrified' rubbery material.

3.6. Possible reasons for the decrease of the glass transition temperature in thin films and surface layers of amorphous polymers

Analysis of the above data indicates that the free-standing thin films and in many cases thin films on the substrates, and also the surface layers of the amorphous polymers are characterised by a large decrease of T_g and that the conditions for large-scale molecular mobility are significantly improved (α-relaxation). Naturally, the effort of many scientists has been directed to the determination of the mechanism of this phenomenon. Regardless of this, at present time there is no united view regarding the mechanism of such a large decrease of T_g in the thin surface layers.

One of the approaches to solving this problem is associated with the examination of the process of glass transition of the polymer in a limited volume (here we are concerned with the situation in which the polymer phase is so small that there are difficulties in the implementation of the equilibrium conformations of the chains and the nature of large-scale molecular motion changes). As is well-known [77], the glass transition process is determined by the cooperative type of molecular motion which requires the presence of a relatively long polymer phase. In [78], the second harmonic generation method was used to measure the mobility in the surface layers of low-molecular glass forming liquids. The results show that the surface layers are characterised by the disruption of the collective dynamics of molecular motion (the rate of molecular motion increases in comparison with the bulk polymer), leading to the glassy state.

A large increase of the rate of molecular motion in the surface layers of the polymers was observed in [79], where the special features of molecular motion in poly-4-methylstyrene, PS and poly-n-tret-butylstyrerne were studied by x-ray photoelectronic spectroscopy. It was shown that the surface energy of activation of the diffusion of the solid polymers at room temperature (92 kJ/mol) is considerably smaller than the activation energy of self-diffusion of polystyrene in the melt at a considerably higher temperature of 170–190°C (167.6 kJ/mol). It was assumed that the result indicates a decrease of the scale of the molecular motion and this reflects the disruption of the collective dynamics of the polymer in the surface layer in relation to the volume. In [80], local thermal analysis was used to determine the T_g of the thin films of PS and PMMA on SiO_2 substrates, treated with hexamethyl disilazane (HMDS). The results show that the T_g of PMMA increases on the SiO_2 substrate (polar) and on the SiO_2–HMDS substrates it decreases (non-polar). In the case of polystyrene the decrease is observed on both substrates.

In the case of the amorphous polymers, this collective dynamics is disrupted at the thickness of the polymer phase smaller than the RMS distance between the ends of the chain. Consequently, the size of the 'devitrified' section of the chain decreases and, as a result, the glass transition temperature is also lower.

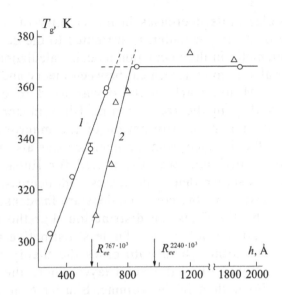

Fig. 3.10. Dependence of the glass transition temperature of the free-standing polystyrene films with the molecular mass $M_w = 767 \cdot 10^3$ (1) and $M_w = 2240 \cdot 10^3$ (2) on the film thickness. The unperturbed dimensions of the coil are indicated by the arrows [27].

Figure 3.10 shows the dependence of the glass transition temperature of the polystyrene films for two values of the molecular mass ($76.7 \cdot 10^4$ and $22.4 \cdot 10^5$) on their thickness h [27]. The same figure shows the appropriate unperturbed dimensions of the coils of the macromolecules for the given molecular masses. It may clearly be seen that the start of decrease of T_g correlates to a certain degree with the unperturbed dimensions of the coils of the macromolecules.

The effect of the spatial restrictions on the nature of the molecular motion of the polymer chains has also been confirmed by some theoretical studies. Using the interaction model, in [81] it was shown theoretically that a large decrease of T_g is observed when the thickness of the free film becomes smaller than the distance between the ends of the unperturbed chains. It is assumed that this circumstance in particular is the reason for the strong dependence of T_g on the molecular mass in the thin polymer films.

It has been proposed that the free films with a larger molecular mass firstly induce the planar orientation of the polymer chains when their distance between the ends of the unperturbed chains becomes comparable with the film thickness and, secondly, reduce the cooperative length of the chain segments in the vicinity of the surface. Consequently, the scale of the local segmental movement of the molecular units decreases in the vicinity of the interface. The decrease of T_g, in particular, is attributed to the decrease of the interaction parameter in the interaction model. Calculation of T_g using this model results in good agreement between theory and experiment.

In [82] the Monte Carlo method was used to examine the molecular mobility in the free-standing films in dependence on their thickness and the conformation of the macromolecules in dependence on the distance from the interface was also investigated. The diffusion coefficient was calculated for linear and cyclic molecules. It was shown that T_g decreases with decrease of the film thickness. In these films the conformations are far denser (compact) than indicated by the Gaussian distribution. For the ring-shaped conformations, all the defects are far less distinctive than for the linear molecules, although in both cases the theory predicts the existence of a liquid-like interfacial layer where the mobility is considerably higher than in the volume. Similar conclusions were also drawn by the authors of the theoretical study [83] in which the molecular dynamics method was used to show that the mobility of amorphous polyethylene increases, and the density decreases on approach from the bulk to the surface. One of the varieties of this viewpoint is the concept, proposed in studies by DeGennes [84] of the possibility of existence of the 'sliding' type of molecular motion of polymer chains in the restricted volume. As shown by calculations, in this case, the glass transition temperature should decrease when the geometrical dimensions of the polymer phase become comparable with the unperturbed dimensions of the molecules.

Another popular viewpoint is the one according to which the reason for the decrease of the glass transition temperature on the surface layers is the segregation of the end groups of macromolecules in these layers [12–15, 17–20]. The concept of segregation of the interfacial surface of the end groups is based on the assumption that, according to the Fox–Flory theory [16], the ends of the chains are in fact a plastification agent decreasing T_g. This is also assumed that the higher concentration of the end groups in the surface layers of the polymers in comparison with the bulk increases the local content of the free volume in these layers and at the same time, decreases the glass transition temperature. To verify the assumption, the method of dynamic secondary ion-mass spectroscopy was used to study deuterised polystyrene with protonised end groups [12]. The data were used to obtain the concentration profile of the H-end groups in the direction normal to the film surface. The increase of the concentration of the H-groups in the surface in comparison with the D-groups was higher and this confirms the hypothesis of the increase of the concentration of the molecules in the surface layer of the polymer.

To explain the mechanism of decrease of T_g in the surface layers and thin films of the amorphous polymers, the authors of [85] synthesised three polystyrene samples with different end groups: conventional PS, and PS2, PS3 – polystyrene with the end groups NH_2 and COOH, respectively and also a large number of the monodispersed fractions of all three polymers. The nature of the end groups of the molecular mass was varied to explain the effect of the end groups and the cooperative length of the vitrified element of the structure on the decrease of T_g on the surface. The values of T_g were determined by atomic force microscopy and scanning viscoelastic microscopy, which make it possible to determine the dynamic mechanical characteristics in the surface layers of the polymers. Photoelectronic x-ray spectroscopy was used to determine the concentration of the end groups at the surface. It was shown that for both the bulk and the surface T_g decreases as a result of the change of the molecular mass. In the entire molecular mass range the surface T_g decreased in comparison with the bulk T_g. Surface T_g started to decrease at smaller molecular masses, and in dependence on the molecular mass its rate of decrease was greater than that of the block. Using the principle of temperature–time superposition, the effective energy of activation of the surface α-relaxation (ΔH) was determined from the expression:

$$\ln a_t = \Delta H/R(1/T - 1/T_0) \qquad (3.2)$$

where a_t is the shear factor, T and T_0 is the measurement temperature and the reduced temperature, respectively, R is the universal gas constant. It was shown that in all cases the activation energy of the α-transition on the surface is equal to 230 ± 10 kJ/mol, whereas for the bulk this value according to different data varied from 367 to 880 kJ/mol [86]. The results obtained in [86] indicate that both factors examined previously may influence the decrease of T_g on the polymer surface.

Some authors [87–89] have proposed a three-layer model for describing the decrease of a_t in the thin polymer layers. The dynamics of each layer was different. In the vicinity of the free surface the mobility was higher than in the bulk and in the centre it was equivalent to the bulk and at the intermediate phase it was limited in comparison with the bulk. The model, based on this approach, resulted in good agreement between theory and experiment.

To conclude this section, it should be mentioned that the spatial restrictions, which the macromolecules experience in the thin-films and surface layers, are reflected not only in the molecular motion, responsible for α-relaxation. In [91], analysis of the fluorescence spectra was used to investigate the association of pyrene groups in PDMS with pyrene end groups. The spectra were used to analyse the relative concentrations of the associated and non-bonded pyrene groups. This relationship does not depend on the film thickness greater than 100–200 nm, but decreases 40 times when the thickness is reduced to 5 nm. The effect of the thickness on association is explained by the presence of two interfacial or surface layers with a thickness of ~10 nm, with each layer having a greatly reduced concentration of the physical cross-links in relation to the rest of the film. It is assumed that the decrease of the extent of association of the groups has the same reasons as the decrease of the glass transition temperature.

Conclusion

In conclusion, should be mentioned that the largest decrease of the glass transition temperature in the thin (nanometric) films and in the surface layers of the amorphous polymers, observed in the middle of the 90s, is their fundamental property. Previously, this phenomenon had not been observed because in the examination of the properties of the bulk polymers the contribution to the properties of the surface layers was negligible and not taken into account. However, in cases in which deformation is accompanied by the development of a high level of the interfacial surface, this contribution must be taken into account. In this section, attention was given only to the small part of the literature data which can be re-examined taking into account the new results of studies of the special features of the polymer surfaces on their structure and the fundamental properties of the polymers as a whole. Evidently, the further development of this important scientific direction offers new possibilities for creating a general theory which would make it possible to describe the adequate relationship between the structure and properties of the polymers.

References

1.   Shchukin E.A., Pertsov A.V., Amelina E.A., Colloid chemistry, 3rd ed., Moscow, Vysshaya shkola, 2004, P. 445.
2.   Voyutsky S.S., Colloidal chemistry course, Moscow, Khimiya, 1976, P. 512.
3.   Schwab A.D., Agra D.M.G., Kim J.H., Kumar S., Dhinojwala A., Macromolecules, 2000, V. 33, No. 13, P. 4903.
4.   Forrest J.A., Svanberg C., Revesz K., Rodahl M., Torell L.M., Kasemo B., Phys. Rev. E, 1998, V. 58, No. 2, P. 1226.
5.   Cao H., Yuan J.P., Zhang R., Sundar C.S., Jean Y.C., Suzuki R., Ohdaira T., Kobayashi Y., Nielsen B., Appl. Surf. Sci., 1999, V. 149, P. 116.
6.   Xie L., DeMaggio G.B., Frieze W.E., DeVries J., Gidley D.W., Hristov H.A., Jee A. F., Phys. Rev. Lett., 1995, V. 74, No. 24, P. 4947.
7.   Jean Y.C., Cao H., Dai G.H., Suzuki R., Ohdaira T., Kobayashi Y., Hirata K., Appl. Surf. Sci., 1997, V. 116, P. 251.
8.   Wolff O., Johannsmann D., J. Appl. Phys., 2000, V. 87, No. 9, P. 4182.
9.   Keddie J.L., Johnes R.A.L., Cory R.A., Europhys. Lett., 1994, V. 27, P. 59.
10. Ferry J. D., Viscoelastic properties of polymers, 2 ed., New York, Wiley, 1970.
11. Hyun J., Aspens D,E., Cuomo J.J., Macromolecules, 2001, V. 34, No. 8, P. 2396.
12. Kajiyama T., Tanaka K., Takahara A., Polymer, 1998, V. 39, No. 19, P. 4665.
13. Kadjiyama T., Tanaka K., Takahara A., Macromolecules, 1997, V. 30, No. 2, P. 280.
14. Satomi N., Takahara A., Kajiyama T., Macromolecules, 1999, V. 32, No. 13, P. 4474.
15. Kajiyama T., Tanaka K., Satomi N., Takahara A., Sci. Technol. Adv. Mater., 2000, V. 1. P. 31.
16. Fox T., Flory P., J. Polym. Sci., 1954, V. 14, P. 315.
17. Tanaka K., Takahara A., Kajiyama T., Macromolecules, 1997, V. 30, No. 21, P. 6626.
18. Kajiama T., Tanaka K., Takahara A., Proc. Japan Acad., 1997, V. 73B, No. 7, P. 132.
19. Tanaka K., Kajiyama T., Takahara A., Acta Polymerica, 1995, V. 46, P. 476.
20. Kajiyama T., Tanaka K., Takahara A., Macromolecules, 1995, V. 28, No. 9, P. 3482.
21. Kajiyama T., Tanaka K., Takahara A., Macromolecules, 1995, V. 28, No. 9, P. 3482.
22. Kim J.H., Jang J., Lee D-Y., Zin W-C., Macromolecules, 2002, V. 35, No. 6, P. 311.
23. Overney R.M., Buenviaje C., Luginbh R., Dinelli F., J. Therm. Anal. Calorim., 2000, V. 59, P. 205-225.
24. Bollinne C., Stone V.W., Carlier V., Jonas A.M., Macromolecules, 1999, V. 32, No. 14, P. 4719.
25. Frank B., Gast A.P., Russel T.R., Brown H.R., Hawker C., Macromolecules, 1996, V. 29, No. 20, P. 6531.
26. Zheng X., Sauer B.B., Van Alsten J.G., Schwartz S.A., Rafailovich M.H., Sokolov J., Rubinstein M., Phys. Rev. Lett., 1995, V. 74, P. 407.
27. Forrest J.A., Dalnoki-Veress K., Dutcher J.R., Phys. Rev. E., 1997, V. 56, No. 5, P. 5705.
28. Pochan J.D., Lin E.K., Satija S.K., Wen-li Wu, Macromolecules, 2001, V. 34, No. 9, P. 3041.
29. Freyer D.S., Peters R.D., Kinm E.J., Tomaszewski J.E., de Pablo J.J., Nealy P. F., Macromolecules, 2001, V. 34, No. 16, P. 5627-5634.
30. Schwab A.D., Agra D.M.G., Kim J.H., Kumar S., Dhinojwala A., Macromolecules, 2000, V. 33, No. 13, P. 4903.
31. Wang J., Tolan M., Seek O.H., Sinha S.K., Bahr O., Rafailovich M.H., Sokolov J., Phys. Rev. Lett., 1999, V. 83, No. 3, P. 564.

32. Van der Lee A., Hammon L., Holl Y., Grohens Y., Langmuir, 2001, V. 17, No. 24, P. 7664.

33. Hartmann L., Gorbatschow W., Hauwede J., Kremer F., Europ. Phys. Journ. E, 2002, V. 8, Issue 2, R. 145-154.

34. Porter C.E., Blum F.D., Macromolecules, 2000, V. 33, No. 19, P. 7016.

35. Forrest J.A., Dalnoki-Veress K., Stevens J.R, Dutcher J.R., Phys. Rev. Lett., 1996, V. 77, P. 2002.

36. Forrest J.A., Dalnoki-Veress K., Adv. Colloid and Interface Sci., 2001, V. 94, P. 167.

37. Stafford C.M., Guo S., Chiang M.Y.M., Harrison C., Rev. Sci. Instrum., 2005, V. 76, No. 6, P. 062207.

38. Stafford, C.M., Vogt B.D., Harrison C., Julthongpiput D., Huang R., Macromolecules, 2006. V. 39, No. 15, P. 5095-5099.

39. Stafford C.M., Harrison C., Beers K.L., Karim A., Amis E.J., Vanlandingham R., Kim H.-C., Volksen W., Miller R.D., Simonyi E.E., Nature materials, 2004, V. 3, P. 545.

40. Huang R., Stafford C.M., Vogt B., Journ. Aerospace Eng., 2007, V. 20, No. 1, P. 38.

41. Bowden N., Brittain S., Evans A.G., Hutchinson J.W., Whitesides G.M., Nature, 1998, V. 393, P. 146.

42. Volynskii A.L., Bazhenov S.L., Lebedeva O.V., Bakeev N.F., J. Mater. Sci. 2000, V. 35, No. 3, P. 547.

43. Lee W.K.., Yoon J.-S., Tanaka K., Satomi N., Jiang X., Takahara A., Ha C.-S., Kajiyama T., Polym. Bull., 1997, V. 39, P. 369.

44. Fischer H., Macromolecules, 2002, V. 35, No. 9, P. 3592.

45. Svorchik V., Rubka V., Jankovskij O., Hnatowicz V., J. Appl. Polym. Sci. 1996, V. 61, P. 1097.

46. Ho P.S., Haight R., Wight R.S., Fapuel F., Fundamentals of adhesion. New York, Plenum, 1991.

47. Helmann A., Hamann C., Prog. Colloid Polym. Sci., 1991, V. 85, P. 102.

48. Mittal K.L. (ed.), Metallized plastics: Fundamentals and applications, New York, Marcel Dekker, 1998.

49. Wilecke R., Fapuel F., J. Polym. Sci., B, 1997, V. 35, P. 1043.

50. Koziol K., Dolgner K., Tsuboi N., Kruse J., Zaporojtchenko V., Deki S., Faupel F., Macromolecules, 2004, V. 37, No. 6, P. 2182-2185.

51. Kim H., Ruhm A., Lurio L.B., Basu J.K., Lal J., Lumma D., Mochrie S.G.J, Sinha S. K., Phys. Rev. Lett., 2003, V. 90, P. 068,302.

52. Teichroeb J.H., Forrest J.A., Proc. Symp. Mat. Res. Soc., 2001, V. 734, P. B3.2.1.

53. Teichroeb J.H., Forrest J.A., Phys. Rev. Lett., 2003, V. 91, P. 016104.

54. Rimai D.S., Schaefer D.M., Bowen R.C., Quesnel D.J., Langmuir, 2002, V. 18, P. 4592.

55. Zaporojtchenko V., Strunskus T., Erichsen S., Fapuel F., Macromolecules, 2001, V. 34, No. 5, P. 1125.

56. Kovacs G.J., Vinsett P.S., J. Colloid. Interface Sci., 1982, V. 90, P. 335.

57. Bechtolsheim C., Zaporotchenco V., Fapuel F., Appl. Surf. Sci. 1999. V. 151, P. 119.

58. Zaporojtchenko V., Behnke K., Strunscus T., Fapuel F., Surf. Interface Anal., 2000, V. 30, P. 439.

59. Sharp J.S., Teichroeb J.H., Forrest J.F., Eur. Phys. J. E, 2004, V. 15, P. 473-487.

60. Zaporojtchenko V., Strunskus T., Behnke K., Von Bechtolsheim C., Kiene V., Fapuel F., J. Adhesion Sci. Technol., 2000, V. 14, No. 3, P. 467.

61. Kiene M., Strunskus T., Peter R., Faupel F., Adv. Mater., 1998, V. 10, P. 1357.

62. Strunskus T., Kiene M., Willecke R., Thran A., von Bechtolsheim C., Faupel F.,

Werkst. Korros., 1998, V. 49, P. 180.
63. Faupel F., Willecke R., Thran A., Mater. Sci. Eng., 1998, V. R22, P. 1.
64. Kovacs G.J., Vincett P.S., Colloid Interface Sci., 1982, V. 90, P. 335.
65. Erichsen J., Kanzow J., Schurmann Ul., Dolgner K., Günther-Schade K., Strunskus T., Zaporojtchenko V., Faupel F., Macromolecules, 2004, V. 37, No. 5. P. 1831-1838.
66. Ogawa T., Masuichi M., J. Appl. Polym. Sci., 1997, V. 30, P. 943-949.
67. Kovacs G.J., Vincett P.S., Tremblay C., Pundsack A.L., Thin Solid Films, 1983, V. 101. P. 21.
68. Rudoi V.M., Yaminsky I.V., Dement'eva O.V., Ogarev V.A., Kolloid. Zh., 1999, V. 61, No. 6, P. 861.
69. Dement'eva O.V., Rudoi V.M., Yaminsky I.V., Sukhov V.M., Stuchebryukov S.D., Ogarev V.A., Proc. Second All-Russian Karginsky Symposium (with international participation), Chemistry and Physics of Polymers at the beginning of the 21st century, Chernogolovka, 2000, V. 41, S1-96S.
70. Rudoi V.M., Dement'eva O.V., Yaminsky I.V., Ogarev V.A., Proc. Second All-Russian Karginsky Symposium (with international participation), Chemistry and Physics of Polymers at the beginning of the 21st century, Chernogolovka, 2000, V. 41, S3-102.
71. Sukhov N.L., Dement'eva O.V., Kartseva M.E., Rudoi V.M., Ogarev VA, Kolloid. Zh., 2004. V. 66, No. 4, P. 539-546.
72. Dement'eva O.V., Kartseva M.E., Bolshakova A.V., Vereshchagin O.F., Ogarev V.A., Kalinin M.A., Rudoi V.M., Kolloid. Zh., 2005. V. 67, No. 2, P. 149.
73. Pochan J.D., Lin E.K., Satija S.K., Wen-li Wu, Macromolecules, 2001, V. 34, No. 9, P. 3041.
74. Boiko Y.M, Prudhomme R.E., Macromolecules, 1998, V. 31, No. 19, P. 6620.
75. Boiko Yu.M., Vysokomolek. Soed., B, 2000, V. 42, No. 3, P. 542.
76. Boiko Yu.M., Vysokomolek. Soed., A, 2002, V. 44, No. 7, P. 1160.
77. Rostiashvili V.G., Irzhak V.I., Rosenberg B.A. Vitrification of polymers, Leningrad, Khimiya, 1987.
78. Jerome B., Commandeur J., Nature, 1997, V. 386, P. 589.
79. Rouse J.S., Twaddle P.L., Ferguson G.S., Macromolecules, 1999, V. 32, No. 5. P. 1665.
80. Freyer D.S., Nealey P.F., de Pablo J., Macromolecules, 2000, V. 33, P. 3376.
81. Ngai K.L., Rizos A.K., Plazek D.J., J. Non-Cryst. Solids, 1998, V. 235–237, P. 435.
82. Jain T.S., de Pablo J.J., Macromolecules, 2002, V. 35, No. 6, P. 2167.
83. Doruker P., Matice W.L., Macromolecules, 1999, V. 32, No. 1, P. 194.
84. De Geness P.G., Eur. Phys. J., 2000, V. 2, P. 201.
85. Satomi N., Tanaka K., Takahara A., Kajiyama T., Ishizone T., Nakahama S., Macromolecules, 2001, V. 34, No. 25, P. 8761.
86. Santagleo P.G., Roland C.M., Macromolecules, 1998, V. 31, No. 12, P. 4581.
87. Forrest J.A., Mattsson J, Phys. Rev. E, 2000, V. 61, P. R53.
88. Fukao K., Miyamoto Y., Phys. Rev. E, 2000, V. 61, P. 1743.
89. Kim J., Jang H., Zin J.W.C., Langmuir, 2000, V. 16, P. 4064.
90. Lebedev D.V., Ivankov E.M., Marikhin V.A., Myasnikov L.P., Saydenwitz V., Fiz. Tverd. Tela, 2009, V. 51, No. 8, P. 1645-1652.
91. Kim S.D., Torkelson J.M., Macromolecules, 2002, V. 35, No. 15, P. 5943

Role of surface phenomena in shear yielding of glassy polymers

The above results indicates that the lowering of the glass transition temperature in thin films and surface layers of amorphous polymers is their fundamental property. Obviously, this feature of the amorphous polymers must affect their macroscopic properties in a temperature range lower than the glass transition temperature in the cases when, for one reason or another, the polymer contains a developed interface

Before proceeding to consider the effect of surface effects on the properties of glassy polymers, it is necessary to briefly recall some basic modern concepts of the mechanism of deformation. All the characteristic properties of polymers are due to the chain structure of their molecules and the fundamental property related to the structure – flexibility. Flexibility of macromolecules, i.e. the ability to change their shape (conformation) under the action of external stresses and thermal motion, determines in particular one of the unique mechanical properties of the amorphous polymers – large reversible deformation. According to this property the polymer stretched many times almost instantaneously restores its size once it is released from the tension device. Note that large reversible deformation is uniform (affine), that is, it develops simultaneously throughout the volume of the deformed polymer, that is, it develops simultaneously throughout the volume of the deformed polymer [1].

In its classic form large reversible deformation occurs only in polymers which are in the rubbery state. Below the glass transition temperature T_g large strains may also be realized, however their reversibility is achieved only when the deformed polymer is heated above T_g.

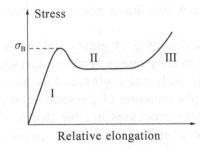

Fig. 4.1. The typical stress–strain curve of a glassy polymer. Explanation in the text.

The mechanical behavior of glassy polymers is usually illustrated by the stress–strain curve (Fig. 4.1). Such a curve contains three main portions: the so-called elastic portion (I), the plateau region, where a neck develops (II), and a portion where there is homogeneous deformation of the polymer transformed to a neck (III). We will examine and analyze the main phenomena and processes taking place in the first two sections of the stress–strain curve of the polymer. A brief description of polymers, subjected to deformation to these values, is given in textbooks as follows: 'In the first section (stress–strain curve) the polymer is deformed elastically. The deformation is achieved by increasing the intermolecular distances. The stretched glassy polymer (up to strains values corresponding to the second portion of the stress–strain curve) does not shorten spontaneously. However, when heated above T_g, once segments gain the ability to thermal movements, the sample will be reduced to a length close to the original' [2].

The above-cited provisions of the glassy state of the polymers as a state in which the large-scale molecular motion is 'frozen', are not completely correct. For more than half a century in the scientific literature experimental data that do not fit in the above frame were accumulated. Let us consider the main results of the anomalous structural and mechanical behaviour of amorphous glassy polymers. In this connection, it is first necessary to investigate the feasibility of certain types of large-scale molecular motion in the temperature within the glassy state of the amorphous polymers.

The structure and properties of glassy polymers over the years have been the subject of extensive research. Interest in the study of the glassy state of the polymers is due to at least two factors. Firstly, the widespread practical use of glassy polymers may provide a better understanding of the relationship between their structure and properties. Secondly, some of the fundamental properties of glassy polymers have a number of features which have not as yet been completely explained.

One of these features is the physical ageing of glassy polymers. Studies on this issue are so numerous that by now a separate section of the science of ageing processes in polymeric glasses has been formed. In the most general form, the phenomenon of physical ageing of glassy polymers included spontaneous processes in time that lead to a noticeable change in the whole complex of their properties (density, enthalpy, permittivity, dynamic and static mechanical properties). These changes occur over time in the conditions of isothermal annealing, and as a result of mechanical action and, very importantly, are carried out in the temperature range of the glassy state of the polymer. On one hand, these phenomena suggest the implementation of large-scale molecular motion in polymers that are below their glass transition temperature. The nature of this relaxation process largely remains controversial. On the other hand, this question has an undoubted practical value, since there is the need for adequate prediction of the long-term properties of numerous products obtained on the basis of glassy polymers.

The purpose of this section is to examine the currently available experimental data on some processes spontaneously taking place in glassy polymers and leading to a noticeable impact on their physical, chemical and physico-mechanical properties.

4.1. Thermal ageing of polymer glasses

For many decades, the authors of numerous studies have noted that the mechanical, thermal, physical and chemical properties of polymer glasses are not stable over time [3]. Such changes over time (physical ageing) are of the general fundamental character. The processes occurring during physical ageing of polymeric glasses are studied by a range of different methods. For example, events that occur during ageing of polycarbonates have been studied by analyzing the mechanical behaviour in uniaxial tension at a constant speed and creep [4, 5], using dynamic thermal analysis [6–10], DSC [11–14], as well as modulated DSC [15–18]. Especially promising seems the use of the positron annihilation technique [19], Doppler spectroscopy [20], as well as dilatometry [21, 22], IR spectroscopy, Fourier transform [23,24], and Raman spectroscopy [25, 26]. A review of these and other methods of investigation of the physical ageing of polymeric glasses is given in [27].

4.2. Main features of the effect of thermal ageing on the properties of glassy polymers

Let us consider the most typical examples of physical ageing phenomena polymer glasses. Figure 4.2 shows the curves of tensile loading of amorphous PET subjected to low-temperature annealing at a temperature of 60°C for various periods time [25]. Note that the above annealing temperature is significantly below the glass transition temperature of PET (75–78°C). Figure 4.2 shows that low-temperature annealing leads to a significant increase in the modulus and yield stress of the glassy polymer. There are similar changes in the dynamic mechanical properties of the polymer. These changes in the mechanical properties in the process of physical ageing of the glassy polymer are also accompanied by a significant change in its thermophysical properties.

This is clearly evidenced by DSC data. Figure 4.3 shows the typical data of this kind. It clearly shows that as annealing occurs a more intense endothermic peak forms in the region of its glass transition temperature. The presence of the peak allows us to estimate the enthalpy of the process occurring in the glassy polymer at its physical ageing [26]. The height of this endothermic peak is also dependent on the time of low-temperature annealing. Figure 4.4 shows the dependence of the dynamic modulus, logarithmic decrement of damping and the enthalpy on the time low-temperature annealing of amorphous PET. It is clearly seen that all these characteristics are clearly correlated with each other and with the data obtained in the mechanical tests, shown in Fig. 4.2.

Importantly, the ageing in the conditions of low-temperature annealing is a thermally reversible process. It is sufficient to heat

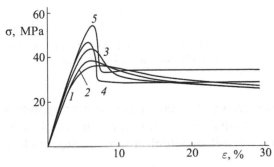

Fig. 4.2. Stress–strain curve of glassy PET at room temperature. Low-temperature annealing time 0.16 (1), 1.5 (2), 16.6 (3), 166.6 (4) and 1166.6 h (5). Annealing temperature 60°C [25].

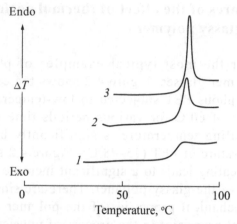

Fig. 4.3. DSC thermograms of amorphous PET, annealed at 65°C for 0 (1), 1 (2) and 45 hours (3) [26].

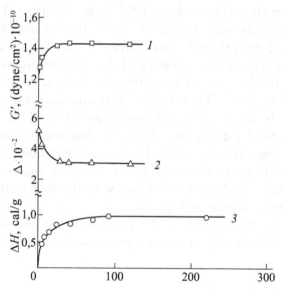

Fig. 4.4. Dependence of the modulus of elasticity (1), the logarithmic decrement of damping (2) and the enthalpy (3) of amorphous PET on annealing time below T_g [28].

the aged polymer above its glass transition temperature to induce its 'rejuvenation' (transition to the state in which it was in prior to annealing), as a result of which a new ageing cycle can be fully implemented

Further studies identified a number of factors that have a significant impact on the ageing process of polymer glasses. For example, in [28]

the authors investigated the mechanical losses and the density of the PMMA quenched after heating to 190°C, depending on the ageing time at low temperature annealing. All the usual changes of the mechanical properties and density with ageing were observed (see Figs. 4.2, 4.3). At the same time it is shown that the rate of change of these properties is not dependent upon the molecular mass of PMMA. The very process of ageing in this study is linked with a change of the conformational set of macromolecules its annealing below T_g.

While the molecular mass of the polymer has no significant effect on the ageing process of the polymer, the latter most strongly depends on its thermal history. This factor has been studied in detail in [29] for the case of amorphous PET by the DSC method. It is shown that for PET the height and the position of the endothermic peak of the polymer (see. Fig. 4.3) is strongly dependent on the thermal prehistory and, in particular, on the cooling rate after heating the polymer to a temperature above T_g. Another factor influencing the kinetics of aging is the chemical structure of a macromolecule and, in particular, the fine structure of its monomer unit. In [30] the authors studied the ageing (relaxation enthalpy) of a number of polystyrenes: PS, poly-4-hydroxystyrene, poly-α-methylstyrene, poly-4-chlorostyrene and poly-methylstyrene. The data obtained were analyzed by the Covey–Fergusson equation:

$$\Delta H(t_a) = \Delta H_\infty \left[1 - \exp\left(-(t_a / \tau)^\beta \right) \right] \qquad (4.1)$$

where $\Delta H(t_a)$ is the isothermal enthalpy of the glass after the aging time t_a, ΔH is the enthalpy of equilibrium glass, $\exp(-(t_a/)/\tau)\beta$ is the structurally sensitive parameter. There was no effect on the aging process for all the *para*-substituted PS. At the same time, the poly-4-hydroxystyrene showed an unexpectedly high value of ΔH_∞. A noticeable impact on the ageing process is also exerted by the structure of the macromolecule. For example, in [31] using DSC, thermally stimulated currents and thermally stimulated creep, the authors studied the effect of the tacticity of the chain on the ageing process of PMMA. Analysis of the data indicates that the molecular mobility depends on the tacticity of PMMA chains. In particular, the greater the amount of the iso-triads, the more flexible the PMMA chain and the more effective the ageing process.

A noticeable impact on the ageing process has the molecular orientation of the polymer. In [32] the authors studied the effect of the drawing ratio of polycarbonate above its glass transition temperature (1600°C) of the ageing rate aging by DSC with the simultaneous evaluation of the dynamic mechanical properties. Surprisingly, the oriented PC samples had the ageing rate 50% greater than the original non-oriented polymer. It is well known that the mobility of the chains decreases with orientation due to more efficient packaging of the chains, and due to the lower percentage of the free volume. It is believed that this effect may be associated with the concept of 'rejuvenation' of the polymer under mechanical stress because the orientation has an activating effect on the chains by transferring them into a higher energy state.

4.4. Physical ageing and the structure of glassy polymers

One of the key issues of the problem of ageing of polymeric glasses is the question of the molecular mechanism of this phenomenon. In this context, it is important to acquire information about structural changes that accompany the ageing process. Amorphous polymers are structurally inhomogeneous solids with the size of the inhomogeneities few to tens of angstroms [33–35]. In numerous studies to explain features of the ageing process, the author used of the inhomogeneous structure of glassy polymers and, in particular, about the existence of ordered domains in them.

The inhomogeneity of the structure of the amorphous polymers is of the fluctuation, non-equilibrium nature, and not of the phase nature [36, 37], and this creates considerable experimental difficulties in investigating the structure and determining the mechanism of the structural rearrangement in the amorphous polymers. These difficulties are caused by the fact that it is not possible to use direct structural methods of investigation, based on the phase contrast of the studied objects (x-ray diffraction analysis, electron diffraction analysis). A large number of the attempts to use [38, 39] direct microscopic methods to solve this problem did not make it possible to create a new universal model of the structure of amorphous polymers

Therefore, the considerations regarding the structure polymer glasses and its evolution in the process of physical ageing are based on the data obtained indirect experimental methods. One of this method is the study of the behaviour low-molecular substances, introduced into the glassy polymers. This is caused by the fact that the characteristics such as diffusion, permittivity, the mobility of low-molecular components in the polymers, are associated with the free volume of the polymer which has a controlling effect on the processes of physical ageing.

For example, in [40] the authors investigated the duration of rotation of the low-molecular probes (tetracene and rubrene) in polystyrene on its ageing time. The results show that the rotation time changes by more than an order of magnitude in the process of isothermal ageing of polystyrene. The time dependence of the variation of the rotation time is completely identical with the time dependences of the relaxation of the volume and the enthalpy. The difference in the evolution of the rotation time of two probes in the polystyrene matrix is explained by its spatial inhomogeneity, as a result of which the ageing process takes place in some small areas of the polymer at a higher rate than in other, denser areas.

Identical results were also obtained by the authors of [41] who investigated the ageing processes in polyetheretherketone at temperatures of 80–1200°C by the method of density measurements, DSC, gas transport, the dynamic mechanical properties and creep. It was shown that in ageing in the vicinity of T_g, in addition to the reduction of the free volume, the extent of sorption of dichloromethane decreases even at low concentrations. The authors of [32–35] explained this phenomenon by the fact that in ageing the polymer shows the formation of domains impermeable for the vapours of the low molecular component.

The physical ageing is accompanied by a large reduction of the permeability of the gases and vapours of the low molecular substances through the glassy polymers. Changes of the permeability (P) are determined by a decrease of the free volume of the polymer as a result of ageing. The permeability coefficient is the product of the equilibrium solubility of the gas (S) and the diffusion coefficient (D) of the gas through the specimen:

$$P = DS \qquad (4.2)$$

Since the solubility of the gases usually changes only slightly in ageing, the variation of the permeability is associated mainly with the diffusion coefficient of the gas. In turn, the diffusion coefficient of the gas is determined by the specific free volume of the polymer (v_f):

$$D = D_0 \exp(-A/v_f) \qquad (4.3)$$

where A is a constant determined by the nature of the gas.

It should be mentioned that the effect of the variation of the permeability is especially strong in ageing of loosely packed hard-chain polymers characterised by the anomalously high value of the specific free volume. For example in ageing for 12 months of the glassy polymer poly-1-trimethylsilyl-1-propane (PTMSP), characterised by a high value of the free volume (29%), the permeability of the gases decreased by more than an order of magnitude [42]. Investigation of the PTMSP film by the positron annihilation method showed that ageing is accompanied by a decrease of the concentration of relatively large elements of the free volume (voids with a diameter greater than 5 Å).

Recently, investigators have been paying special attention to the effect of the scale factor (the thickness of the specimen) on the variation of permeability during ageing. Understanding the ageing processes in the thin films is especially important because in the industry gas separation is carried out using very thin membranes. In early studies [43] it was reported that the rate of physical ageing is independent of the sample thickness. At the same time, in a number of later investigations, the authors reported dramatic changes of the permeability of the thin films of the glassy polymers during ageing in a relatively short period of time. On the example of the polyimide and polysulfone films it was shown that the rate of variation of permeability during ageing depends on the thickness of the samples [44].

In [45] the authors studied the permeability of oxygen through polyarylate films (produced on the basis of biphenol-A-benzophenone dicarboxylic acid) of different thickness. It was shown that the permeability of oxygen decreases with ageing time and this is accompanied by relatively complicated relationships governing the variation of gas permeability with the ageing time for films of different thickness. The authors specified two ranges of the thickness of the polyarylate films. For thick films (thickness range 2.5–20 μm), the loss of permeability as a result of ageing was relatively small and the ageing rate was independent of the film thickness. For thinner films (thickness range 0.25–2.5 μm) the loss of permeability of oxygen as a result of ageing was greater. The ageing rate depends on the thickness of the film: as the film thickness increases, the reduction of permeability becomes greater.

Two mechanisms of evolution of the free volume in ageing was investigated in the study – the natural relaxation of the free volume and the relaxation of the system according to the 'lattice contraction' theory.

According to the authors, the dependence of the ageing rate (the rate of variation of permeability) on the film thickness can be explained on the basis of the assumptions according to which the loss of the free volume takes place as a result of diffusion of holes to the surface of the specimen. The second mechanism – the 'lattice contraction' theory – assumes that the decrease of the intermolecular spacing during ageing takes place simultaneously throughout the entire volume of the sample. According to this mechanism, the rate of the ageing process should not depend on the specimen thickness. The authors assume that the entire set of the experimental data for the ageing of the thick and thin polyarylate films can be explained without contradiction if it is assumed that both mechanisms of the relaxation of the free volume take place at the same time.

At the same time, the authors of [46] assume that the accelerated ageing of the thin films (thickness smaller than 1 μm) can be qualitatively explained on the basis of the well-known assumptions according to which the glass transition temperature of the thin films (thickness smaller than 1 μm) rapidly decreases with a decrease of the thickness of the films [47, 48]. This circumstance increases, with other conditions being equal, the speed of molecular motion in the thin films and this also determines the increase of the rate of physical ageing in these films.

Important information on the free volume of the polymer and its evolution during ageing is provided by the positron annihilation method. For example, in [49] this method was used to study the variation of the free volume of the polycarbonate at three different ageing temperatures: 25, 100 and 120°C. The dependences of the decrease of the free volume on ageing time were obtained in all cases. The results are interpreted as a decrease of the number of holes of the free volume during ageing and this is in good agreement with the Struik free volume model [3]. At the same time, in [50] the positron annihilation method was also used to study the change of the free volume in epoxy resins during ageing. The variation of the free volume is explained by the formation in the ageing polymer of a certain type of the local order as a result of segmental regrouping. In [51], DSC, thermally stimulated currents and thermal stability creep were used to study the effect of the heating rate on the relaxation processes in amorphous PET. The ageing process was linked with the displacement of the α-transition in dependence on the heating rate. The temperature dependences of the relaxation time and the retardation time were obtained, and these dependences were used to calculate the effective energy of activation of the α-transition. On the basis of the shift of the T_g, determined by different methods, in relation to the heating rate it is possible to evaluate the effect of the ageing process on molecular mobility. According to the authors, the magnitude of the displacement of the peak of α-relaxation indicates the change (increase) of the dimensions of the domains of the inhomogeneous structure of amorphous PET in dependence on the heating rate and the change of the cooperative nature of this process as a result of physical ageing. It was assumed that as the size of the domains increases, the cooperative nature of segmental motion becomes more distinctive.

A more complicated mechanism of physical ageing was found by the authors of [52] when studying the processes of physical ageing in polycarbonates. Using the indentation procedure combined with DSC, they studied the changes of the modulus, rigidity and creep of the PC during ageing at room and elevated temperatures. The results show abrupt changes in the modulus and rigidity of the PC after approximately the same ageing time at different temperatures. It was assumed that the free volume has a gradient in the volume of the polymer and is distributed in the PC with a higher concentration and (or) larger holes closer to the surface than in the bulk. Consequently, densening of the PC in annealing takes place non-uniformly with a larger decrease of the free volume closer to the surface and a smaller or no decrease in the bulk. Comparison of the results of the mechanical tests and DSC data showed that there is no direct (linear) relationship between the change of the mechanical properties and the change of the enthalpy taking place during annealing. This observation is in good agreement with the results in [53]. It was assumed that the changed of the mechanical properties are caused by changes of the volume during annealing, whereas the shift of the position and change of the size of the endothermic DSC peaks are associated with the changes of the internal energy of the polymer.

The ageing processes were also observed in partially crystallised polymers. In [54] investigations were carried out into the ageing of polytrimethylene terephthalate. All the classic changes of the properties (modulus, endothermic peak of DSC, and others) were observed in the process of low-temperature annealing. X-ray diffraction studies did not show any changes in the crystallinity or the structure of the crystals of the polymer during ageing. To explain the resultant data, it was assumed that the polymer contains some third phase, in addition to the crystalline and amorphous phases, which includes mutually oriented chains of macromolecules with limited mobility in the vicinity of the crystal surfaces.

In [55] DSC was used to study the effect of ageing of amorphous polyethylene terephthalate on the process of its subsequent cold crystallisation. Ageing at 40°C for 24 hours resulted in the displacement of the peak of cold crystallisation by 17°C. It was assumed that this reflects the movement to a more regular system with the formation of ordered domains, acting as the nuclei for subsequent crystallisation. This is in agreement with the data published in [56]. Ageing at 60°C resulted in the same effect but already after 1 h. However, after 24 hour annealing, the crystallisation peak started increase and reached 150°C after annealing for a week. This effect is explained by the fact that an even larger order, formed in such long-term annealing, decreases the mobility of the chains because T_g increases by 8°C in this case. It was assumed that the ageing of polyethylene terephthalate is a complicated process in which two competing processes take place, one of which facilitates crystallisation (at short ageing times) and the other one complicates this process – at longer times

It is also important to mention the attempts to detect inhomogeneities in the structure (domains) in the amorphous polymers by the direct microscopic method. In [38] investigations were carried out into the effect of low-temperature annealing (below T_g) on the mechanical properties in tensile loading of the polycarbonate. The DSC showed the formation of an endothermic peak instead of a step, usually observed at T_g, with the size of the peak increasing with increasing annealing temperature and time. The position of the peak was displaced into the high-temperature range (by 160°C). Electron microscopic studies (replicas) showed the presence of grains which became larger with increasing annealing temperature and time (the size of the grains changed during annealing from 300 to 700 Å). The increase of the inhomogeneity was also recorded by electron diffraction. It was assumed that all these changes cannot be explained only in terms of the change of the free volume. It was suggested that ageing is accompanied by morphological changes in the structure, but the relationship of these changes with the properties has not been determined.

4.4. Molecular mechanism of thermal ageing of glassy polymers

The formation of the non-equilibrium structure of the polymer glass and its relaxation in the low-temperature annealing conditions are usually associated with the evolution of the free volume during the transition of the polymer from the rubbery to glassy state. The decrease of the fraction of the free volume during cooling of the rubber-like polymer corresponds to the equilibrium conditions only away from the glass transition temperature of the polymer. On approaching the glass transition temperature, the viscosity of the polymer starts to increase rapidly and, consequently, the structure of the polymer, characteristic of the rubber-like state, is 'frozen in' during the transition of the polymer to the glassy state. The process of physical ageing is in fact a spontaneous transition of the polymer glass to the thermodynamic equilibrium state. Evolution of the free volume plays the main role in the process of physical ageing. On the basis of this pattern, several models [3, 57, 58] have been proposed for describing the physical ageing mechanism. All the models are based on the experimental results which make it possible to separate the ageing process into two components: the thermally activated process, described by the Arrhenius equation, and the process, in which the driving force is the excess of the free volume in the system, determining the distance of the system from the thermodynamic equilibrium. In other words, the process of physical ageing of the polymer depends, firstly, on some type of molecular motion in the temperature range of the glassy state and, secondly, on the migration of the free volume. Regardless of the large number of investigations, concerned with the explanation of the physical ageing mechanism, there are still some questions in this problem which have not been answered.

In this section, it is necessary to discuss briefly the question of the nature of molecular motion, responsible for the ageing of polymer glasses. As mentioned previously, the cooling of the polymer to temperatures higher than T_g results in 'freezing' of the non-equilibrium conformations of the molecules. The transition to the equilibrium conformations is attributed to the physical ageing process. Since the ageing process is interrupted at temperatures lower than the temperature of the β-transition, it is rational to assume that in particular the β-relaxation is responsible for the physical ageing processes. The activation energy of the β-transition is usually

~30–50 kJ/mol and increases to 200–300 kJ/mol on reaching T_g [59]. This is explained by the fact that the β-transition is accompanied by the movement of only 1–2 bonds, whereas the α-transition is more cooperative and requires 8–20 bonds to occur. The authors of [60] assume that the physical ageing process is associated with the redistribution of the free volume and not simply with the shift of the relaxation processes on the temperature scale.

The most detailed analysis of this problem was published in [61]. The authors concluded that the observed effects of the physical ageing of the polymer glasses are caused by the β-molecular motion. It should be shown that the β-relaxation is, like the α-relaxation, of the segmental origin, but this is the non-cooperative motion of the segment which in the case of the vitrified solid may take place in the areas of low density molecular packing (free volume). The disruption of the cooperative nature of the molecular motion results in the structural inhomogeneity of the glassy polymer. The latter means that the polymer glass has a non-uniform structure. In some areas the structure of the glassy polymers is such that the cooperative nature of the large-scale molecular motion is disrupted. Therefore, this is the reason why molecular motion is also observed in the temperature ranges below the glass transition temperature. In may be seen that the viewpoint of the authors of [61] is in full agreement with the previously discussed results of a large number of investigations were the authors assume that the structure of the glassy amorphous polymer is non-uniform.

To conclude this section, it should be mentioned that the problem of determination of the physical ageing mechanism of the polymer glasses is far from being solved. In this connection, it is important to mention a number of investigations with the results of these investigations fitting in the framework of the assumptions made about. It is important to find a clear answer to the main question: what is the main reason for the physical ageing of the glassy polymer – the evolution of the free volume or some type of molecular motion, accompanied by the change of the conformation state of the glassy polymer? We shall discuss several examples showing that there is no united view regarding this question.

In [64], the positron annihilation method was used to study the change of the free volume in the polycarbonate at three different ageing temperatures: 25, 100 and 120°C. In all cases, the dependence of the decrease of the free volume on the ageing time was determined. The results are interpreted as a decrease of the number of holes of the free volume during ageing and this is in good agreement with the Struik free volume model [3].

Completely contradicting results were obtained in [63]. In this study, the authors also used the positron annihilation method to investigate the content of the free volume in polyethylene terephthalate and a number of related polymers. The results show that the effect of physical ageing in annealing below T_g was sufficiently distinctive in all cases. Nevertheless, the free volume did not change in these cases. In [64] investigations were carried out into the process of ageing of the polycarbonate and the effect of drawing (to fracture at the room temperature) by Raman and infrared spectroscopy. It was observed that the Raman spectra show a shift of several bands at the orientation of the polymer, which indicates the sensitivity of this method to the conformation rearrangement during tensile loading of the polycarbonate. At the same time, ageing was accompanied by a very small shift of these bands indicating that ageing resulted in quite insignificant conformation changes.

In many studies the authors reported differences in the kinetics of the process determined by different methods. In [65] it was established that there is no direct (linear) relationship between the variation of the mechanical properties and the changes in the enthalpy during ageing of the polycarbonate. This observation is in good agreement with the data in [53]. It was assumed that the changes in the mechanical properties and enthalpy are due to different reasons. The changes of the mechanical properties are associated with the changes of volume during annealing, whereas the shift of the position and the change of the size of the endothermic peak are explained by the change of the internal energy of the polymer.

In [66] the physical ageing of the amorphous polyimide (T_g = 238°C) was studied by DSC and tests of the mechanical properties in the annealing conditions from 15 to 65°C below T_g. The relaxation rate of the enthalpy and the rate of mechanical ageing during creep were determined as a function of the ageing temperature. The relaxation rate of the enthalpy decreased with the decrease of the ageing temperature, whereas the ageing rate of creep remained relatively constant. The thermal rheological complex nature of the phenomenon prevents the application of the principle of temperature–time superposition to the dynamic mechanical data, and this problem is caused by the overlapping of the primary and secondary relaxation processes inside the frequency range used during isothermal ageing. The tensile loading curves at room temperature as a function of the ageing and at 204°C showed no embrittlement, associated with the physical ageing, regardless of the increase of the yield stress and modulus. The attempts to predict the rate of relaxation of enthalpy by means of the mechanical data (by means of the appropriate shift) show that the relaxation time, responsible for the relaxation of enthalpy, is considerably longer than the corresponding time responsible for the relaxation of mechanical behaviour.

It is assumed that ageing takes place by the transition of the quenched glass to a more equilibrium state. This is accompanied by the increase of the density of the glass and, correspondingly, a decrease of the free volume, the configuration entropy, and consequently the molecular mobility. In turn, these processes limit the capacity of the material for further relaxation and densening. Such a process can be referred to as self-limitation or self-retardation. However, recently, the authors of [67] detected that the polycarbonate, aged at 20°C, shows the self-retardation of bulk relaxation only up to the ageing time of $\sim 10^7$ s. Subsequently, the rate of bulk relaxation rapidly increases. This phenomenon contradicts the studies [3, 68] dealing with the self-retardation of the process. In [69] the authors carried out comparative studies of the long-term aged (bulk relaxation) polycarbonate and polystyrene at room temperature. The results show that the polycarbonate is in fact characterised by the self-acceleration of the process in accordance with the data in [67]. At the same time, the polystyrene showed no acceleration of ageing and showed the self-retardation process in the same conditions. It was assumed that since the polycarbonate, in contrast to the polystyrene, can crystallise, the observed acceleration of densening is associated with its crystallisation, or more accurately with the formation of some microregions as domains–precursors of possible crystallisation. However, there has been no confirmation of any crystallisation by the DSC method in the aged specimens and, therefore, this explanation is speculative.

Thus, physical (thermal) ageing of the glassy polymers is their fundamental property. The consequence of this process is the spontaneous structural rearrangement with time leading to a large change of the entire set of the physical properties. It is important to note that the aged polymer can be 'rejuvenated' by heating it above T_g. Subsequently, the ageing process starts again. As shown above, many details of this process have not as yet been explained but important conclusions can already be made. The process of physical (thermal) ageing of the polymer glasses is an affine process, i.e., takes place simultaneously throughout the entire volume of the amorphous glassy polymer. In this sense, the physical ageing of the polymer can be regarded as a volume process. This circumstance must mentioned in the context of the overall subject of this monograph.

4.5. Effect of mechanical action on the process of physical ageing of polymer glasses

Previously, it was mentioned that the process of physical ageing can be easily 'restarted'. For this purpose, it is sufficient to heat the aged polymer glass to a temperature higher than its T_g so that the polymer acquires all the properties which it had prior to the start of thermal ageing. Thus, subsequent low-temperature annealing will again be accompanied by all the special features of thermal ageing, discussed previously.

At the same time, it appears that there is another factor also capable of 'starting up' the physical ageing mechanism. This factor is the mechanical effect on the glassy polymer. This effect can be easily realised by, for example, uniaxial compression or rolling in the temperature range below the T_g of the polymer film. A typical example of this type is shown in Fig. 4.5 [70]. The graph indicates that the preliminary mechanical effect (rolling in this case) results in the effect identical the effect of 'rejuvenation' of the aged polymer by heating the polymer above the glass transition temperature followed by rapid cooling (quenching). In fact, the mechanically 'rejuvenated' the polymer has no yield stress or subsequent decrease of the stress.

It is well-known that in the conditions of uniaxial tensile loading at room temperature the polystyrene is a brittle material and its fracture takes place at strains of the order of 1–2% (in the so-called

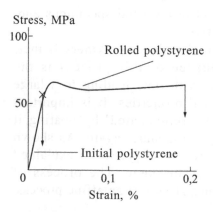

Fig. 4.5. Effect of rolling on the stress–strain curve of polystyrene at room temperature. The arrows indicate the fracture elongation of polystyrene [70].

Hookean section of the stress–strain curve). At the same time, after the mechanical effect, the polystyrene shows the plastic behaviour and its fracture elongation reaches 30% at room temperature. The effect of mechanical action on the glassy polymer can also be detected visually. Figure 4.6 shows the photographs of two polymer samples and their mechanical behaviour is shown in Fig. 4.5. It may be seen that the initial polystyrene is a brittle material and fractures at a tensile strain of 1–2%. Fracture is preceded by the rapid development of crazes which act as the nuclei of the main crack [71]. However, if the polystyrene is subjected to preliminary deformation in the rolls (rolling), it becomes turbid, opaque and can be deformed in any conditions, up to folding to a pipe (Fig. 4.6 *b*).

Annealing of the deformed polymer above its glass transition temperature completely restores all the properties of the initial polymer and eliminates the effect of mechanical 'rejuvenation' [72, 73]. In addition, the mechanically 'rejuvenated' polymer restores spontaneously with time all its initial mechanical properties without any annealing. This effect is shown in Fig. 4.7 [74]. It is clearly seen that the yield stress appears with time and starts to increase, and already after 48 hours after rolling (Fig. 4.7) restores at room temperature all its initial properties, including the capacity to fracture at low strains. As indicated by the data, presented in Figs. 4.5–4.7, the 'rejuvenation of the polymer by means of mechanical effects also 'starts' some type of physical ageing, taking place spontaneously with time.

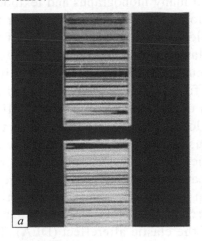

Fig. 4.6. External appearance of the polystyrene film after deformation in air at room temperature: a) initial polystyrene, b) polystyrene sample after preliminary rolling [70].

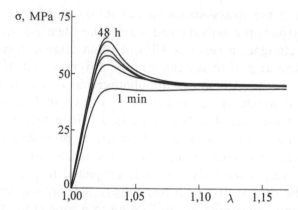

Fig. 4.7. Stress–strain curves of polystyrene samples at room temperature different periods of time after rolling [74].

4.6. Properties of glassy polymers subjected to mechanical effects

Thus, the mechanical effect on the glassy polymer has a very strong impact on the mechanical properties of the polymer, even in the range of low strains. We will examine in detail the effect of the mechanical loading of the glassy polymer in different conditions on its fundamental properties.

Stress relaxation
Previously, it was mentioned that in many monographs and textbooks it is concluded that the deformation of the glassy polymers in the first deformation stage (the so-called elastic Hookean section of the tension–compression curve) is elastic and takes place as a result of the increase of the intermolecular spacing [2]. However, it is clear that Yu.S. Lazurkin was the first scientists [75] who noted the inelastic nature of the deformation of the glassy polymer in the first, elastic Hooke stage of the tensile loading curve. He established that in the temperature range of the glassy state and at strains below the yield stress the amorphous polymer undergoes 'comparatively fast relaxation processes. These strains after unloading gradually decrease and at the end they convert to 0 in a relatively short period of time without heating above T_g'. The processes of this type were named by Lazurkin as the deformation of the elastic aftereffect (DEA). The consequence of the inelastic processes is, for example, the extensive relaxation of the stresses in the glassy polymer in the Hookean section of its tensile loading (compression) curve. Figure 4.8 shows

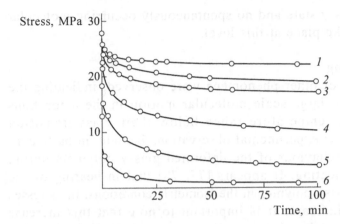

Fig. 4.8. Stress relaxation curves of PMMA in the glassy state. Test temperature 18 (1), 25 (2), 40 (3), 54 (4), 71 (5) and 83°C (6). Initial stress approximately 0.5 of the yield stress [75].

a series of such stress relaxation curves of glassy PMMA [75]. It can be seen clearly that the glassy polymer shows in these conditions the mechanical behaviour completely different in comparison with the elastic, Hookean solid. It should also be mentioned that the stress relaxation is also reflected in both the force and temperature serious below T_g and below the yield stress.

Lazurkin also found that the loading of the glassy polymer to stresses corresponding to the Hookean section of the tensile loading curve is accompanied by the formation in the polymer of strong birefringence (BRF) which relaxes with time. The digits of the BRF are always the same in DEA and forced plastic deformation, i.e. in the polymers oriented 'in a neck'. At the same time, the sign of the BRF for elastic deformation differs from the digit which polymer acquires as a result of its molecular orientation. Consequently, Lazurkin concluded that there is no large difference between the BRF and the forced elastic deformation. These two types of deformation are linked with the large scale structural rearrangements caused by the conformation transformations in the polymer chains. Lazurkin concluded that, firstly, the deformed amorphous polymer has molecular mobility in the glassy state and, secondly, 'both types of deformation – forced elastic and elastic aftereffect – are of the orientation nature. They are linked with the regrouping of the molecular chains, leading to the formation of preferential orientation'. The results have not been convincingly explained because it is assumed that the large scale molecular motion is 'frozen' within the limits of the glassy state and no spontaneously occurring molecular processes can take place at this level.

Isometric heating

Later, a number of other phenomena were observed, indicating the possibility of the large-scale molecular motion in the amorphous polymers in the temperature range below their glass transition temperature. These experimental observations include, in particular, the mechanical behaviour of the deformed glassy polymers during their isometric heating. It appears [77, 78] that in heating of the tensile loaded glassy polymer in the isometric conditions, the stresses in the polymer increase. It is important to note that this increase occurs long before the glass transition temperature of the polymer is reached. The result also confirms unambiguously that the large-scale molecular motion can take place in the glassy state of the amorphous polymers. An example of such mechanical behaviour is shown in Fig. 4.9. It may clearly be seen that the deformed polymer (PMMA) shows a large increase of the stress in the temperature range considerably lower than the glass transition temperature (115°C). It should also be noted that these experimental results confirm the entropy (polymeric) nature of this phenomenon because only the oriented polymer can show the increase of stresses during heating in the isometric conditions [1].

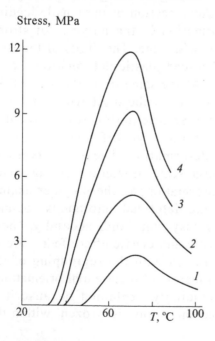

Fig. 4.9. Curves of isometric heating of PMMA samples, deformed to $\lambda = 1.5$ (1), 2.5 (2), 3.25 (3) and 3.9 (4) at 105°C [77].

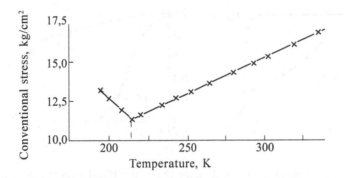

Fig. 4.10. Dependence of the equilibrium stress at a constant length on the temperature of natural rubber, tensile loaded at −50°C to 350% [79].

A typical example of this type of thermomechanical behaviour is shown in Fig. 4.10 [79]. Although this study was published a long time ago, the data illustrate efficiently the fundamental special features of the thermomechanical behaviour of the oriented polymers. Figure 4.10 shows that the stressed oriented polymer in the conditions of isometric heating shows a decrease of the stress, like all other solids, as a result of its thermal expansion. However, as soon as the glass transition temperature of the polymer is reached, the temperature dependence of the thermal expansion coefficient changes its digit and the stress in the polymer starts to increase with increasing temperature. The reasons for this thermomechanical behaviour are clear. In the oriented and rubbery polymer the stress will always increase in the isometric heating conditions as a result of the entropic nature of its elasticity [1]. The data presented in Fig. 4.9 indicate the increase of stress during isometric heating in the temperature range well below the glass transition temperature of the polymer (120°C) which is anomalous and requires explanation.

Thermally stimulated restoration of the dimensions of deformed glassy polymers
The possibility of the large-scale molecular motion below T_g is even more distinctive when examining the thermally stimulated restoration of the dimensions of the deformed glassy polymers. It was shown in a large number of studies [80–84] that the deformed glassy polymer restores its dimensions in annealing in a unique manner.

Figure 4.11 *a, b* shows the typical temperature dependences of the restoration of the dimensions of PMMA, deformed in the uniaxial

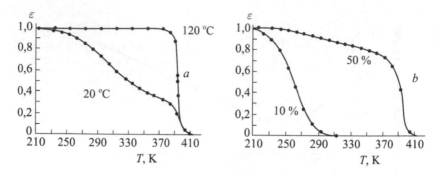

Fig. 4.11. Effect of temperature (a) and strain (b) of the PMMA on the thermally stimulated restoration of its dimensions [82].

compression conditions [82]. It may clearly be seen that this process is controlled by two factors: temperature and the magnitude of strain in the initial polymer. As indicated by Fig. 4.11 *a*, the restoration of the dimensions of the polymer, deformed at 120°C (above T_g) takes place in the same manner as described in textbooks, i.e., the deformed polymer restores its dimensions when its temperature reaches T_g. If the specimen is deformed at room temperature (below T_g), the specimen fully restores its dimensions in the temperature range below the appropriate T_g.

The magnitude of strain also has the strongest effect on the nature of relaxation of the deformed polymer (Fig. 4.11 *b*). This factor is observed for the polymer deformed at room temperature (below T_g). If the polymer is deformed to a certain value below the yield stress or in the region of the yield stress of the polymer (less than ~15%), the specimen completely restores its dimensions in the temperature range below T_g. Deformation of the polymer to the values higher than the yield stress results in the appearance of the high-temperature component of the thermally stimulated restoration of the dimensions of the polymer.

The effect of the magnitude of strain of the glassy polymer on the nature of its thermally stimulated restoration is shown in Fig. 4.12. In the figure, the curves of uniaxial compression of the polymer is compared with the high-temperature (ε_1) and low-temperature (ε_1) contributions to the thermally this stimulated restoration of the dimensions of the polymer. The graph shows that in the glassy polymer is deformed below the yield stress, the polymer restores its dimensions completely in the temperature range below the glass transition temperature. When the strain is increase of the values higher than the yield stress, the polymer starts to accumulate the

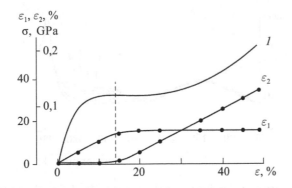

Fig. 4.12. The curve of uniaxial compression of PMMA at room temperature (1), low-temperature (ε_1) and high-temperature (ε_2) contributions to the thermally stimulated shrinkage in annealing [81].

temperature components of its thermally stimulated restoration of the dimensions, i.e., the part which relaxes in the range of the glass transition temperature of the polymer. The low-temperature contribution to the restoration of the dimensions of the specimen does not disappear but simply stops to change in value.

It is important to mention another special feature of the thermomechanical behaviour of the deformed glassy polymers. After deformation of the polymer there is a part of the strain which relaxes at the experiment temperature immediately after removing the load. It appears that this contribution to the thermally stimulated restoration dimensions of the specimen is not elastic in the true meaning of the word and can be 'fixed' by deep cooling of the polymer in the deformed state [80]. After removing the load and annealing, this part of the residual strain relaxes in the temperature range below the deformation temperature. This result again stresses the assumption made for the first time in [75] according to which the deformation of the glassy polymer in the so-called Hookean section is not truly elastic (reversible).

Dynamic mechanical properties
The investigation of the dynamic mechanical properties of the amorphous polymers also indicates a number of anomalies showing the possibility of large-scale (segmental) molecular motion in the amorphous polymers below their T_g which would appear to contradict the accepted assumptions regarding the nature of the glassy state [2]. The molecular motion of this type was observed in experiments in a number of studies [85–89].

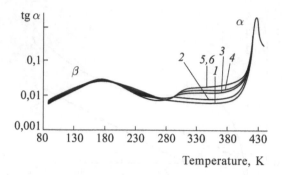

Fig. 4.13. Temperature dependence of the tangent of the angle of shift of the phases of the polycarbonate samples, deformed at room temperature in the uniaxial compression conditions by 0 (1), 6 (2), 10 (3), 20 (4), 30 (5) and 40% (6) [90].

This phenomenon will be discussed in greater detail on the example of the data published in [90]. In this study, the amorphous polycarbonate (PC) was deformed in the uniaxial compression conditions at room temperature to different values and subsequently the samples were cooled with liquid nitrogen and this was followed by determining the temperature dependence of their mechanical characteristics. Figure 4.13 shows the evolution of tg φ at a loading frequency of 1 Hz and a heating rate of 1 K·min^{-1} for the sample deformed in the plane by 0, 6, 10, 20, 30 and 40%.

For the non-deformed polycarbonate, the peaks of the β- and α-relaxation are separated by more than 200°C. The α-peak of the polycarbonate at the given frequency corresponds to 420 K, whereas the β-peak is situated at 170 K and is greatly stretched along the temperature axis because it is associated with the wide distribution of the relaxation time [91, 92]. It can be clearly seen that when studying the deformed polycarbonate samples there is a very wide relaxation transition between the α- and β-transitions, with the relaxation transition spreading from 280 K up to T_g. This peak is completely eliminated during annealing of the polymer above T_g; its position does not depend on the strain. As indicated by Fig. 4.13, this transition has the form of a wide relaxation peak which then transforms to a plateau spreading up to the α-transition. The height of this plateau increases with increasing strain and reaches saturation at strains higher than 30%. The authors assume that the deformed glassy polymer contains zones in which the molecular mobility is much higher in comparison with the initial bulk polymer. The formation in the glassy polymer of the 'zones' with higher

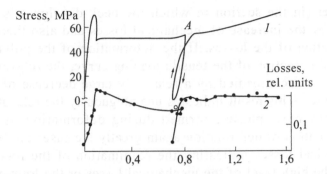

Fig. 4.14. Stress–strain curve (1) and the appropriate mechanical losses (2) in the polycarbonate at room temperature. After reaching the elongation, corresponding to the point A on the curve, the stress was reduced to zero and the specimen was held in this condition for 1 s. After recovery, tensile loading was continued [95].

compliance results in the completely unusual mechanical properties of the polymer as a whole. The presented experimental data (Fig. 4.12) show unambiguously that the inelastic deformation of the polymer glass is accompanied by the formation of a second 'phase' in the polymer which has greatly different properties. It should be mentioned that the formation of the relaxation peak, caused by the deformation of the glassy polymer, was also observed previously in [93, 94].

The structural inhomogeneity of the deformed polymer glasses was also observed when examining the dynamic characteristics of the polycarbonate in tensile loading at a constant rate. In [95] investigations were carried out into the inelastic deformation of the glassy polymer (polycarbonate) in the uniaxial tensile loading conditions and internal friction processes were measured at the same time.

Figure 4.14 compares the stress–strain curve of the polycarbonate with the corresponding mechanical losses, measured in parallel with the mechanical response of the polymer in the conditions of tensile loading at a constant rate. It may be seen that initially (prior to the application of mechanical stress) the losses are small, as expected for the polymer glasses. After applying fracture load, long before reaching the yield stress of the polymer (starting at ~2% elongation) a new 'phase' forms in the polymer which has a considerably higher compliance and, correspondingly, higher mechanical losses. The amount of this phase continuously increases since there is a continuous increase of internal friction. This increase is interrupted quite rapidly after reaching the yield stress. Further deformation of the polymer (in the section in which the neck develops) is not accompanied by the increase of mechanical losses and also there is no large reduction of the losses. If the deformation of the polymer is arrested in this section of the tensile loading curve, the relaxation of the stresses is accompanied by a relatively large decrease of the mechanical losses. This result indicates unambiguously the relaxation (healing) of the new 'phase', formed during deformation of the polymer. Restoration of deformation again greatly increases the level of the mechanical losses indicating the reanimation of the relaxed regions with the high level of the mechanical losses or the formation of new regions in the amount that is the same as that in the initial condition, formed after reaching the yield stress by the polymer. This 'phase' forms already in the Hookean section of the loading curve and its amount continuously increases up to the moment when the polymer reaches the yield stress. As soon as the yield stress is reached, i.e., the stress in the polymer ceases to increase, the amount of the nucleated new 'phase' also stops increasing. Consequently, the deformed glass has a different structure in comparison with the initial glass, and it may be assumed that the initial non-equilibrium structure of the polymer glasses is not responsible for the physical–chemical and physical–mechanical anomalies shown by the deformed polymer glass.

It is important to note that the processes taking place in the deformed glassy polymer are not identical with the physical ageing processes, discussed previously. In fact, we have noted that the physical ageing is of the affine nature, i.e. it develops uniformly and simultaneously in the entire volume of the glassy polymer. At the same time, the data in [90, 95] indicate that the deformed polymer glass contains discrete zones in which the large-scale molecular motion is greatly facilitated.

Thermophysical properties
In a large number of thermodynamic investigations, carried out using deformation calorimetry [82–84], it has been shown that the plastic deformation of the polymer glasses differs drastically from the appropriate process in the low-molecular solids. It appears that a large part of the deformation work (mostly prior to and in the range of the yield stress) is stored by the polymer and is not transformed to heat (Fig. 4.15) [82].

The results indicate that deformation of the glassy polymer is accompanied by structural rearrangements in the polymer, responsible

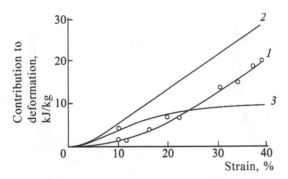

Fig. 4.15. Dependence of the heat (1), mechanical work (2) and stored internal energy (3) on the magnitude of deformation in uniaxial compression of polystyrene at 30°C, strain rate $5 \cdot 10^2$ min^{-1} [82].

for the buildup of internal energy. Figure 4.15 also shows that these structures form in the initial stages of deformation (up to and in the region of the yield stress of the polymer), i.e., in the area where the anomalies in the thermally stimulated recovery of the dimensions and in the dynamic mechanical properties appear (compare Figs. 4.12 and 4.15).

In [96] examination by DSC showed that the deformed polymer contains measurable amounts of internal energy. It has also been noted that the stored energy starts to be released during annealing at temperatures 30–40°C lower than T_g. Studies of polystyrene and some of its copolymers with polymetacrylic acid showed [61] that the samples deformed by 40% also store internal energy (1.5 cal/g). The authors attributed the origin of this excess internal energy to the weakening of the intensity of the intermolecular interaction in the polymer associated with its inelastic deformation.

In analysis of the DSC data it should be noted that this method records not only the effect of deformation on the thermophysical properties of the glassy polymer but also special types of molecular motion, determined by the processes of its physical ageing (Fig. 4.3). These changes of the mechanical properties in physical ageing are also accompanied by large changes of its thermophysical properties. Figure 4.16 a shows DSC thermograms characterising the process of physical ageing of the amorphous glassy polystyrene. The polystyrene was produced by quenching from the high elasticity state, followed by low-temperature annealing for various periods of time. It is clearly seen that annealing is accompanied by the above described phenomena during physical ageing of the amorphous glassy polymer (compare Figs. 4.16 a and 4.3) [61]. During annealing, the glass

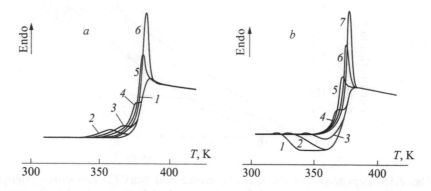

Fig. 4.16. DSC curves of the quenched (a) and compression pre-deformed polystyrene (b) after annealing at 60 and 70°C respectively: a) annealing for 0 (1), 0.2 (2), 1 (3), 10 (4), 90 (5) and 170 (6) hours; b) annealing for 0 (1), 0.003 (2), 0.08 (3), 1.5 (4), 6 (5), 11 (6) and 50 (7) hours [61].

transition 'step' transforms to the endothermic effect (Fig. 4.16 *a*) and the intensity of the effect increases with increase of the annealing time. Detailed analysis of the effect of the physical ageing of the polymer glasses on the evolution of the DSC thermograms was carried out in [3, 61, 97]. Here it should only be mentioned that the effect of the ageing process on the thermophysical behaviour of the polymer does not prevent to use the same specimens for the independent investigation of the processes taking place during their deformation.

Actually, the deformation of the polymer in its glassy state (below T_g) also has the strongest effect on its thermophysical properties (Fig. 4.16 *b*), but this effect is completely different. Figure 4.16 *b* shows that the deformed polymer contains a wide exothermic DSC peak, situated below the glass transition temperature, whereas the non-deformed polymer contains only the endothermic peak in the glass transition temperature range (compare Figs. 4.16 *a* and 4.16 *b*). It is important to note the evolution of these peaks during low-temperature annealing. It appears that the exothermic peaks (curves 1–3) gradually decreases during annealing and completely disappears (curves 4–7) at long annealing time. At the same time, the endothermic peak, responsible for the physical ageing of the glassy polymer increases during annealing. In other words, the physical ageing and relaxation of the structure of the deformed glassy polymer during low-temperature annealing is accompanied by the thermal effects of the opposite peaks. In addition, comparison of Figs. 4.16 *a* and Fig. 4.16 *b* shows that the previously mentioned physical processes appear not 'to notice' each other and evolve independently during annealing.

Thus, the presented experimental data indicate that in the range of the glassy state of the amorphous polymer there should be at least two types of large-scale molecular motion. Firstly, the molecular motion responsible for the process of physical thermal ageing which results in a more equilibrium state of the polymer glass. It is very important to note that the molecular motion of this type is of the affine nature and takes place simultaneously throughout the entire volume of the polymer.

Secondly, there is some type of large-scale molecular motion which is induced in the glassy polymer by mechanical deformation below the T_g of the bulk polymer and also takes place below its glass transition temperature. An important special feature is the localisation of this type of motion in special zones, formed in the unchanged polymer matrix [90, 95]. This means that the deformed glassy polymers are inhomogeneous as regards their structure so that it is important to determine the structural rearrangements accompanying their deformation.

4.7. The spatial inhomogeneity of deformation of polymer glasses

The investigations of the structure of the deformed glassy polymers are numerous and started a long time ago. These investigations determined a very important experimental fact. It appears that the affinity (homogeneity) of deformation is characteristic only of the amorphous polymers which are in the rubbery state (in the temperature range above the corresponding glass transition temperature) [1]; the results of a large number of direct microscopic studies show that the inelastic deformation of the glassy polymer is always inhomogeneous throughout the volume of the polymer. In particular, the inhomogeneity of the deformation in the volume is evidently manifested by the fact that a neck forms and grows in the conditions of uniaxial tensile loading in the glassy polymer. However, the structural inhomogeneity of deformation is also detected prior to the formation of the neck, at strains not exceeding the yield stress of the polymer, i.e., in the areas where all the previously mentioned anomalies in the properties are detected.

For example, Lazurkin in his classic study [75] noted, obviously for the first time, that the deformation of the polymer in the region

Fig. 4.17. Light micrograph of a polyethylene terephthalate sample deformed at room temperature with the formation of a neck. On the right is a section of the sample transformed into a neck, on the left the non-deformed area of the sample. The photograph was produced in crossed polaroids [98].

of the yield stress is inhomogeneous. The polymer is characterised by the formation of a system of inhomogeneities which can easily be recorded by light microscopy and observed even with the naked eye. These inhomogeneities are expressed by straight lines intersecting the polymer under the angle of 45–55° in relation to the axis of tensile loading. By analogy with the low-molecular solids, Lazurkin referred as shear bands and noted that their formation usually requires the presence of microscopic inhomogeneities – stress concentrators.

These shear bands can be easily detected in direct microscopic studies. Figure 4.17 shows the light micrograph of a polyethylene terephthalate sample deformed at room temperature with the formation of a neck. It may be clearly seen that the part of the specimen which has not as yet transformed to the oriented state (to a neck) is penetrated by shear bands. The material of the neck is at first sight free from shear bands [98].

In [99] the authors used the method of decoration of samples for electron microscopic studies which made it possible to study, in particular, the evolution of the shear bands, included in the structure of the growing neck. The method is based on the deposition of a thin (nanometric) metal layer on the surface of the polymer prior to deformation. Subsequent deformation of the volume of makes it possible to visualise by means of the deposited coating the structural rearrangements taking place in the polymers. The method was described in greater detail in the first chapter of this book.

Figure 4.18 *a* shows a micrograph of the section of the polymer localized and the interface between the neck at the non-oriented part. In this area, the thickness of the specimen decreases and this results in a relatively distinctive bending of the initially straight surface. Naturally, this bending is accompanied by the cracking of the coating. At the same time, the method can be used to visualise the formation of shear bands in this area. It may be seen that these

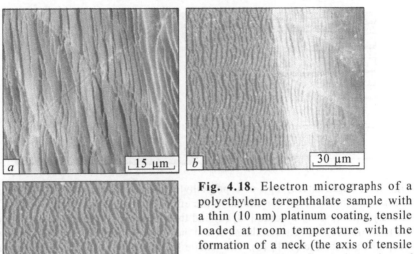

Fig. 4.18. Electron micrographs of a polyethylene terephthalate sample with a thin (10 nm) platinum coating, tensile loaded at room temperature with the formation of a neck (the axis of tensile loading is horizontal): a) the region of transition of the polymer to the neck; b) the section of the specimen in which the transition layer (right) and the region of the produced neck (left) coexist; c) the region of the produced neck, away from the transition layer [99].

bands propagate through the given transition zone in a straight direction and approximately under the angle of ~45° in relation to the tensile loading axis.

The transition layer between the non-oriented part of the polymer and the growing neck during deformation of the specimen move through the specimen until it is transformed completely to the neck. Figure 4.18 *b* shows the section of the specimen with a fragment of the transition layer, and also part of the specimen transformed to a neck. It may be seen that the shear bands, formed in the transition layer reach the region of the neck. At exit into the neck, the shear bands lose their 45° orientation in relation to the direction to the drawing axis. They unfold almost completely along the axis of tensile loading and are included in the structure of the neck. These shear bands, included in the structure of the neck, are clearly visible in the sections situated at any distance from the zone of transition of the polymer to the neck (Fig. 4.18 *c*). After all, after inclusion of the shear bands in the structure of the neck they can no longer be referred to as the shear bands. As indicated by the data, the resultant neck does not have a homogeneous structure. Although the shear bands are included in the neck material this material still

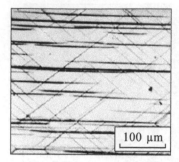 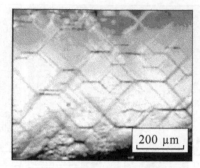

Fig. 4.19. Light micrograph of a polyethylene terephthalate sample, deformed at room temperature in the conditions of the effect of the tensile constant load of ~0.7 of the yield stress for 2 hours [98] (a) and the electron micrograph of the section of the transition region between the neck and the non-deformed part of the polyethylene terephthalate sample [101] (b). The tensile loading axis is vertical in both cases.

'remembers' that the transition to the neck is implemented by the polymer containing the shear bands. If the polyethylene terephthalate neck is subjected to shrinking, for example, using a swelling solvent, the resultant material will contain a system of shears, recorded in a light microscope [100].

The inhomogeneity of plastic deformation of the glassy polymer is also clearly visible in the strain range not exceeding the yield stress. Figure 4.19 a shows the light micrograph of a sample of glassy polyethylene terephthalate, subjected to the effect of a constant load at a stress equalling ~0.7 of the yield stress. Nevertheless, it can be seen that clearly visible inhomogeneities form in the polymer. In addition to a system of shear bands, oriented under the angle of ~45° in relation to the tensile stress, the polymer also contains other zones of the plastically oriented polymer propagating in the direction normal to the tensile loading axis. These are the so-called crazes – the zones of the deformed polymer which contain a large number of microscopic cavities [102–104]. Figure 4.19 *b* shows the electron micrograph of the transition zone between neck and the non-deformed part of the polyethylene terephthalate sample. Both the shear bands and the crazes are clearly visible. The structure and properties of the crazes will be discussed in greater detail later and here we only mention that both types of formations (shear bands and crazes) often co-exist during mechanical loading of the hard polymers.

The structural inhomogeneity of the deformed glassy polymer is unambiguously detected by the method of small angle x-ray scattering. For example, in [105] it was reported that at 3–5% strain (lower than the corresponding yield stress) polystyrene and polyvinyl acetate (PVA) show the high intensity diffusion x-ray scattering in the temperature range 10–20°C below the appropriate glass transition temperature. The result confirms unambiguously that in these conditions the Hookean section of the stress–strain curves shows the disruption of the continuity of the glassy polymers in the formation of interfaces which are also sources of small angle x-ray scattering.

The formation and growth of structural inhomogeneities take place not only during deformation of polymer films discussed above. In the uniaxial compression of the bulk polymers which was accompanied by many anomalies in the mechanical behaviour of the glassy polymers [80–84], the development of the zones of inelastic deformation was much more difficult to study. In this case, a thick cylindrical sample, subjected to uniaxial compression, usually acquires the barrel shape. Recording of any inhomogeneities in such samples is very difficult.

Nevertheless, the application of the well-known methods of preparation of the samples for direct microscopic studies makes it possible to solve this problem. For example, in [106] the authors produced microsections from the bulk of a number of glassy polymers, deformed in uniaxial compression. The resultant sections were investigated in a polarization optical microscope. The results show that the deformed glassy polymers (PS, PC, PET, PMMA, and others) contain straight bands with a width of ~1 μm characterised by strong birefringence and separated by the blocks of the initial non-deformed material.

Figure 4.20 show is the micrograph of a thin section taken from a block sample of polystyrene, deformed in the uniaxial compression. In other words, the polymer is deformed in these conditions by the development of shear bands. It should be mentioned that the strong

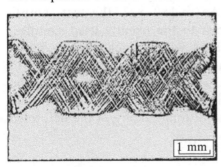

Fig. 4.20. Light micrograph of a thin section of a polystyrene sample, subjected to uniaxial compression. The axis of compressive loading is vertical [106].

birefringence is detected not only in the shear bands but also in the crazes [103], and this indicates the molecular orientation of the material, filling these formations.

Regardless of the above general features of the mechanisms of inelastic deformation of different glassy polymers, there are certain differences in the methods of producing samples for visualisation of structural rearrangement of the deformed polymer and a large number of investigations carried out in the past. Usually, the formation of the polymer in the uniaxial compression conditions is carried out using cylindrical samples [80–84]. The reverse process of thermally stimulated restoration of the dimensions is studied by measuring the height of the deformed samples during annealing.

The uniaxial compressive loading is used to study the deformation mechanism of the glassy polymers because of the easy and convenient procedure. The point is that uniaxial tensile loading is accompanied by the formation of a neck in which the orientation of the molecule changes abruptly. The magnitude of this orientation is determined by the so-called degree of natural drawing of the polymer which cannot be changed or regulated in a wide range. At the same time, uniaxial compression does not lead to the formation of the neck and deformation of the polymer can be smoothly regulated in wide ranges. In this case, it is assumed that the degree of deformation is of the affine (homogeneous) nature throughout the entire volume of the deformed polymer.

As indicated by the microscopic data (Fig. 4.20), in this case the deformation of the glassy polymer is also inhomogeneous throughout the volume of the sample as a result of the development of shear bands. In order to characterise the deformation of the polymer in the uniaxial compression conditions, the authors of [107] developed a simple and efficient procedure. A complete analogue of the sample was produced for this purpose. The analogue is usually used for examining the mechanical behaviour of polymers in the uniaxial compression loading conditions. As in [80–84], the test sample was cylindrical. However, in contrast to the usually used samples, the cylinder was assembled from discs with a thickness of ~1 mm (Fig. 4.21). After deformation, the sample can be dismantled to individual elements each of which in turn can be studied using the above method of direct microscopic evaluation. From the physical viewpoint, it is the method of structural tomography of the deformed bulk sample of the polymer.

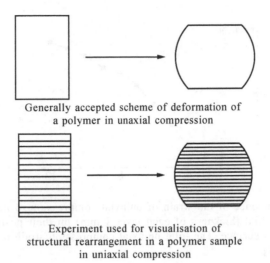

Generally accepted scheme of deformation of
a polymer in unaxial compression

Experiment used for visualisation of
structural rearrangement in a polymer sample
in uniaxial compression

Fig. 4.21. Diagram of the method of structural tomography used for studying the structure of the polymer deformed in the uniaxial compression conditions [107].

In [107] it was shown that the tomographic procedure does not cause any significant changes in the deformation mechanism of the polymer. Firstly, under the effect of uniaxial compression the multilayer sample (the structure of the sample is shown in Fig. 4.21) is deformed as an integral unit so that its shape changes from cylindrical to barrel, in full agreement with the change of the shape reported in [80–84] in deformation of monolithic specimens. Secondly, the uniaxial compression loading curve of the multilayer sample corresponds fully to the uniaxial compression curve of the monolithic polymer.

The proposed approach can be used to obtain new important information on the deformed polymer which can be used for examining the structural–mechanical behaviour of the polymer inside thick barrel-shaped samples. Figure 4.22 shows the magnitude of compressive deformation of each layer of the polymer in dependence on their location (distance) in relation to the compressed surface.

This indicates that the polymer, deformed in the uniaxial compression conditions, has a highly inhomogeneous structure. In fact, the PMMA sample, deformed by 30% in the uniaxial compression, contains sections which greatly differ from each other by the magnitude of deformation. The first 3–4 layers, adjacent to the compressed surfaces, are almost free from residual strain. When the other hand, in the central part, formed as a result of compression of the 'barrel', the strain in the polymer exceeded 70%.

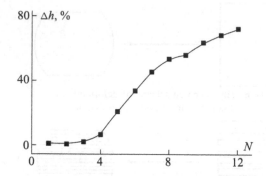

Fig. 4.22. Dependence of the strain of uniaxial compression of individual layers of the PMMA (initial thickness of each layer 1 mm) on their position in relation to the compressed surface. The total uniaxial compression strain of the multilayer sample 30% [107].

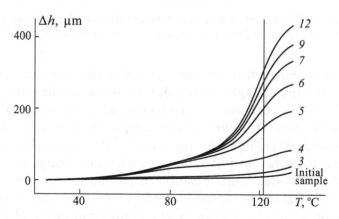

Fig. 4.23. Dependence of the absolute change of the height of the individual layers of PMMA after deformation of a multilayer sample by 30% at room temperature. The numbers at the curves correspond to the position of each layer in relation to the compressed surface. The vertical line indicates the glass transition temperature of the PMMA.

The proposed method can be used to investigate the thermally stimulated restoration of the dimensions of each layer forming the 'barrel'. In Fig. 4.23 the data are presented in the form of the dependence of the absolute change of the height of each layer, forming the 'barrel', on annealing temperature. Depending on the position of each layer in the overall structure of the sample, there are three types of the thermomechanical behaviour. The layers 1–3 change only slightly their dimensions during annealing in relation to their position related to the compressed surface (the layer 2 coincides with the layer 3). The thermomechanical behaviour is almost identical with that of the initial non-deformed PMMA and, therefore, they are not shown in the figure. Sample 4 shows a distinctive low-temperature component of the thermally stimulated restoration of the dimensions. This component appears starting at an annealing temperature of ~50°C. This temperature also indicates the start of the restoration of the dimensions of all other fragments of the multilayer sample. In addition, in the samples 5–9 the high-temperature (in the T_g range of the PMMA) component of the thermally stimulated restoration of the dimensions forms and becomes more and more significant. With increase of the distance from the surface of the 'barrel' this contribution to the thermally stimulated shrinkage continuously increases and gradually its absolute value becomes greater than that of the low-temperature component.

Thus, the cylindrical sample of the polymer, which is usually used in the uniaxial compression experiments, is deformed highly non-uniformly. Inside a specimen, subjected to uniaxial compression, there are zones deformed to different degrees. In Figs. 4.22 and 4.23 these layers not only differ in the strain but also show different response to the subsequent heating in the annealing experiments. These data make it possible to improve the accuracy of the experimental results obtained in the thermally stimulated restoration of the block, six samples. Since the barrel-shaped polymer, deformed in the uniaxial compression conditions, contains regions with different compressive stress, the sample cannot react uniformly to the acting load and, correspondingly, cannot restore uniformly its dimensions during annealing.

The above anomalies in the thermophysical properties of the deformed glassy polymers become evident only in the cases in which the discrete zones of the deformed material (shear and/or crazes) form in them during deformation. For example, in a polymer deformed in the hydrostatic compression conditions resulting in the true elastic deformation, there are no significant anomalies in the DSC thermograms. At the same time, the deformation by hydrostatic compression combined with shear, i.e., in the conditions in which the shear bands form, lead to the appearance of an exo-effect on the DSC curves in the temperature range below the T_g of the amorphous polymer [108].

It is important to mention that both the crazes [103] and the shear bands are completely healed when the polymer is annealed. Later, this structural special feature of the interfaces formed during deformation of the glassy polymers will be discussed in greater detail, and here we only mention that the healing of this type is their fundamental property. This property will be described on the example of thermally stimulated healing of the shear bands in amorphous polyethylene terephthalate. In [109] investigations were carried out into the effect of the annealing temperature on the state of the shear bands produced in the polymer in the conditions close to uniaxial compression. A spherical indentor was pushed into the film of the amorphous polyethylene terephthalate at room temperature and the deformation pattern was studied in an optical microscope. Figure 4.24 *a* shows the pattern after annealing of the deformed polymer at 40°C, i.e., well below its T_g temperature. It may be seen that in these conditions the deformation of the polymer is accompanied by the formation and development of a system of shear bands. With the increase of the annealing temperature to 70°C (Figure 4.24 *b*), the shear bands become 'blurry' and less distinctive and, finally, at the glass transition temperature (75°C) the shear bands are completely healed so that they cannot be observed by the optical microscope (Fig. 4.24 *c*).

.It may be seen that the annealing of the polymer, containing the shear bands, results in the complete healing of the interfaces in the process takes place in the temperature range below is T_g. The low-temperature (below T_g of the bulk polymer) the relaxation is also characteristic of the crazes [98].

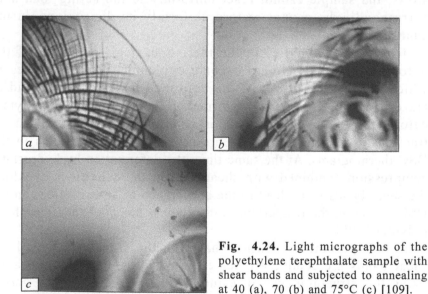

Fig. 4.24. Light micrographs of the polyethylene terephthalate sample with shear bands and subjected to annealing at 40 (a), 70 (b) and 75°C (c) [109].

Thus, the discussed experimental data can be used to assume that the previously described anomalies in the mechanical and thermophysical behaviour of the deformed glassy polymers are connected in some way with the processes taking place in the zones of the plastically deformed polymer with developed interfaces (in the crazes and shear bands). This is clearly evident because the blocks of the non-deformed polymers, separated by the shear bands and/or crazes, do not differ at all from the initial polymer, i.e., they do not contain the excess of internal energy, do not show any low-temperature shrinkage or any other anomalies in the physical-mechanical behaviour examined in detail previously.

The deformation of the glassy polymer even in the early stages of its development is highly inhomogeneous. The above data indicate that both the shear bands and the crazes, formed during deformation of the amorphous glassy polymer, have very unusual structural–mechanical properties. It is rational to assume that all the previously mentioned anomalies in the mechanical behaviour of the glassy polymers are associated with the formation and development of these structural formations. In the context of the investigated problems this means that an important role in the investigated structural–mechanical behaviour of the glassy polymers is played by the characteristic developed interfaces, easily recorded in these discrete zones of the deformed polymer glasses.

4.8. Structure of shear bands formed during deformation of glassy polymers

It is now important to discuss in greater detail the structure and properties of the previously described in homogeneity is (shear bands and crazes) forms during inelastic plastic deformation of the polymer glasses. We will pay special attention to examining the problems associated with the structure and properties of the crazes and here we will discuss the nucleation, a volume should and structure of the shear bands, which appear to accompany the well studied phenomenon of plastic deformation of the glassy polymers.

Figure 4.24 *a* shows the photograph of a polymer sample in which a shear band is formed in the initial stage of tensile loading [110]. It may be clearly seen that, as in the low-molecular solids, this band grows through the specimen under the angle of 45–50° in relation to the direction of tensile stress. However, here the similarity of the shear bands in the polymers and low-molecular solids ends. It appears

Fig. 4.25. External appearance of a polystyrene sample containing a shear band (a) and the electron micrograph of the band (b) showing its internal structure 110].

[110, 111] that the shear bands in the polymers have a complicated structure. They are filled with a highly dispersed oriented, fibrillised material, completely identical with the material filling the volume of the crazes [102, 103].

This important property is confirmed by direct microscopic studies. When the material containing shear bands is subjected to low tensile stresses in the direction normal to the plane of the shear bands, the latter open (Fig. 4.25 *b*) exposing a craze-like structure. The similarity with the craze structure is so great that these open shear bands were referred to as shear band craze in [110]. Some morphological differences of the shear band and the crazes in comparison with classic crazes are caused by the fact that the fibrils in its structure are inclined in relation to the plane of the shear band (Fig. 4.25 *b*).

The complicated structure of the shear bands in the glassy polymers has a specific effect on the morphology of the fracture surfaces of the bands. Figure 4.26 *a* shows the photograph of a polystyrene sample in which a shear band was initiated. In further deformation, fracture took place through this band. It can be seen that the volume of the shear band is characterised by the formation of a large amount of the plastically deformed polymer. Under large magnification (Fig. 4.26 *b*) it may be seen that the material consists of parallel plates formed by the connected fibrils. The structure of this type is morphologically very similar to the craze material

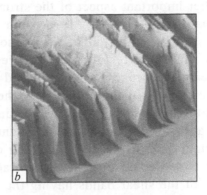

Fig. 4.26. External appearance of a polystyrene sample fractured through the shear band (a) and the electron micrograph of the fracture surface in the shear band (b) [110].

detected on brittle cleavage areas of the crazed polymers [98]. As shown in [110], this material can be easily separated from the sample using a pair of tweezers and studied by the DSC method. It was found that it is the material of the shear band that is responsible for the exothermic low-temperature (below T_g) effect observed by DSC in the deformed glassy polymer (Fig. 4.16 b). At the same time, the material distributed between the shear bands is not deformed and, consequently, does not differ at all from the initial block polymer. In particular, it does not show the low-temperature endothermic effect in annealing.

In [110] it was possible to produce both a shear band and a craze in the same polystyrene sample. Figure 4.27 shows the electron micrograph of such a sample and it can be seen that the difference between the shear craze band and the classic craze is only that the fibrils in its structure are inclined in relation to the plane of the shear band. Otherwise, the morphological features of these two structures are completely identical.

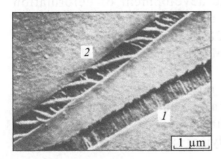

Fig. 4.27. Electron micrograph of a polystyrene sample, containing a craze (1) and a shear band (2) [110].

Another important aspect of the structure of microheterogeneities (shear bands and crazes), formed in deformation of the glassy polymers, is the question of the presence in them of real micro-cavities (actual interfaces). The presence of the microporous structure in the crazes is not doubted at the moment [102, 103]. The development of the polymer by the crazing mechanism during deformation in adsorption-active media takes place at least in the initial stages, by the growth of the general surface of the polymer. At the same time, in the shear yielding conditions the tensile loaded polymer is subjected to general contraction and, regardless of the presence of the shear bands having the real interfaces, the problem of the development of the real general porosity of the polymer is not so evident.

It was shown on [112] that the transport of methanol in PMMA samples subjected to 23–24% compressive strain greatly differs from its transport to the initial non-deformed PMMA. The diffusion rate at 40°C is twice as high and at 25°C it is five times higher than in the non-deformed PMMA. These data show that the shear bands in the glassy polymers contain regions in which the feeling material is so porous that it greatly facilitates the transport of low-molecular substances.

It is quite easy to show how the polymer is deformed in the same conditions either by the crazing mechanism or by means of the developed shear bands [113]. In tensile loading of the thick specimens of PET in the AAM their surface shows the nucleation of crazes which start to propagate into the bulk of the specimen. Naturally, as they grow, the hydrodynamic resistance to the flow of the liquid to the tip increases. Finally, a situation forms in which the liquid cannot effectively and in sufficient quantities travel to the areas of orientation transformation of the polymer (the tips of the crazes). Since the development of the crazes requires the presence of the AAM in the areas of active deformation, the polymer 'selects' an alternative path of the development of deformation – by the mechanism of growth of the shear bands. In particular, this situation is shown in Fig. 4.28. The figure shows the light micrograph of a thin section of a polyethylene terephthalate sample with a thickness of 0.7 mm, tensile loaded at room temperature in the AAM (n-hexanol) by 50% at a rate of ~100%/min. In may be seen that in these conditions the crazes cannot intersect the entire cross-section of the polymer and start the stage of their widening (Fig. 4.28 a). Nevertheless, the polymer continues to be deformed

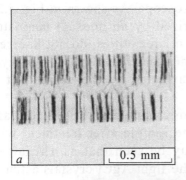
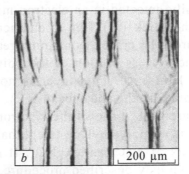

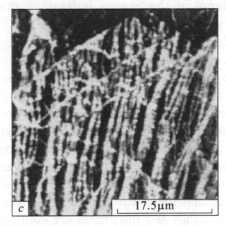

Fig. 4.28. Light micrograph of the thin section of a polyethylene terephthalate sample, containing crazes and shear bands (a) and (b); scanning electron micrograph of the low temperature cleavage of the polyethylene terephthalate sample, shown in Fig. 27 a, b after using it as a membrane in a dialysis cell, containing aqueous solutions of NaCl and AgNO$_3$ (c) [113]. Explanation in the text.

in its centre by means of the development of the shear bands which do not require the presence of the AAM. The micrographs (Fig. 4.28 *b*) shows a number of important special features of this type of deformation of the polymer. The shear bands, like the crazes, are clearly seen in the light microscope indicating evidently that they have distinctive interfaces.

The film (the structure is shown in Fig. 4.28 *a*, *b*) was used as a membrane in a dialysis cell with the chambers of the cell filed with an aqueous solution of NaCl on one side and AgNO$_3$ on the other side. It was shown previously [114] that if the crazes penetrate the polymer from one side to the other, the solutions of NaCl and AgNO$_3$, diffusing against each other through a system of interconnected pores in their structure, meet in their volume and form AgCl crystals which

are easily detected in an electron microscope. As shown in Fig. 4.28
a, b, the thick PET film is characterised by an unusual laminated
structure. The crazes, containing real microcavities, do not intersect
the whole cross section of the polymer and leave a polymer layer
in the film core. This layer is not affected by the crazes but is
penetrated by the shear bands.

Figure 4.28 *c* shows the electron micrograph of brittle cleavage
in the same polyethylene terephthalate sample after treatment with
solutions of NaCl and $AgNO_3$. It may be seen that as a result of
the previously described procedure, the light AgCl crystals actually
lay aside in the volume of the crazes indicating the continuous
penetration of the solutions of NaCl and $AgNO_3$ through the
polyethylene terephthalate film. In addition to this, Fig. 4.28 *c* also
shows that the AgCl crystals lay aside not only in the microcavities of
the crazes but also in the shear bands (straight channels, intersecting
the limit under the angle of ~45°), increasing greatly their contrast.
The results show unambiguously that the shear bands, even if they do
not contain the actual microcavities, have such low density that the
low-molecular liquid can diffuse through them like through channels.

Summing up the material in this section, it is relevant to mention
the important identical feature of the two previously examined
structures – the crazes and the shear bands. It is expressed by the
fact that both the shear bands and the crazes are continuous channels
in the polymer film with the highly dispersed fibrillised material.
As shown later, this similarity is the basis of the many anomalies
of the physical–mechanical, physical–chemical and other properties
of the deformed glassy polymers.

4.9. The nature of structural–mechanical anomalies in the properties of deformed glassy polymers

To explain the entire set of the special features of the structural-
mechanical behaviour of the deformed polymers, the experimental
results have been used to propose several mechanisms which,
however, do not take into account or explain the role of the
previously mentioned structural formations (shear and crazes,
characterised by the highly developed surface) in the development
of high inelastic strains in them.

The first approach in this group (in the chronological order) is
represented by a series of studies carried out to examine the thermally
stimulated restoration of the dimensions of the deformed polymer

glasses (Figs. 4.11 and 4.12) [80, 81, 115, 116]. In these studies, the anomalies in the mechanical behaviour of the polymer glasses (low-temperature restoration of the dimensions during annealing) are attributed to their initial structural inhomogeneity. It is proposed that the mechanism of high-temperature restoration is associated with the entropic relaxation of the excited elongated macromolecule at angles and with the transition to the initial state as a result of 'unfreezing' of the segmental mobility at T_g of the polymer. The nature of the low-temperature component of restoration of the dimensions is explained by the conformation rearrangement of the macromolecules, stimulated by the combined effect of temperature and of the internal stresses, stored during deformation [80, 81, 115, 116].

By analogy with the term 'forced elastic deformation', this phenomenon is referred to as 'forced elasticity relaxation' indicating the general nature of the mechanisms of deformation in subsequent restoration of the dimensions of the deformed polymer according to which the deformation of the polymer is activated by the external stress and restoration by internal stress. Later [117, 198], within the framework of this model it was assumed that the low-temperature relaxation of the deformed polymer glass is determined by the inhomogeneity of the structure of the initial polymer glasses, and the mechanism of this phenomenon is associated with the consecutive devitrification of the local structural regions with increasing temperature, with the set of these regions in the initial deformed material.

The main special features of this approach are: 1) the assumption of the entropic nature of the observed molecular motion, responsible for the anomalies in the low-temperature restoration of the residual strain of the deformed glassy polymers, and 2) the assumption according to which these anomalies exist in the structure of the initial polymer glass which has an inhomogeneous structure and a set of glass transition temperatures.

It should be mentioned that the most extensively discussed item in the framework of this model is the assumption according to which the initial non-deformed polymer contains regions with greatly different glass transition temperatures. This assumption contradicts the results of a large number of experimental studies in which only one relatively narrow range of the glass transition temperature was recorded in the measurement of the glass transition temperatures of the amorphous polymers by different methods.

Another approach to explaining the same structural–mechanical special features of the polymer glass was proposed in [82–84]. The considerations, developed in these studies, are based on the concept according to which the entire inelastic strain and the stationary plastic yielding do not take place in the initial structure of the polymer and occur instead in the polymer structure saturated with the fine-scale plastic shear transformations (PST). This structure is excited and metastable. Formation of the PST starts in the very early stages of loading and reaches the stationary mode at low strains (20–35%). The PST is the main source of macroscopic deformation of the polymer. The conformation rearrangements in the chains at $T < T_g$ do not take place directly under the effect of the stress and are the product of the annihilation of the PST – non-confirmation, 'volumeless' shear formations, surrounded by the elastic stress fields. The entire energy, stored by the sample in deformation, is concentrated in these fields. The processes of relaxation, physical ageing and molecular mobility in the deformed glass are closely connected with the formation and annihilation of the PST. Mass transfer in deformation of the glass takes place by means of the fine-scale motion of the γ, β and, possibly δ-types, and not by segmental motion.

The main special features of this approach are the assumptions that: 1) all the anomalies in the structural–mechanical behaviour of the polymer glasses are associated with structural changes which the polymer undergoes in the process of its inelastic deformation, and 2) the low-temperature restoration of the deformed polymer glasses is caused by the structural transitions in shear transformations not connected with the entropic elasticity of the molecules.

A shortcoming of this approach is the assumption according to which the reversibility of the high strains in the deformed polymer is caused by some energy transitions in the PST and the elastic mechanical stresses around them. It is now agreed that the high reversible strains is the 'privilege' of the polymer solids and the nature of this elasticity is determined by the entropic shortening of the polymer chains during their transition to the most probable state [1]. For example, the PST cannot be used to explain the increase of stress in the deformed glassy polymer during its isometric heating in the temperature range below T_g (Fig. 4.9). The existence of the PST has not been confirmed by any experimental method.

In some studies [61, 119, 120] the ductility of the polymers below their glass transition temperature is attributed to the β-molecular motion. However, the data in Fig. 4.9 indicate that the molecular motion, generated in the glassy polymer by deformation, does not affect the β-peak, and is situated on the temperature scale above the β-molecular motion, but below the α-transition.

It is also important to mention the studies [121] in which the process of plastic deformation of the hard polymer is described by the free volume concept.

In all the previously cited studies, the mechanism of plastic deformation is regarded on the microscopic, submolecular level. It is assumed that plastic deformation is delocalized throughout the volume of the polymer, and its elementary act takes place in the volumes with the size of tens or hundreds of angströms.

Finally, in a number of published studies, the authors assume that the anomalous low-temperature molecular motion, generated in the glassy polymer by mechanical deformation, is localized in some 'zones' dispersed in the glassy matrix [122–126]. Unfortunately, the authors of these studies did not specify what these 'zones' are, how they form and what is there morphology and properties.

The coexistence of several viewpoints regarding the mechanism of deformation of the polymer glasses is explained, in our opinion, by the absence of direct experimental data concerned with explaining the structural rearrangements, accompanying this process.

It is important to note that there is another aspect of the problem of the deformation of polymer glasses which until recently was not studied or taken into account. Any deformation of the solid and, in particular, the polymer is accompanied by the change of its surface area, in addition to the changes of the geometrical dimensions. The volume of the deformed polymer may remain constant [1] but its surface area always changes.

The main reason why there is no reliable information on such an important phenomenon as the mass transfer in deformation of the polymers and the associated change of the interfacial area is the fact that there is no reliable investigation method. The problem of obtaining information on the mass transfer phenomena in the deformed polymer is very complicated. It is therefore very important to carry out direct microscopic studies capable of casting light on this problem. Chapter 1 described and substantiated a method of visualisation of the structural rearrangements, accompanying the deformation of the amorphous polymers. This method was used in [127] to described structural rearrangements responsible for the above-mentioned anomalies in the structural–mechanical behaviour of the deformed glassy polymers.

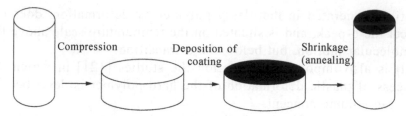

Fig. 4.29. Diagram of the experiment to visualise structural rearrangement in thermally stimulated shrinkage of the polymer deformed in the conditions of uniaxial compression (planar tension) [127].

In [127] a cylindrical sample (Fig. 4.29) was subjected to uniaxial compression in the conditions in which is surfaces, adjacent to the compressed services, could freely slide in relation to the latter. As a result of this compression the polymer acquired the form of a cylinder with a small height but a large diameter (in area) of circular bases. Subsequently, a thin metallic coating was deposited on these bases. The resultant sample was then annealed and the metal-coated surfaces were investigated.

Evidently, in annealing, which is accompanied by the restoration of the initial dimensions of the sample (Fig. 4.29), the bases of the cylinder reduced their surface areas. This is possible only by migration of part of the material of this surface to the volume of the polymer. Microscopic studies of the surface of the polymer with the coating provides important information on the migration of this type.

Figure 4.30 *a, b* shows two micrographs produced by the methods described above when examining the mechanism of thermally stimulated restoration of the dimensions of deformed amorphous polyethylene terephthalate. The figure shows the results of examination of two polyethylene terephthalate samples after preliminary deformation in the uniaxial compression conditions to approximately the same value (22–24%). The only difference between the samples *a* and *b* was that one of them (*a*) was deformed above the glass transition temperature (100°C) and the other one (*b*) below the glass transition temperature, at room temperature. It should be mentioned that in the restoration of the dimensions of the investigated specimens during annealing in the absence of the coating the surface in both cases remains smooth in all stages, irrespective of the preliminary deformation temperature.

In particular, the presence of the coating makes it possible to observe and characterise the structural rearrangements of the

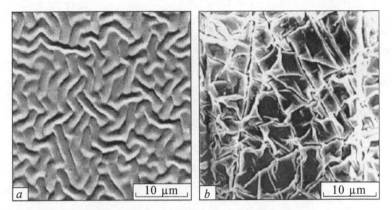

Fig. 4.30. Electron micrographs of polyethylene terephthalate samples deformed by uniaxial compression at 100°C (a) and at room temperature (b); after deformation, thin (10 nm) platinum layers were deposited on the sample surfaces and the samples were annealed at 105°C [127].

polymer during thermally stimulated restoration of its dimensions. It may be seen (Fig. 4.30 *a*) that the reduction of the area of the sample (shrinkage), with the coating deposited on the sample, in the deformation of the polymer above the glass transition temperature produces a regular and distinctive microrelief on the platinum coating. The mechanism of formation and development of this type of relief can be regarded as a special type of the loss of mechanical stability of the hard coating during planar compression on a compliant base. This process was discussed in detail in the first chapter. Annealing of the deformed polymer is accompanied by a decrease of the surface area of the investigated surface on which the metallic coating was deposited, by diffusion of the polymer from its surface to the volume (Fig. 4.29). In this case, the process is quite uniform so that the coating on the surface of volume is uniformly compressed, loses stability and acquires the microrelief, shown in Fig. 4.30 *a*.

It is important to mention here that this microrelief is uniformly distributed on the entire surface of the sample, indicating the general homogeneity (affinity) of the planar deformation and, correspondingly, of the process of restoration of the dimensions of polyethylene terephthalate in annealing above the glass transition temperature. The latter result is quite evident because uniaxial compression loading was applied to the polymer at a temperature higher than T_g, i.e., the polymer in the rubbery state, and the homogeneity (affinity) of deformation is an integral property of the rubber-like polymers [1].

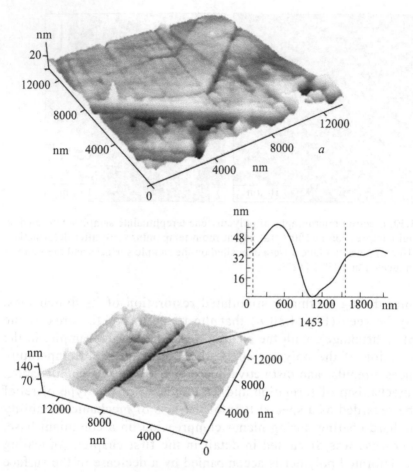

Fig. 4.31. Three-dimensional reconstruction of the atomic-force image of the surface of a polyethylene terephthalate sample which restored its dimensions during annealing. The sample was deformed to 18% as shown in Fig. 4.29. After depositing a thin (10 nm) platinum layer on the sample surface, the sample was heated to 105°C and, consequently, its initial dimensions were restored: a) general view of the surface; b) the image of the individual shear band [127].

We will now examine the reaction of the deposited metallic coating on the restoration of the dimensions in annealing of the polyethylene terephthalate deformed at a temperature below the glass transition temperature. As indicated by Fig. 4.30 *b*, the thermally stimulated restoration of the dimensions of the polymer is accompanied by completely different structural rearrangements in the surface layer in comparison with the polymer deformed above its T_g. It may clearly be seen that in annealing the entire sample surface is covered with straight bands which intersect the entire surface of the sample and each other under different angles. The data obtained in atomic force microscopy (Fig. 4.31 *a*) show that these bands form depressions in the surface layer of the polymer. Since the detected bands intersect the entire studied surface of the samples, it may be assumed that they also intersect the entire cross-section of the deformed polymer. These bands are straight channels (Fig. 4.31 *b*) in which the polymer is 'pulled' from the surface into the volume during annealing.

The procedure (developed in [99, 127] (see chapter 1)) for visualisation of the structural rearrangement, taking place during the restoration of the dimensions of the deformed glassy polymers, can be used to characterise the structural rearrangement accompanying this process in individual stages. Figure 4.32 shows the results of investigation of the evolution of the surface structure in the process of restoration of the dimensions of another deformed glassy polymer (polystyrene). In this case, the experimental procedure was slightly different. All the samples were deformed to the same compression strain (~25%) and then a thin metallic layer was deposited on the surface of the samples. Different degrees of restoration of the dimensions (shrinkage) were obtained by changing the annealing temperature [98].

The results show that the method used to obtain a specific degree of shrinkage did not have any controlling effect on the structural rearrangement of the polymer. For example, at a relatively low annealing temperature of 70°C the shrinkage of the specimen was small (1.7%) and the surface of the sample was covered with a network of shear bands (Fig. 4.32 *a*). Increase of the annealing temperature to 80°C (Fig. 4.32 *b*) results in the shrinkage of 7.5%. In this case, the shear bands have a more distinctive relief and

Fig. 4.32. Scanning electron micrographs of polystyrene samples, subjected to uniaxial compression deformation by ~25% at room temperatures and subjected to shrinkage by 1.2 (a), 7.5 (b) and 25% (c) as a result of annealing at 70, 80 and 90°C, respectively [98].

their density is higher. Finally, annealing in the range of the glass transition temperature (90°C) results in the overall shrinkage of the entire deformed material. It may be seen that the shear bands, formed in the early stages of deformation (shrinkage), are retained in this case (Fig. 4.32 c). Shrinkage takes place on the entire surface of the polymer. In addition, it is important to note the clear similarity of the relief formed in thermally stimulated shrinkage of polystyrene and polyethylene terephthalate (compare Figs. 4.30 b and 4.30 c), indicating the general nature of these phenomena. The folded relief, localized between the shear bands, is very similar to the microrelief which was described previously (its formation mechanism). This relief, detected by the microscopic procedure, is externally identical with the relief formed in shrinkage of the polymer deformed at a temperature higher than the glass transition temperature (Fig. 4.30 a). It should be remembered that the shrinkage of the polymer, deformed at temperatures higher than the glass transition temperature, is uniform so that the surface of the polymer acquires a regular microrelief uniformly distributed over the entire surface of the specimen.

Thus, the data obtained in direct visualisation of the process of thermally stimulated restoration of the initial dimensions of the deformed glassy polymer can be used to draw two conclusions. Firstly, the thermally stimulated restoration of the initial dimensions of the polymer, deformed at a temperature lower than the glass transition temperature, also takes place in the temperature range below the T_g of the polymer in complete agreement with the results of other investigations, examined previously [80–84], and secondly the process is concentrated in special zones (shear bands).

The existence in the deformed glassy polymers of special zones, containing a highly dispersed oriented material (shear bands and/or crazes) results on the whole in the very unusual thermomechanical properties of the polymer. Figure 4.33 compares the curves of restoration of the dimensions of the polycarbonate deformed in air with the formation of a neck (2) and in an adsorption-active medium by the crazing mechanism (1). It may be seen that in annealing of the specimen, deformed in air, there is a distinctive low-temperature contribution to the restoration of the dimensions of the polymer.

This contribution is ~20%, which corresponds approximately to the yield stress of the polycarbonate in tensile loading. This result is in complete agreement with the data obtained in other studies [80–84]. At the same time, the polymer, deformed by the

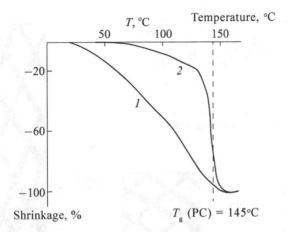

Fig. 4.33. Dependence of the lattice change of the dimensions on the temperature of the polycarbonate specimens, deformed in the AAM by the crazing mechanism (1) and in air with the formation of a neck (2) [98].

crazing mechanism restores its dimensions almost completely in the temperature range below the T_g of the bulk polymer. Evidently, this difference is caused by the fact that in deformation of the polymer in the AAM the free surface of the fibrillised material of the crazes develops in a wide range of the degrees of tensile loading of the polymer, and is not restricted by the strain corresponding to the yield stress (Figs. 4.22, 4.23). In the case of shear yielding, this process occurs only in the low strain range – up to and in the vicinity of the yield stress.

It is interesting to note that the low-temperature contribution to the restoration of the dimensions of the polymer deformed in air has been the subject of a large number of investigations over many years [32–36], whereas the considerably stronger effect of the low-temperature restoration of the dimensions of the crazed polymers (Fig. 4.34) did not attract much attention. The properties of the crazed polymers will be investigated in greater detail below.

Mechanism of low-temperature restoration of the dimensions of deformed glassy polymers in annealing
All the above data show convincingly that the anomalous thermomechanical behaviour of the deformed glassy polymers is associated with the processes taking place in special zones of the plastically deformed polymer, 'implanted' in the initial polymer matrix. Nevertheless, the following important question should be answered: what are the physical reasons for the low-temperature

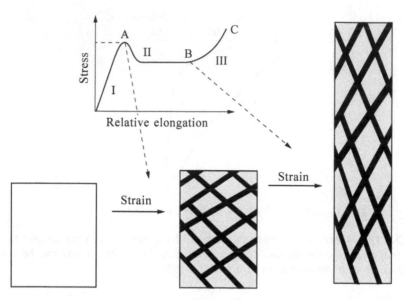

Fig. 4.34. Structural rearrangements accompanying the deformation of the glassy polymer (explanation in the text).

(below T_g) restoration of the dimensions of the deformed polymer glasses during annealing or, in other words, what types of molecular motion in the polymers are responsible for these spontaneously occurring structural rearrangement?

It should be mentioned that in the middle of the 90s of the previous century a new scientific direction was formed and rapidly developed. This direction is based on the study of the properties of amorphous glassy polymers in thin (nanometric) films and surface layers. The fundamental dependence of T_g of the amorphous polymer films on thickness was found [128]. This question was examined in detail in chapter 3. Here, it is important to note the very fact of the reduction of T_g of the polymer with a decrease of the dimension of the polymer phase. The decrease of this type equal to many tens or even hundreds of degrees. We believe that this phenomenon forms the basis of all anomalies in the structural–mechanical behaviour of the glassy polymers. It should be mentioned that the inelastic deformation of the glassy polymer by the development of shear bands or crazes is in fact the process of the dispersion of the polymer into the finest aggregates of the oriented macromolecules – fibrils – with the size equal to units or tens of nanometres. This type of fibrillisation of the block polymer denotes the transition of the polymer to the thinnest surface layer because the fibrillar aggregate of the macromolecules of this diameter containing several tens of oriented macromolecules [129]. Taking into account the data published in [103, 110], it may be assumed convincingly that the material of the shear bands and the crazes has a considerably lower T_g than that of the initial block polymer. The latter circumstance is the actual physical basis for explaining the existence of the large-scale molecular mobility in the deformed glassy polymers in the temperature range considerably lower than the T_g of the block polymer.

In deformation of the polymer in air the polymer appears to be deformed in two stages. In the first stage (up to the yield stress) the inelastic deformation of the polymer is localized in the shear bands which separate the blocks of the non-oriented initial polymer.

In the second stage (the plateau region on the tensile or compression curve) the polymer is transformed completely to the oriented state. It is very important to know that, according to direct microscopic observations (Fig. 4.18), the shear bands, nucleated in the initial stages of deformation, are included in the oriented structure of the polymer which forms in the second stage of deformation. This inclusion takes place in such a manner that the shear bands retain their individuality and the properties. In the case of crazing, this dispersion of the polymer in the discrete zones (crazes) is even more intensive because it is not limited by the strain, corresponding to the yield stress of the polymer [103].

The data for the structure of the shear bands and the crazes, on the one hand [103, 110], and the fundamental relationships between the glass transition temperature and the dimension of the polymer phase on the other hand, described in [98, 130], can be used to make justified assumptions regarding the structure of the deformed glassy polymers. As a result of deformation, the glassy polymer acquires the 'two-phase' structure. The shear bands and/or crazes, containing the highly dispersed fibrillised material, appear to be 'implanted' into the block glassy matrix. It is important to note that these zones of the plastically deformed polymer penetrate (percolate) throughout the entire cross-section of the polymer.

The detailed examination of the structural rearrangement, accompanying the deformation of the glassy polymer, makes it possible to propose the following pattern of inelastic deformation of the polymer and the thermally stimulated restoration of its dimensions. Deformation of the glassy polymer in the initial stages (up to and in the region of the yield stress) produces the structure consisting of two mutually connected components: the part of the oriented polymer, localized in the shear bands and/or crazes, and blocks of non-deformed polymer, distributed between them (Figs. 4.16–4.20). As shown previously, the shear bands and crazes are filled with a highly dispersed fibrillised polymer with the T_g temperature considerably lower than that of the surrounding block polymer. Further inelastic deformation of the polymer in air (the plateau region on the stress–strain curves) results in the molecular orientation of the polymer blocks situated between the shear bands. The orientation of this part of the polymer takes place without the formation of new interfaces and, in fact, does not differ at all from the orientation of the polymer which is in the rubbery state.

Annealing of this two-component system, formed in inelastic deformation of the glassy polymer, is accompanied by the following structural rearrangements. Heating of the deformed polymer in the temperature range below its glass transition temperature causes devitrification and shrinkage of the material in the shear bands which has a lower glass transition temperature in a wide temperature range (low-temperature contribution to the thermally stimulated restoration). As a physical phenomenon, this part of the thermally stimulated shrinkage of the deformed polymer is associated with the entropic shortening of the oriented material filling the shear bands and characterised by a reduced glass transition temperature.

Further heating of the polymer leads to the relaxation of the main part of the oriented polymer (volume component) which, in fact, does not differ from the shrinkage of the block-like polymer and, therefore, takes place in the glass transition temperature range of the block polymer (high-temperature contribution to thermally stimulated restoration). In the case of the crazed glassy polymer the high-temperature contribution to the thermally stimulated restoration of the dimensions is fully degenerated (Fig. 4.33). The latter circumstance is associated with the fact that in the crazing conditions, the development of the zones, containing the highly dispersed oriented material (crazes), takes place in the entire strain range and, therefore, the transition of the polymer to the oriented state as it takes place in the block (during tensile loading by the development of a neck) does not take place.

Thus, all the anomalies in the mechanical behaviour of the deformed glassy polymers (stress relaxation in the initial section of the tension (compression) curve, the increase of stress during isometric heating of the deformed glassy polymer, the existence of the low-temperature contribution to the thermally stimulated restoration of the deformed polymer glasses, the appearance of a soft, compliant 'phase' in deformation of the glassy polymer, detected by examining the dynamic mechanical properties), can be easily explained if two fundamental circumstances are taken into account. Firstly, the inhomogeneity of the inelastic deformation of the glassy polymers as a result of which the zones (shear bands and/or crazes) form in the initial polymer and, secondly, a large decrease of the T_g of the highly dispersed oriented material filling the shear bands and crazes. Usually, all the previously described anomalies in the mechanical behaviour of the glassy polymers are determined by their polymer nature so that they are not found in the low-molecular solids.

Physical reasons for anomalies in the thermophysical properties of deformed glassy polymers

Thus, the analysis of the results of structural studies makes it possible to formulate the physical reasons for the anomalies in the mechanical behaviour of the deformed polymer glasses. Attention will be given to the available data relating to the anomalies in the thermophysical properties of the deformed glassy polymers. It should be mentioned that these anomalies relate, firstly, to the data for deformation calorimetry according to which the first stages of the deformation of the glassy polymers are characterised by the accumulation in the polymer of a large amount of internal energy of unknown nature [82–84] (Fig. 4.15). Secondly, deformation of the glassy polymer results in the formation of a wide exothermic peak on the DSC diagrams in the temperature range below T_g (Fig. 4.1 *b*) [61].

The energy balance of the deformation of the glassy polymer differs greatly from the corresponding characteristics of the low-molecular ductile solids. At the same time, as in the deformation of the low molecular ductile solids (for example, metallic copper) the all expended mechanical energy is converted to heat [131], in deformation of the glassy polymer part of the mechanical energy builds up in the polymer in the form of internal energy (Fig. 4.15). Evidently, this effect is associated with irreversible changes in the structure of the deformed polymer. Previously, it was shown that the structure of the glassy polymer undergoes extensive changes during its deformation. These changes are associated with the formation in the polymer of discrete zones (shear bands and crazes), containing a highly dispersed fibrillised material. According to some estimates

[103], this material has the specific surface of several hundreds of square metres per gram. Evidently, the presence of the excess free surface must be reflected in the internal energy of the polymer. In this connection, it is important to evaluate the principal possibility of buildup of the internal energy by the deformed polymer as a result of increase of its surface energy.

It is necessary to answer the question whether the experimentally determined buildup of such large amounts of internal energy in deformation of the polymer can be associated with the development of the interfaces, characteristic of the shear bands and/or crazes? To answer this question, in [132] the authors carried out the quantitative evaluation of the internal energy which the polymer can store exclusively as a result of the development of the interfacial surface which forms during its deformation by the development of shear bands and/or crazes in the polymer. This estimate is based on the assumption according to which in tensile loading of the glassy polymer up to (approximately) its yield stress the entire deformation takes place by the nucleation and development of discrete zones (shear bands and/or crazes), containing the plastically deformed fibrillised polymer. This assumption was based on the large number of the experimental data available at the time. Firstly, the experimental data, relating to the mechanism of deformation of the glassy polymer in air, indicate that in deformation of the polymer up to and in the region of the yield stress deformation takes place in fact by the development of shear bands (Figs. 4.44–4.28). In the deformation of the glassy polymer in the adsorption-active medium (by the crazing mechanism), the development of the crazes, having a highly developed interfacial surface, does not cause any doubt even in a wider strain range [103]. Secondly, all the anomalies in the mechanical behaviour of the glassy polymers, deformed in air (Figs. 4.12, 4.15), are observed in the strain range corresponding to the values up to and in the range of the yield stress).

As mentioned several times previously, there is no principal difference between the structure of the crazes and shear bands. Both cases are characterised by the development of structural formations with a fibrillar morphology and, consequently, a large interfacial surface.

It is assumed that the entire inelastic strain in the first stage (up to and in the range of the yield stress of the polymer) develops by the nucleation and growth of the zones containing the plastically deformed fibrillised material. Taking into account this mechanism of deformation of the polymer, it is possible to make a substantiated evaluation of the interfacial surface area which the polymer acquires during deformation. In fact, the strain, corresponding to the yield stress of the glassy polymers (ε) is known from the experiments (~10%). The degree of drawing of the fibrillised polymer in the shear bands and crazes (λ_s) is assumed to be equal to 2. This corresponds to direct microscopic evaluations. According to the currently available experimental data [102–104], the polymer filling the zones of the shear bands and the crazes is divided into a system of fibrils with a diameter D_f which is equal to ~10^{-6} cm. In this case, the total surface area of the fibrils (S_f) of the polymer in the shear bands (crazes) is equal to [132]:

$$S_f = 4\varepsilon/(\lambda_s - 1)D_f \qquad (4.4)$$

This equation can be used to determined the surface energy U stored by the glassy polymer during deformation in the first stage when the inelastic deformation takes place mainly by the development of shear bands or crazes (up to and in the range of the yield stress):

$$U = S_f \gamma = 4\gamma\varepsilon/ (\lambda_s - 1)D_f \qquad (4.5)$$

where γ is the specific surface energy of the polymer.

The specific surface energy γ of the most widely used glassy polymers, for example, PS, PET, PMMA, equals 0.04–0.05 J/m^2. If the shear strain of the polymer ε is assumed to be equal to 10% (approximate strain at the yield stress of the majority of glassy polymers), the general surface area of the fibrillised materials in shear bands is ~40 m^2/cm^3, and the corresponding surface energy is 1.6–2.0 J/cm^3. It should be mentioned that the internal energy, stored by the polymer during its deformation to the yield stress is, according to the deformation calorimetry data [82–84], equal to 2–3 J/cm^2 for polymers such sa PS, PC, PMMA.

Regardless of their approximate nature, the results can be used to justifiably assume that there is another contribution in addition to the contributions [61] to the internal energy, stored by the glassy polymer during its deformation (change of the conformation of the chains, i.e. intramolecular energy as a result of T–G transitions, change of molecular packing and intermolecular interaction energy, distortion of the valence angles and bonds and rupture of chemical bonds). In addition to the above-mentioned factors, the excess internal energy may be associated with the development of highly developed interfacial surfaces. In addition, as shown previously, this contribution to the internal energy of the deformed polymer can be so large that if completely ensures the experimentally measured internal energy stored by the deformed polymer.

We now examine what takes place during annealing of the deformed polymer glasses. As shown previously, the deformed glassy polymer is inhomogeneous in the structural aspect and contains discrete bands (year bands and crazes), characterised by the highly developed interfacial surface. Annealing of the deformed polymer not only leads to the restoration of its dimensions but is also accompanied by the complete disappearance (healing) of these interfaces (Fig. 4.24) [109]. Annealing of this type is also carried out in DSC experiments [61]. The data presented in Fig. 4.24 show that the healing of the interfaces during low-temperature annealing (below T_g) in the deformed glassy polymer is accompanied by the formation on DSC thermograms of a wide exothermic peak, spreading virtually from the β-transition to T_g. It should be noted that the length of this DSC peak is almost completely identical with the temperature range of the anomalies in the mechanical behaviour of the deformed polymers.

Taking into account the above considerations regarding the role of the excess interfacial surface in the discrete zones (shear bands and crazes), formed during the inelastic deformation of the polymer glasses, we can explain the existence of this low-temperature exothermic DSC peak (Fig. 4.16) [61]. In heating of the deformed glassy polymer in the temperature range below T_g it is possible to reach gradually the local T_g of the highly dispersed oriented material of the shear bands and the crazes.

This circumstance has two consequences. Firstly, it is the shrinkage of the oriented material of the shear bands and the crazes which is accompanied by the restoration of the dimensions of the polymer by the conventional entropic mechanism, as was the case in the previous section. Secondly, since the devitrified material of the shear bands and the crazes, subjected to shrinkage, has the temperature higher than its local T_g (but lower than T_g of the block polymer), the process of healing of the excess interfacial surface starts to take place. The point is that the healing (self-adhesion) of the interfacial surface can take place only if the polymer is in the rubbery state (see chapter 2). The mechanism of the healing process consists of the mutual diffusion of the segments of the chains through the interface, and the process can take place if the segmental mobility of this type takes place in the polymer. The processes of the decrease (healing) of the interfacial surface are thermodynamically advantageous and, therefore, they take place spontaneously when the required molecular mobility is reached. Evidently, the healing of the interfacial surface is always an exothermic process because it is accompanied by a general decrease of the free energy of the system. This effect is also detected on the DSC thermograms (Fig. 4.16 b) in the form of the exothermic peaks situated between the β- and α-transitions of the block polymer.

The internal energy, stored by the polymer in the initial stages of deformation, is released in the form of heat during its subsequent annealing as a result of healing the interfacial surface. It is not surprising that the thermal energy, determined by the DSC method, is 2–3 J/cm^3, i.e., completely corresponds to the amount of the internal energy, stored during deformation of the glassy polymer (Fig. 4.15) [82].

Thus, all the known special features in the mechanical and thermophysical behaviour of the glassy polymers, which are not included in the conventional considerations regarding their properties, can be explained as follows. Firstly, deformation of the glassy polymer is highly inhomogeneous so that the deformed glassy polymer represents a structure consisting of two mutually connected 'phases': the block polymer, which includes the zones of the plastically deformed material (shear bands and/or crazes). Secondly, the zones (shear bands or crazes) are filled with the highly dispersed oriented material whose glass transition temperature is lower than that of the block polymer and has a highly developed interface. The latter two circumstances can be used to explain without contradiction all the anomalies in the mechanical and thermophysical behaviour of the deformed glassy polymers from a single viewpoint. It is important to note that the above considerations regarding the structural–mechanical behaviour of the glassy polymers are based on the results of direct structural studies. In fact, all the previously mentioned anomalies in the mechanical and thermophysical behaviour of the glassy polymers are associated with the processes taking place at the interfaces of the polymer and, therefore, can be regarded as a complicated surface phenomenon.

Conclusions
Taking into account the above considerations, the process of deformation of the glassy polymer in the tensile loading conditions

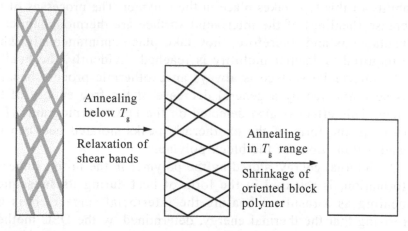

Fig. 4.35. Diagram of structural rearrangements accompanying the thermally stimulated restoration of the dimensions of the deformed glassy polymer (explanation in the text).

can be described as follows (Fig. 4.34). The initial stages of deformation (up to and in the range of the yield stress) are characterised by the nucleation and growth of the shear bands characterised by a specific structure. The polymer, situated between the shear bands, is not deformed and the structure and properties along is identical with those of the initial non-oriented polymer.

Further inelastic deformation (the plateau region on the stress-strain curves) results in the molecular orientation of the polymer blocks, distributed between the shear bands. The orientation of this part of the polymer can take place without the formation of new interfaces so that there is almost no buildup of the internal energy in this deformation stage. It is important to note that the shear bands, formed in the initial stages of deformation of the polymer, are included in the structure of the polymer and retain their individuality, interfaces and properties even after complete transition of the polymer to the oriented state.

The reverse process of restoration of the dimensions and structure of the deformed glassy polymer can also be described graphically taking the above considerations into account (Fig. 4.35).

Heating of the deformed polymer in the temperature range below its glass transition temperature results initially in the shrinkage of the material in the shear bands which has a reduced glass transition temperature in a wide temperature range (low-temperature contribution to the thermally stimulated restoration of the dimensions). As a physical phenomenon, this part of the thermally stimulated shrinkage of the deformed polymer is associated not only with the entropic shortening of the oriented material, filling the shear bands and having a lower glass transition temperature, but also with the healing of the interfaces in the structure of the shear bands. The process of healing of the interfaces in annealing of this system is characterised by the exothermic effect and complete relaxation of the stored internal energy.

Further heating of the polymer results in the relaxation of the main part of the oriented polymer which, in fact, does not differ from the shrinkage of the block rubber-like polymer and, therefore, takes place in the range of its glass transition temperature (high-temperature contribution to the thermally stimulated restoration of the dimensions). The above analysis showed the important role of the surface phenomena in the structural–mechanical behaviour of the glassy polymers.

References

1.   Treloar L.R.G., Physics of rubber elasticity, Moscow, Mir, 1975.
2.   Kuleznev V.N., Shershnev V.A., Chemistry and physics of polymers. Moscow, Ko-losS, 2007.
3.   Struik L.C.E., Physical aging in amorphous glassy polymers and other materials, Amsterdam, Elsevier, 1978.
4.   Shelby M.D., Wilkes G.L, Polymer, 1998, V. 39, P. 6767.
5.   Jamieson A.M., Simha R., Mcgervey J.D., J. Polym. Sci., Polym. Phys., Ed. 1995, V. 33, P. 2295.
6.   Shelby M.D., Hill A.J., Burgar M.I., Wilkes G.L., J. Polym. Sci., Polym. Phys., Ed. 2001, V. 39, P. 32.
7.   Brauwens-Crowet C., J. Mater. Sci., 1999, V. 34, P. 1701.
8.   Guest M.J., van Daele R., J. Appl. Polym. Sci., 1995, V. 55, P. 1417.
9.   Bendler J.T., Comput. Theor. Polym. Sci., 1998, V. 8, P. 83.
10.   Othmezouri-Decerf J., J. Mater. Sci., 1999, V. 34, P. 2351.
11.   Higuchi H., Yu Z., Jamieson A.M., Simha R., Mcgervey J.D., J. Polym. Sci., Polym. Phys., Ed. 1995, V. 33, P. 2295.
12.   Orreindy S., Rincon G.A., J. Appl. Polym. Sci., 1999, V. 74, P. 1646.
13.   Robertson G.C., Wilkes G.L., Macromolecules, 2000, V. 33, P. 3954.
14.   Hutchinson J.M., Tong A.B., Jiang Z., Thermochim. Acta, 1999, V. 335, P. 27.
15.   Greer R.W., Wilkes G.L., Polymer, 1998, V. 18, P. 4205.
16.   Davis W.J., Pethrick R.A., Eur. Polym. J., 1998, V. 34, P. 1747.
17.   Shelby M.D., Wilkes G.L., Polymer, 1998, V. 39. P. 6767.
18.   Hill A.J., Heater K.J., Agrawal C.M., J. Polym. Sci., Polym. Phys., Ed. 1990, V. 28, P. 387.
19.   Heymans N., Polymer, 1997, V. 38, P. 3435.
20.   Dybal J., Schmidt P., Baldrian J., Kratochvil J., Macromolecules, 1998, V. 31, P. 6611.
21.   Lee S.-N., Stolarski V., Letton A., Laane J, J. Mol. Struct., 2000, V. 521, P. 19.

22. Hutchinson J.M., Prog. Polym. Sci., 1995, V. 20, P. 703.
23. Hourston D.J., Song M., Hammiche A., Pollock H.M., Reading M., Polymer, 1996, V. 37, P. 243.
24. Reading M., Trends Polym. Sci., 1993, V. 1, P. 248.
25. Tant M.R., Wilkes G.L, Polym. Eng. Sci., 1981, V. 21, No. 14, P. 874.
26. Petrie S.E.B, J. Macromol. Sci., Phys., 1976, V. 12, No. 2, P. 225.
27. Wang Y., Song R., Shen D., J. Macromol. Sci., Phys., 1998, V. B37, No. 5, P. 709.
28. Shelby M.D., Wilkes G.L, J. Polym. Sci. B: Polym. Phys., 1998, V. 38, P. 495.
29. Wang Y., Song R., Shen D, J. Macromol. Sci. Phys., 1998, V. B37, No. 5, P. 709.
30. Brunacci A., Cowie J.M.G., Fergusson R., McEwen I.J., Polymer, 1997, V. 38, No. 4, P. 865.
31. Doulut S., Bacharan C., Demont P., Bernes A., Lacabanne C., J. Non Cryst. Solids, 1998, V. 235-237, P. 6645.
32. Shelby M.D., Wilkes G.L., J. Polym. Sci. B: Polym. Phys., 1998, V. 38, P. 495.
33. Kargin V.A., Slonimsky G.L., Short essays on the physical chemistry of polymers, Moscow, Khimiya, 1967.
34. Shlyapnikov Yu.A., Usp. Khimii, 1997, V. 66, No. 1, P. 1.
35. Belousov V.N., Kozlov G.V., Mikitaev A.L., Lipatov Yu.S., Dokl. AN SSSR, 1990, V. 313, No. 3, S. 630.
36. Bove L., D'Aniello C., Gorrasi G., Guadagno L., Vittoria V., Polym. Bull., 1997, V. 3S, P. 579.
37. Balta-Calleja F.J., Santa Cruz C., Asano T.J., Polym. Sci. Phys. Ed., 1993, V. 31, P. 557.
38. Neki K., Geil P.H, J. Macromol. Sci., Phys. 1973. V. 8, No. 12. P. 295.
39. Yeh G.S.Y., J. Macromol. Sci. Phys., 1973, V. 7, No. 4, P. 729.
40. Hwang Y., Inoue T., Wagner P.A., Ediger M.D., J. Polym. Sci. B: Polym. Phys., 2000, V. 38, P. 68.
41. Capodanno V., Petrillo E., Romano G., Russo R., Vittoria V., J. Appl. Polym. Sci., 1997, V. 65, P. 2635.
42. Nagai K., Higuchi A., Nakagawa T., J. Polym. Sci., Polym. Phys., 1995, V. 33, No. 2, P. 289.
43. Kovacs A.J, J. Polymer Sci., 1958, V. 29, No. 30, P. 131.
44. Pfromm P.H., Koros W.J., Polymer, 1995, V. 50, P. 2379.
45. VcCaig M.S., Paul D.R, Polymer, 2000, V. 41, P. 629.
46. Dorkenoo K.D., Pfromm P.H, J. Polym. Sci. Polym. Phys., 1999, V. 37, No. 16, P. 2239.
47. Forrest J.A., Dalnoki-Veress K., Adv. Colloid Interface Sci., 2001, V. 94, No. 1-3, P. 167.53.
48. Forrest J A, Eur. Phys. J. E, 2002, V. 8, No. 3, P. 261.
49. Cangialosi D., Schut H., van Veen A., Picken S.J., Macromolecules, 2003, V. 36, No. 1, P. 143.
50. Wang B., Gong W., Liu W.H., Wang Z.F., Qi N., Li X.W., Liu M.J., Li S.J., Polymer, 2003, V. 44, No. 14, P. 4047.
51. Doulut S., Bacharan C., Demont P., Bernes A., Lacabanne C, J. Non-Cryst. Solids, 1998, V. 235–237, P. 6645.
52. Soloukhin V.A., Brokken-Zijp J.C.M., van Asselen O.L.J, de With G., Macromolecules, 2003, V. 36, No. 20, P. 7585.
53. Hutchinson J.M., Smith S., Horne B., Gourlay G.M., Macromolecules, 1999, V. 32, No. 15, P. 5046.
54. Shafee E. El, Polymer, 2003, V. 44, No. 13, P. 3727.

55.   McGonigle E.A., Daly J.H., Gallagher S., Jenkins S.D., Liggat J.J., Olsson I., Pethrick R.A., Polymer, 1999, V. 40, No. 17, P. 4977.

56.   Vittoria V., Petrillo E., Russo R., J. Macromol. Sci., Phys., 1996, V. 35, No. 1, P. 147.

57.   Moynihan C.T., Easteal A.J., DeBolt M.A., Tucker J, J. Am. Ceram. Soc., 1976, V. 59, No. 1, P. 12.63.

58.   Williams G., Watts D.C., Trans. Faraday Soc., 1970, V. 66, P. 80.

59.   Bailey R.T., North A.M., Pethrick R.A., Molecular motions in high polymers, Oxford, Clarendon, 1981.

60.   Davis J.W., Pethrick R.A., Polymer, 1998, V. 39, No. 2, P. 255.

61.   Bernstein V.A., Egorov V.M., Differential scanning calorimetry in the physical chemistry of polymers, Leningrad, Khimiya, 1990.

62.   Yampolskii Y.P., Shishatskii S.M., Shantarovich Y.P., Antipov E.M., Kuzmin N.N., Rykov S.V., Khodjaeva V.L., Plate N.A., J. Appl. Polym. Sci., 1993, V. 48, No. 11, P. 1935.

63.   McGonigle E.-A., Daly J.H., Jenkins S.D., Liggat J.J., Pethrick R.A., Macromolecules, 2000, V. 33, No. 2. P. 480.

64.   Lee S.N., Stolarski V., Letton A., Laane J., J. Molec. Struct., 2000, V. 521, No. 1, P. 19.

65.   Soloukhin V.A., Brokken-Zijp J.C.M., van Asselen O.L.J., de With G., Macromolecules, 2003, V. 36, No. 20, P. 7585.

66.   Robertson C.G., Monat J.E., Wilkes G.L., J. Polym. Sci., Polym. Phys. 1999, V. 37, No. 15, P. 1931.

67.   Wimberger-Friedl R., de Bruin J.G., Macromolecules, 1996, V. 29, No. 14, P. 4992.

68.   Kovacs A.J., Fortschr. Hochpolym.-Forsch., 1964, V. 3, P. 394.

69.   Scherer G.W., Relaxation in glass and composites. New York, Wiley, 1986.

70.   Robertson C.G., Wilkes G.L., Macromolecules, 2000, V. 33, No. 11, P. 3954.

71.   Govaert L,E,, van Melick H.G.H., Meijer H.E.H., Polymer, 2001, V. 42, P. 1271.

72.   Narisawa J., Jee A.F., Material Science and Technology. Structure and Properties of Polymers. 1993. V. 12, P. 701.

73.   Gurevich G., Kobeko P., Rubber. Chem. Tech., 1940, V. 13, P. 904.

74.   Haward R.N., Trans. Faraday Soc., 1942, V. 38, P. 394.

75.   Van Melick H.G.H., Govaert L.E., Meijer H.E.H., Localization phenomena in glassy polymers: influence of thermal and mechanical history, submitted for publication, and van Melick H.G.H., Deformation and failure of polymer glasses, PhD Thesis, Eindhoven University of Technology, 2002.

76.   Lazurkin Yu.S., Dissertation for the title of Doctor of physical-mathematical sciences, Moscow, S.I. Vavilov Institute of Physical Problems, USSR Academy of Sciences, 1954.

77.   Laius L.A., Kuvshinskii E.V., Vysokomolek. Soed., 1964, V. 6, No. 1, P. 52.

78.   Shoshina V.I., Nikanovich G.V., Tashpulatov Yu.T., Isometric method of studying polymer materials, FAN Tashkent, Uzb. SSR, 1989.

79.   Meyer M., Ferry D., Rubber Chem. Technol., 1935, No. 8, P. 319.

80.   Arzhakov S.A., Dissertation Dr. chem. Sciences, Moscow, Karpov Institute, 1975.

81.   Arzhakov M.S., Arzhakov S.A., Structural and mechanical behavior of glassy polymers, Zaikov G.E., (ed.) New York: Nova Science Publishers, Inc., Commack, 1997, P. 275.

82.   Oleynik E.F., High performance polymers, Baer E., Moet S. (ed.) Berlin, Hauser Verlag, 1991, P. 79.

83.   Oleinik E.F., Salamatina O.B., Rudnev S.N., Shenogin S.V., Vysokomolek. Soed.,

A, 1993, V. 35, No. 11, P. 1819.

84.   Shenogin S.V., Homme G.W.H., Salamatina O.B., Rudnev S.N., Oleinik E.F.,
       Vysokomolek. Soed., A, 2004, V. 46, No. 1, P. 30.
85.   Gauthier C., Pelletier J.-M., David L., Vigier G., Perez J., J. Non-Cryst. Solids,
       2000, V. 274, P. 181-187.
86.   Trzadnel M., Pakula T., Kryszewski M., Polymer, 1988, V. 29, P. 619.
87.   Munch E., Pelletier J. B., Vigier G., Phys. Rev. Lett., 2006, V. 97, P. 207, 801.
88.   Wallace M.L., Jous B., Phys. Rev. Lett., 2006, V. 96, P. 025, 501.
89.   Warren M., Rottler J., Journ. Chem. Phys., 2010, V. 133, P. 164, 513.
90.   Munch E., Pelletier J.-M., Vigier G., J. Polym. Sci., Part B: Polym. Phys., 2008, V.
       46, P. 497-505.
91.   Havriliak S. Jr., Shortridge T.J.,, Polymer, 1990, V. 31, P. 1782-1786.
92.   Starkweather J., Polymer, 1991, V. 32, P. 2443-2448.
93.   Bauwens-Crowet O., J. Mater. Sci., 1999, V. 34, P. 1701-1709.
94.   Berensa R., Hodge I.M, Macromolecules, 1982, V. 15, P. 756-761.
95.   Parisot J., Rafi O., Choi W., J. Polymer. Eng. Sci., 1984, V. 24, No. 11, P. 886.
96.   Shu-Sing Chang, J. Chem. Therm., 1977, V. 9, P. 189.
97.   Volynskii A.L., Efimov A.V., Bakeev N.F., Vysokomolek. Soed., 2007, V. 49, No.
       7, P. 1317-1343.
98.   Volynskii A.L., Bakeev N.F., Structural self-organization of polymers, Moscow, Fiz-
       matlit, 2005.
99.   Volynskii A.L., Kulebyakina A.I., Panchuck A., Moiseeva S.V., BolshakovA.V.,
       Grokhovskaya T.E., Yarysheva L.M., Kechekyan A., Bazhenov S.L., Bakeev N.F.,
       Vysokomolek. Soed., A, 2007, Vol. 49, No. 12, S. 2063-2084.
100.   Kechek'yan A., Vysokomolek. Soed., B, 1987, V. 29, No. 11, P. 804.
101.   Sell C.G., Hiver J.M., Dahoun A., International Journal of Solids and Structures,
       2002, V. 39, P. 3857.
102.   Kambour R.P., J. Polym. Sci., Macromol. Rev., 1973, V. 7, P. 1.
103.   Volynskii A.L., Bakeev N.F., Solvent crazing of polymers, Amsterdam, New York,
       Elsevier, 1996. P. 410.
104.   Passaglia E., J. Phys. Chem. Solids, 1987, V. 48, No. 11, P. 1075.
105.   Nadezhin Yu.S., Sidorovich A.V., Usher B.A., Vysokomolek. Soed., A, 1976, V. 18,
       No. 12, S. 2626.
106.   Bowden P.B., Raha S., Phil. Mag., 1970, V. 22, P. 463.
107. Volynskii A.L., Grokhovskaya T.E., Lyulevich V.V., Yarysheva L.M., Bol'shakova A.V.,
       Kechek'yan A.S., Bakeev N.F., Vysokomolek. Soed., A, 2004, V. 46, No. 2, P. 247.
108.   Oleinik E.F., Rudnev S.N., Salamatina O.B., DAN SSSR, 1986, V. 286, No. 1, P.
       1235-1300.
109.   Bazhenov S.L., Rodionova Yu.A., Kechek'yan A.S., Vysokomolek. Soed., A, 2005,
       Vol 47, No. 2. P. 255.
110.   Li J. C. M., Polym. Eng Sci., 1984, V. 24, No. 10, P. 750.
111.   Friedrich K., Adv. Polym. Sci., 1983, V. 52/53, P. 266.
112.   Harmon J.P., Lee S., Li J.C.V., Polymer, 1988, V. 29, No. 7, P. 1221.
113.   Volynskii A.L., Yarysheva L.M., Mironov A.A., Arzhakova O.V., Kechek'yan A.S.,
       Ozerin A.N., Rebrov A.V., Bakeev N.F., Vysokomolek. Soed. A, 1996, V. 38, No. 2,
       P. 269.
114.   Volynskii A.L., Yarysheva L.M., Arzhakova O.V., Bakeev N.F., Vysokomolek.
       Soed.. A. 1991, V. 33, No. 2, P. 418.
115.   Arzhakov S.A., Kabanov V.A., Vysokomolek. Soed., B, 1971, V. 13, No. 5, P. 318.

116. Arzhakov S.A., Bakeev N.F., Kabanov V.A., Vysokomolek. Soed., A, 1973, V. 15, No. 10, S. 1154.
117. Arzhakov M.S., Lukovkin G.M., Arzhakov S.A., Dokl. RAN, V. 1999, 369, No. 5, P. 629.
118. Arzhakov M.S., Lukovkin G.M., Arzhakov S.A., Dokl. RAN, 2000, V. 371, No. 4, P. 484.
119. Brady T.E., Yeh G.S.Y., J. Macromol. Sci. Phys., 1974, V. 9, No. 4, P. 659.
120. Bershtein V.A., Peschanskaya N.N., Halary J.L., Monnerie L., Polymer, 1999, V. 40, P. 6687.
121. Sanditov D.S., Sangadiev S.Sh., Vysokomolek. Soed., A, 1999, V. 41, No. 6, P. 977.
122. Johari J.P., J. Chem. Phys., 1973, V. 58, P. 1766-1770.
123. Hasan O.A., Boyce M.S., Polym. Eng. Sci., 1995, V. 35, P. 331-344.
124. G'Sell C., El Bari H., Perez J., Cavaille J,Y,, Johari G.P., Mater Sci. Eng. A, 1989, V. 110, P. 223-229.
125. Perez J., Cavaille J.-Y., David L., J. Mol. Struct., 1999, V. 479, P. 183-194.
126. David L., Quinson R., Gauthier C., Perez J., J. Polym. Eng. Sci., 1997, V. 37, P. 1633-1640.
127. Volynskii A.L., Yarysheva L.M., Bakeev N.F., Vysokomolek. Soed., A, 2011, V. 53, No. 10, P. 871-898.
128. Volynskii A.L., Bakeev N.F., Vysokomolek. Soed., B, 2003, V. 45, No.7, P. 1209-1231.
129. Yang A.S.M., Kramer E.J., J. Polym. Sci. A, 1985, V. 23, P. 1353.
130. Forrest J.A., Dalnoki-Veress K., Adv. Colloid and Interface Sci., 2001, V. 94, P. 167.
131. Shenogin S.V., Hone G.W.H., Oleinik E.F., Thermochim. Acta, 2002, V. 391, P. 13-23
132. Volynskii A.L., Efimov A.V., Grokhovskaya T.E., Yarysheva L.M., Bakeev N.F., Vysokomolek. Soed., A. 2004, V. 46, No. 7, P. 1158-1167.

Role of surface phenomena in the strain softening of glassy and crystalline polymers

In chapter 4, we described the important role of surface phenomena in, it would appear, the well-examined process such as the deformation of glassy polymers. The present chapter is concerned with the analysis of another aspect of the structural–mechanical behaviour of polymers of general nature – strain softening of glassy and crystalline polymers.

In the most general form, the phenomenon of strain (force) softening of solids takes place if the solid is not deformed in the brittle manner. Actually, as soon as the yield stress of the solid body is reached, its actual modulus starts to decrease (softening) and the deformation curve shows a maximum (the yielding 'tooth'). This type of softening is general, it is typical not only of polymers and is not the subject of the present monograph.

A typical case of the strain softening of polymer systems is the force softening of filled rubbers. This effect is based on the fact that the repeated deformation of the filled rubber requires a considerably lower force in comparison with the first deformation. This effect is irreversible, and its appearance is attributed to the special features of the interaction of polymer chains with the surface of the solid filler in the volume of the rubber. As a result of its practical significance, the force softening of the filled resins has been the subject of special attention of a large number of investigators and, consequently, it is also not the subject of the present book.

In this chapter, we confine ourselves to the examination and generalisation of the problems related to the structural rearrangement, taking place in the amorphous and crystalline polymers under the effect of mechanical stresses. The main special feature of these systems is the structural rearrangement which leads not only to a large change of the structural–mechanical behaviour of the initial polymer, but also the fact that the polymer becomes capable of anomalously high reversible deformation in the temperature range below the appropriate glass transition temperature and/or melting point. It is important to mention that to explain the physical reasons for the formation of this type of high reversible strains, we have several alternative approaches, i.e., at the present time, there is no united view regarding this phenomenon.

5.1. Strain softening of polymer systems, taking place without the formation of porosity

Crystallising rubber

The phenomenon of strain softening in polymers was detected a relatively long time ago. Figure 5.1 shows the curves of tensile loading and restoration of the dimensions (shrinkage) of the gutta-percha [3]. The gutta-percha is a polymer which is a geometrical isomer of natural rubber, but in contrast to the latter it is present at room temperature in the crystalline state [4]. Figure 5.1 shows that this polymer in tensile loading in the first cycle shows the stress–strain curve characteristic of the crystalline polymer (the relatively high modulus, yield stress, the plateau region). This deformation

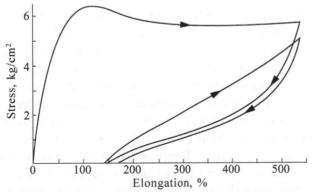

Fig. 5.1. Curves of tensile loading and shortening (shrinkage) of gutta-percha at room temperature [3].

may be carried out to a very high value (to 500 or more percent). However, after interrupting tensile loading and removing the stresses, the tensile loaded polymer restores its dimensions almost completely and spontaneously (Fig. 5.1). In other words, this is accompanied by extensive shrinkage, characteristic of the polymer, which is in the rubbery state. In repeated tensile loading this polymer shows a mechanical behaviour characteristic of the polymer in the rubbery state (low modulus, absence of the yield stress and the plateau region, very low residual strains).

The deformation behaviour of this type is associated with the fact that the first deformation cycle is characterised by the melting of the continuous crystalline frame of the polymer under the effect of mechanical stress [5]. Consequently, is continuous phase in the material is the amorphous component of the polymer which is in the rubbery state. Naturally, in repeated tensile loading such a polymer shows a mechanical behaviour characteristic of the rubber-like polymer.

Block copolymers
The block copolymers are multiphase systems constructed from incompatible fragments of the chains. The incompatibility of this type results in the phase separation of the block copolymer and the formation of distinctive regular structures. There are several varieties of these systems in which the continuous phase is either the rigid component, which is in the glassy state at the deformation temperature and is dispersed in the continuous rubbery phase, or conversely, a system in which the continuous phase is the rubbery component. There are also block copolymers with two continuous phases in the structure [6, 7]. These special features of the structure of the block copolymers are determined in particular by the molecular mass of the individual blocks, their content, the affinity of the blocks for the solvent in the case of separation of the blocks in the solid phase from the solutions, and by some other factors [8].

Figure 5.2 provides information on the structure of the styrene–butadiene–styrene (SBS) block copolymer. As a result of the contrast, the rubber (polybutadiene(PB)) phase forms the dark bands, and the polystyrene phase the light bands [6]. It is clearly seen that the resultant film has a distinctive laminar structure, constructed from alternating lamellae of PS and PB of the same size. It is important to note that in this case the hard glassy phase (PS) is continuous in the volume of the block copolymer.

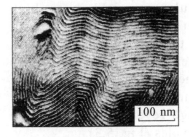

Fig. 5.2. Electron micrograph of a film of ternary styrene–butadiene–styrene copolymer cast from a solution [6].

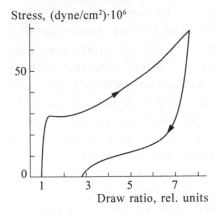

Fig. 5.3. Curves of tensile loading (1) and shrinkage (2) of the film of the styrene–butadiene–styrene block copolymer [6].

Figure 5.3 shows the curves of tensile loading and shrinkage of the block polymer whose structure is shown in Fig. 5.2. It is clearly seen that the deformation of this block copolymer is characterised by the distinctive effect of strain softening of the material. As in the case of the gutta-percha (compare Figs. 5.1 and 5.3), in the first deformation cycle the material shows the mechanical behaviour typical of the glassy polymer (the relatively high modulus, yield stress, plateau region). However, in the process of the first tensile loading cycle the material shows the formation of the 'phase' with considerably lower mechanical parameters in comparison with the initial material and showing high reversible deformation. The detailed analysis of the structural–mechanical behaviour of the SBS block copolymer, carried out in [6], shows that in the case in which the morphology represents a set of continuous alternating layers of the glassy and rubbery polymers, the material shows a high modulus and strain softening (the glass–rubber transition). However, if the block copolymer is constructed on the basis of the 'glassy domains in the rubber matrix' principle, this material shows the mechanical behaviour characterised of the filled rubbers. In this case, the effect of force softening is also detected but its mechanism is different. The initial tensile loading curve is identical with the tensile loading curves of the filled rubbers (low modulus, no yield stress, high reversible strain), and the softening effect is expressed in the general displacement of the tensile loading curve to low stresses [1, 2]. Thus, the strain softening of the block polymer, containing a continuous glassy phase, consists of the failure in the first cycle of deformation of this continuous structure so that the rubbery phase becomes continuous and also determines the high reversible strain in the second and subsequent tensile loading cycles (Fig. 5.3).

Thus, in the investigated systems, the mechanism of strain softening of the material can be described as follows. These materials have the distinctive two-phase structure. In the first case, they contain a continuous crystalline frame, 'filled' with the rubbery component, and in the second case – a continuous glassy frame, also filled with the rubbery polymer. In the first deformation cycle the rigid frame fails in both cases so that the continuous phase becomes the rubber-like polymer. It is not surprising that in subsequent deformation cycles this component shows the mechanical behaviour typical of the polymer in the rubbery state (low modulus, high reversible strain). In the context of the subject of this book the surface phenomena in the investigated cases of force softening are very important. This is obviously due to the fact that the initial systems (the crystalline polymer and the block copolymer) are highly dispersed and consists of several phases. Naturally, the deformation of the system is accompanied by structural rearrangement on the nanometre scale. However, in this book, we will not examine in detail the structural–mechanical behaviour of these systems. In this case, we want to demonstrate the effect of force softening of the glassy and crystalline polymers and its main special feature – formation of the high reversible strain in the material as a result of the first tensile loading cycle.

5.2. Strain softening of glassy polymers

Strain softening is typical not only of the two-phase systems examined previously. The structural–mechanical behaviour of this type is also shown by the single-phase polymers. One of the fundamental conclusions of inelastic plastic deformation of the glassy polymers is crazing [9–11]. An important feature of the crazing of the polymers is the fact that the development of this type of plastic deformation is accompanied by the formation and growth of special zones containing the oriented polymer – crazes. Figure 5.4 shows a

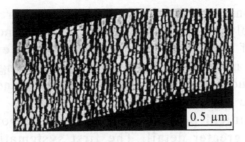

0.5 µm

Fig. 5.4. Transmission electron micrograph of a single craze in a deformed thin polystyrene film. The tensile loading axis is vertical.

transmission electron micrograph of a single craze in the deformed thin-film of the PS.

It may easily be seen that the craze represents the destruction of continuity in the polymer. The principal special feature of the craze structure is that the blocks of the initial non-deformed material are connected by the thinnest strands (fibrils) of the oriented polymer, separated in space. The size (diameter) of these fibrils and the distance between them equal units or tens of nanometres. It is important to know that the highest rate of crazing is detected in the tensile loading of glassy polymers in adsorption-active media (AAM) [11].

The crazing phenomenon in the AAM will be examined in detail in chapter 6, and here we only mention that it was found a long time ago that the polymers deformed in the AAM by the crazing mechanism show the mechanical properties not typical of the glassy state. The most characteristic manifestation of this type is the reversibility of the deformation of the crazed polymer [11]. It is well-known that the amorphous polymer, deformed below the glass transition temperature, shows the phenomenon of the so-called cold drawing, manifested in the formation and growth of a neck. The material of the neck is completely stable at the deformation temperature. To induce shrinkage of the cold drawn polymer, the polymer must be heated to the appropriate glass transition temperature [12, 13].

As shown for the first time in [14, 15], the glassy polymers, deformed in the AAM by the crazing mechanism, show high reversible strains. After tensile loading, removal from the clamps of the tensile loading system and the removal of the active liquid from the volume of the crazes, the polymer shrinks by almost 100%. In other words, the deformation of the glassy polymer by the crazing mechanism is characterised by some type of force softening of the polymer, accompanied by the formation of high reversible strains. The glassy polymers are single-phase systems and, therefore, the mechanism of their force softening can not be explained by the reasons given in section 5.1 for the force softening of the polymer systems, produced from the rigid frame (crystalline or glassy) and the rubber-like 'filler'.

The mechanical properties of the crazed polymers will be discussed in greater detail. The first systematic study of the mechanical properties of the 'dry', i.e. not containing the low-molecular liquid, crazes of the polycarbonate (PC) was carried out by Kambour and Kopp [16]. In this study, the single crazes were produced by deformation of the glassy PC in the AAM (ethanol). Using a special device, the authors investigated the change of the distance between the edges of the single craze in relation to the force, applied to the end of the specimen. The tensile curves obtained in this manner for individual crazes have the characteristic form shown in Fig. 5.5. It may be seen that the yield stress of the crazed material is several times smaller than the yield stress of the initial non-deformed polymer.

The application of the several consecutive deformation–restoration cycles shows that the initial modulus of elasticity and yield stress decrease with the increase of the width of the initial craze. The elasticity modulus is approximately four times smaller than the elasticity modulus of the initial non-deformed polymer. The same study showed that the 'dry' crazes of the PC, deformed by 50–60%, are characterised by the capacity for the spontaneous slow restoration of their dimensions after removing the load. It should be

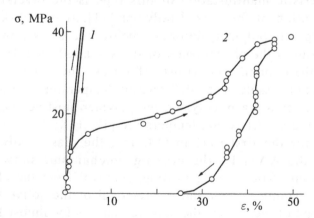

Fig. 5.5. Curves of tensile loading and shrinkage at a constant rate of the PC (1) and of the individual craze in the PC (2), to a stress of 42 MPa [16].

mentioned that this phenomenon is not characteristic at all of the glassy polymer, oriented in the cold drawing mode. It may be seen that the mechanical properties of the 'dry' material of the individual craze differ dramatically from the properties of the initial block polymer.

A similar mechanical behaviour of the crazed polymers was detected in [15, 17]. The results show that the polyethylene terephthalate (PET), crazed in the AAM, shows the tensile curve with two yield stresss, with the first limit considerably (~3 times) lower than the yield stress of the initial glassy polymer. In the same studies, it was found that the crazed polyethylene terephthalate which does not contain any fluid shows the spontaneous reduction of the dimensions after tensile loading. The shrinkage in this case is relatively large and reaches 60%.

The unusual mechanical behaviour of the crazed polyethylene terephthalate was also demonstrated in [18] where it was attempted to explain the role of capillary forces in the observed shrinkage of the crazed polymer in the process of evaporation of the active fluid from the porous structure of the craze. Generally speaking, the capillary forces, which form unavoidably in the evaporation of the AAM from the volume of the craze after tensile loading of the polymer, may be the driving force of the observed irreversible deformation. To eliminate the capillary forces, formed during evaporation of the AAM from the volume of the craze, immediately after tensile loading the polymer in the AAM it was frozen by deep cooling and, subsequently, the active fluid was removed from the volume of the crazes by the method of lyophilic drying. Thus, the removal of the low-molecular component took place without the formation of bend menisci at the polymer–liquid– air interface and, consequently, no capillary forces, which could cause 'closure' of the crazes, formed. Nevertheless, the results show that the shrinkage of the dry crazed polymer at room temperature is up to 30% after removing the sample from the clamps of the tensile loading device. This mechanical behaviour of the PET, which does not contain any liquid component, is also unusual because its glass transition temperature is considerably higher than room temperature (75°C).

In the context of the discussed subject, it is useful to mention, although briefly, the structural–mechanical behaviour of the polymer glasses, reinforced with rubber dispersions. The typical representative of this group of materials is the high-impact polystyrene (HIPS). The methods of production of these materials are well-known [19]. Here

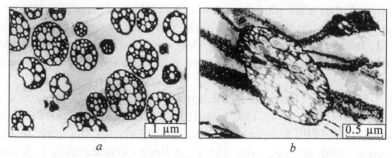

a   *b*

Fig. 5.6. Transmission electron micrographs of specimens of high-impact polystyrene prior to (*a*) [22] and after (*b*) deformation [23].

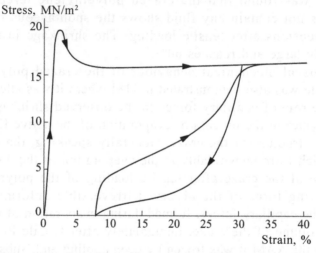

Fig. 5.7. The curves of tensile loading and shrinkage of high-impact polystyrene at 21°C [21].

it is important to mention that the structure of the HIPS and other plastics, modified with rubbers, is characterised by isolated rubber particles, enclosed in the continuous glass matrix [20, 22]. A typical structure of this type is shown in Fig. 5.6 *a*. Dark inclusions are the rubber-like component (polybutadiene), dispersed in the continuous glass matrix (PS). The deformation of these materials is accompanied by the formation and growth of the crazes between the particles of the rubber phase in the glassy matrix (Fig. 5.6 *b*). The formation of the crazes results in the properties of the material unusual on the whole for the glassy state.

Figure 5.7 shows the curves of tensile loading and shrinkage of the high-impact polystyrene at room temperature [21]. It may be seen that in the first deformation cycle the material shows the mechanical behaviour the characteristic of the glassy state (initial straight section, yield stress, the plateau region). However, the removal of stress results in large shrinkage of the polymer and subsequent deformation of this material indicates its strain softening, characteristic of the materials subjected to crazing (compare Figs. 5.5 and 5.7). It is important to note that the strain softening of the high-impact polystyrene is not associated with the presence of the rubber phase in this material and it is due to the material of developed crazes, because the continuous phase in the high-impact plastics is the glassy component.

Similarly, the high susceptibility of the material of the crazes to the healing of the interfacial surface in the temperature range below the appropriate glass transition temperature is also not characteristic of the glassy state. In [16] it was noted that if the crazed polymer is stored for a long period of time, the material gradually becomes monolithic and transparent, i.e., there is not only simple closure of the walls of the crazes but also healing of their porous structure. This phenomenon was studied in greater detail on the example of the PS in [24]. In this study, the formation of crazes in the PS was initiated at different temperatures, and the time to the formation of the first craze and its growth rate were measured. This was followed by removing the stress and relaxation of the polymer containing the crazes for different periods of time at a different temperatures. The same stress was then again applied and the duration of initiation and rate of growth of the same crazes were measured. With increasing recovery time at the given temperature the initiation time increased, and the rate of repeated increase decreased to 0 and other crazes could form in the sample. In particular, the healing process was expressed in the fact that the craze, selected for the investigation, disappeared visually during the healing cycle. These results were used to construct the temperature–time diagram indicating (determining) the time required at the given temperature for complete healing. For example, according to these data at a temperature of 70°C (T_g of polystyrene is 100°C), the total closure (healing) time is longer than 10^5 s. These data indicate the high molecular mobility in the fibrils of the crazes in the temperature range not exceeding the glass transition temperature of the polymer.

The possibility of the large-scale molecular motion in the crazes below the glass transition temperature of the bulk polycarbonate (PC) is also confirmed by the data in [25]. In this study, the methods of measuring thermally stimulated polarization currents were used to examine the molecular mobility in the crazed and bulk specimens of the polycarbonate and brominated polycarbonate. The results show that in the highly crazed samples the dielectric signal becomes considerably stronger at temperatures almost 100° below the glass transition temperature. In low-temperature annealing (10-20° below the glass transition temperature) the low-temperature relaxation becomes more difficult indicating the healing of the highly dispersed material of the crazes (coalescence of the interfaces) and the restoration of the bulk structure of the polymer. Evidently, this effect is associated with the higher molecular mobility in the highly developed surface of the crazes in the temperature range of the glassy state of the bulk polycarbonate. Although no experimental data are yet available at the present time relating to the special features of the molecular mobility in the thin surface layers of the polymers, the authors of [25] explain the easy molecular mobility of the polymer molecules in the crazes by their proximity to the free surface.

Finally, we can present the data obtained in [26] where the crazes were initiated in thin polystyrene films, followed by recording of the dimensions of the crazes and measurement of the distance between the fibrils and the diameter of the fibrils by electron diffraction directly in the video camera of the transmission electron microscope. The results show that the distance between the fibrils increases from 25 to 350 nm over a period of 250 h at room temperature. The diameter of the fibrils increases in the same fashion. These changes are accompanied by the coalescence of the fibrils, and after 750 h the fibrils could not be separated visually.

The development of atomic force microscopy has made it possible in recent years to evaluate directly the mechanical characteristics of small (nanometric) zones on the surface of the polymers [27–29]. In [30] these methods were used for direct measurements of the elasticity modulus on the surface of the polymer polystyrene film, including in individual crazes. The modulus of the crazed materials was very low, from 3 to 10% of the modulus of the surrounding block polymer. In other words, the transition of the polymer to the 'matter' of the craze is accompanied by a large change of its physical and mechanical properties. In particular, there is a large reduction of the modulus of the polymer, transferred into the 'matter' of the craze, i.e., there is a distinctive effect of strain softening.

Taking these considerations into account, it may be concluded that the large number of the experimental data, concerned with the examination of the structure and properties of the crazed polymers, indicate the special state of the polymer, dispersed in the structure of the craze. There is no doubt that the process of crazing in the glassy polymer is accompanied by the formation of some 'phase' with the physical and mechanical properties not characteristic at all of the glassy state. This 'phase' has a low modulus and is not capable of high reversible strains. Evidently, this 'phase' is responsible for the strain softening and formation of high reversible strains in the glassy polymers, subjected to crazing.

5.3. Factors causing force softening of glassy polymers during crazing

The above described mechanical behaviour is so very unusual for the oriented glassy polymer that some new approach needs to be taken to explain its mechanism. Several mechanisms have been proposed for explaining the structural–mechanical behaviour of the glassy crazed polymers. Examining the experimental data obtained in the restoration of the dimensions of the crazes, the authors of [9] assume that the driving force, causing shrinkage, is the entropic elasticity and, most importantly, the excess of the free surface energy, determined by the presence of a large specific surface of the crazed material. The entropic force was evaluated on the basis of the kinetic theory of high elasticity and it was assumed that the elasticity modulus of the material of the craze of the polycarbonate is equal to its modulus at 160°C (15° higher than T_g). Calculations show that in this case the stresses, causing shrinkage, equal approximately 3–4 MPa. This value is considerably lower than the yield stress of the polycarbonate at room temperature and, consequently, can not be the reason for the shrinkage of the polymer which takes place. In addition to this, the authors did not explain the mechanism by which the entropic mechanism of reversibility of operation can operate at temperatures lower than the glass transition temperature of the polymer.

Therefore, it was attempted to calculate the compressive stresses, determined by the change of the free surface energy. It was assumed that the polymer in the craze is dispersed in the form of long cylindrical filaments with radius r and length H which connect the opposite walls of the craze. The authors also assumed that shrinkage is accompanied by a reduction of length H and increase of thickness r of the filaments, connecting the walls of the craze. In fact, this means the shortening of the interfacial surface which takes place in a liquid droplet under the effect of surface forces. Evidently, for this type of shortening of the surface of the cylindrical fibrils, the external forces (surface forces in this case) must be capable of deforming the glassy polymer, i.e., overcome its yield stress. The authors, assuming that the volume of the filaments is constant, showed that the stress, acting in the direction of the tensile loading (shrinkage) axis of the polymer, is equal to $\sigma = \gamma/r$ (where γ is the surface tension of the polymer). The value of the restoring force, calculated from this equation, in the case of the polycarbonate was equal to approximately 4–9 MPa.

Thus, the total shortening stress is approximately 7–14 MPa which is also considerably lower than the limit of forced elasticity of even non-oriented polycarbonate at room temperature, equalling approximately 63 MPa. Consequently, the surface forces cannot be the physical reason for the deformation of the polymer fibrils, capable of causing its shrinkage. Nevertheless, the authors assume that the calculated stress may be sufficient to cause 'collapse' of the crazes, if the following circumstance is taken into account. Firstly, the restoration rate in this case is considerably lower than the rate at which the yield stress, which depends on the test rate, is usually measured. Secondly, the test may be accompanied by a reduction of the viscosity of the craze material as a result of its large free specific surface and high dispersion.

Consequently, it may be concluded that the large number of experimental data, concerned with the investigation of the structure and properties of crazed polymers, indicate the special state of the polymer, dispersed in the structure of the craze. In particular, to develop the crazing theory, based on the phenomenon of instability of the meniscus [31], Kramer had to postulate that the wall of the growing craze is coated with a thin layer of a devitrified, rubbery material. It should be remembered that the theory of widening of the craze, developed by Kramer, describes satisfactorily the available experimental data.

Generally speaking, when examining the physical–mechanical behaviour of the crazed glassy polymers, it is important to answer one question: what is the reason for the high reversible strains of the crazed polymers at temperatures lower than the glass transition temperature? The answer to this question is given by the investigations carried out in recent years [32–36]. The authors of the studies detected and confirmed that the glass transition temperature of the thin films and surface layers of the glassy polymers is greatly reduced. Although the physics of this fundamental phenomenon is not the completely understood and is still the subject of extensive studies

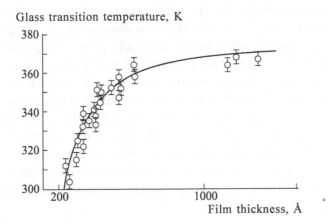

Fig. 5.8. Dependence of the glass transition temperature of polystyrene free-standing films on their thickness [40].

[37–39], the very fact of the large reduction of the glass transition temperature in the thin films and surface layers of the amorphous polymers is not doubted at all.

Figure 5.8 shows the generalised dependence of the glass transition temperature of polystyrene on the thickness of the phase, published in [40]. It is clearly seen that, starting at approximately 80 nm, the glass transition temperature of the polymer starts to decrease with the reduction of the length of the phase of the polymer. Depending on the molecular mass of the polymer, this reduction may reach many tens or even hundreds of degrees.

Since the crazing is in fact the processes of dispersion of the polymer into aggregates with the size of tens or hundreds of angstroms, it is no surprise that the craze material has a greatly reduced glass transition temperature. This fact makes it possible to explain easily and without contradiction the entire set of the examined and anomalous physical–mechanical properties of the craze polymers. The high reversible strains and the low-temperature thermally stimulated shrinkage are associated with the entropic elasticity of the crazed polymer with a reduced glass transition temperature.

In fact, the estimates show that at a diameter of 10 nm the fibril does not contain more than 10 chains. It follows from here that all the macromolecules in the fibrils of the crazes are actually located in the surface layer or, in other words, are characterised by a considerably lower glass transition temperature than the surrounding block polymer. Since the macromolecules inside the fibrils, characteristic for the structure of the crazes, are mutually orientated, the crazing results in a situation in which the glassy block polymer is 'stuffed' with unique mutually penetrating zones (crazes) filled with the orientated amorphous polymer having a reduced (below the experiment temperature) glass transition temperature. Naturally, this rubber-like material should undergo shrinkage (high reversible strain), and this is also detected in the experiments (Fig. 5.7). In other words, the physical reason for the high reversible strains of the craze polymers in the temperature range of the glassy state of the block polymer is the reduction of the glass transition temperature of the amorphous polymers in the thin films and surface, nanometric layers.

5.4. Strain softening of crystalline polymers

Hard elastic crystalline polymers
At the beginning of the 70s of the previous century it was reported that the crystalline polymers, such as polypropylene (PP), high-density polyethylene (HDPE) [42, 43] polyvinylidene fluoride (PVF) [44, 45], polyamide 6,6 [46], poly-4-methylpentane-1 [47], polyamide-11 [48] and others, can show completely unexpected properties.

In [49–54] it was shown that the crystalline polymers oriented in special conditions and having a relatively high elasticity modulus, can be stretched by tens or even hundreds of percent. However, most surprising is the fact which shows that after such tensile loading and removing the stress the tensile loaded crystalline polymers restore almost completely and spontaneously their dimensions at the experiment temperature; in other words, they show high reversible deformation. A typical example of this type is shown in Fig. 5.9 [52], which shows that in the first deformation cycle the polymer has a tensile loading curve characteristic of the crystalline polymers (initial straight section, yield stress, plateau region). However, after removing the stress, the specimens spontaneously shorten to almost the initial dimensions (high reversible strain) and its mechanical behaviour in subsequent tensile loading cycles is characteristic of the rubber-like polymer (low modulus, monotonically increasing stress). In other words, the crystalline polymer shows the distinctive effect of strain softening, completely identical with that discussed for the amorphous polymers. In further stages, the materials, combining the properties of rigid crystalline polymers and the rubber-like materials

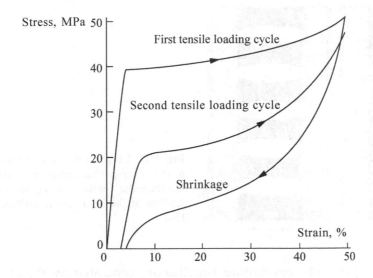

Fig. 5.9. Typical curve of tensile loading and shrinkage of lamellar polypropylene at room temperature [52].

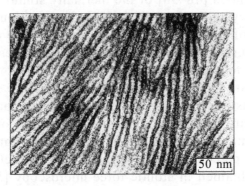

Fig. 5.10. Transmission electron micrograph of a thin polypropylene film with a lamellar structure, contrasted with RuO_4 (dark bands – amorphous interlayers; light bands – crystalline lamellae) [56].

were regarded as hard elastic fibres because their properties are in contrast with the well-known properties of the highly oriented crystalline polymers which have a high modulus, strength and low deformability [55].

The formation of these properties, unusual for the oriented crystalline polymers, is determined by the specific structure of these polymers. The structure can be regarded as a set of regularly distributed (parallel to each other) crystalline lamellae, in which the c axis of the macromolecules is oriented normal to their plane

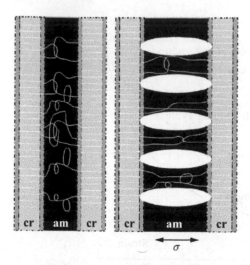

Fig. 5.11. Diagram of the formation of the fibril–porous structure in the interlamellar regions during tensile loading of the hard elastic polymer [50].

(Fig. 5.10) [56]. The crystalline lamellae are separated by the also regularly distributed amorphous interlayers. The structure of this type forms in the conditions of the force effect on the solidifying melts of the polymers [48–50] or the thermally stimulated shrinkage of the oriented crystalline polymers [57]. The previously described effect of strain softening (Fig. 5.9) is observed in the case in which the tensile loading of the rigid elastic material takes place in the direction normal to the plane of the lamellae.

These materials have been subjected to detailed studies and a number of exhaustive reviews of the problem have been published [51–54]. As mentioned previously, the crystalline oriented (hard elastic) polymer may show high reversible strain and can be deformed by many tens of percent (up to 100%). X-ray diffraction data [58] and the results obtained in atomic force microscopy [59] show that the high deformation of these materials is accompanied by the rapid development of porosity and the specific surface. Porosity develops in the first tensile loading cycle in the inter-lamellar regions of the polymer during simultaneous separation (moving away) of the lamellae during the tensile loading of the polymer in the direction normal to the plane of the crystalline lamellae. The structural rearrangement of this type is shown schematically at least in the initial stages of deformation in Fig. 5.11 [50].

Important results have been obtained in the investigation of the effect of the temperature of tensile loading on the deformation response of the hard elastic polymer. Figure 5.12 shows the tensile loading curves of the hard elastic PP, obtained at different temperatures [60].

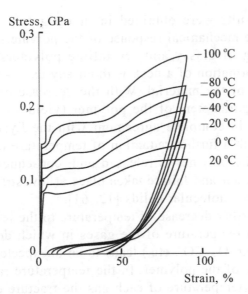

Fig. 5.12. The curves of tensile loading and shrinkage of the hard elastic PP obtained at different temperatures [60].

It may be seen that as the temperature is lowered the tensile loading curve of the initial hard elastic PP is displaced, as expected in the range of higher stresses. However, the temperature has almost no influence on the effect of strain softening. In all cases, the deformed hard elastic PP shows high reversible strains and rubber-like behaviour under repeated tensile loading. In addition, as the deformation temperature of the hard elastic polymer decreases, its reversible strain (shrinkage) becomes higher. This fact indicates that the driving force of the reversibility is the entropy factor, and the reversible deformation itself is associated in some way with the conformation rearrangement in the chains of the polymer.

Crazing of crystalline polymers
The data, presented in the previous section, indicate that the unusual properties of the hard elastic polymers, including strain softening, are associated in particular with the development in them of nanosized porosity and highly developed surface during deformation. At the same time, the crystalline polymers, like amorphous polymers, are capable of deformation with the development of porosity by the crazing mechanism. In this case, the nanosized porosity develops in the polymer even if it does not contain anisotropic lamellar structures, characteristic of the hard elastic materials [11].

Important results were obtained in investigating the effect of gas media on the mechanical response of the polymers. It is known that the drawing of glassy and crystalline polymers in air takes place with the formation of a neck without any extensive disruption of the solidness of the material. With the decrease of temperature the mechanical parameters of the polymer (yield stress, strength) usually increase in complete agreement with the Eyring–Lazurkin concept [12]. With a further reduction of temperature the brittleness temperature of the polymer is released and, consequently, no neck forms in the polymer and failure takes place at low strains, as in the case of brittle low-molecular solids [12, 61].

However, a further decrease of temperature to the level similar to the condensation temperature of the gases in which deformation is carried out (N_2, Ar, O_2, CO_2, etc.) leads to an unexpected appearance of high plasticity of the polymer. In the temperature range close to the condensation temperature of each gas the fracture elongation of the polymer greatly increases and the yield stress of the polymer decreases. Consequently, a maximum appears in the region of very low temperatures on the temperature dependence of the fracture elongation and yield stress. At the same time, deformation of the same polymers at such low temperatures in vacuum or helium does not result in any significant plasticity.

The low-temperature plasticity of the crystalline polymer (polypropylene) was described for the first time by V.A. Kargin et al [62–65] and it was shown that the polypropylene may show extra high strains (hundreds of percent) in the low-temperature range, down to the liquid nitrogen temperature. In addition to this, such strains are reversible when the tensile loaded polymer is heated to room temperature.

The polymer, deformed at a low temperature, is characterised by the restoration of the initial mechanical properties and structure in annealing. Figure 5.13 shows the tensile loading curves of polypropylene at a low temperature (−50°C) which is considerably lower than the glass transition temperature. It may be seen that the tensile loading of the polymer immediately after the first tensile loading cycle is accompanied by the strong effect of strain softening. The entire tensile loading curve is displaced into the range of low stresses and the yield stress disappears. Such a tensile loading curve resembles the tensile loading curve of the rubber-like polymer and fully demonstrates the effect of force softening, characteristic of the hard elastic materials (compare Figs. 5.9 and 5.13). Annealing of

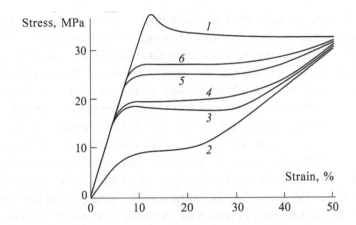

Fig. 5.13. Stress–strain curves of polypropylene at – 50°C: 1) first tensile loading; 2) repeated tensile loading immediately after the first tensile loading cycle; the tensile loading curves after annealing of the tensile loaded specimens at 25°C for 15 min (3), 30 min (4), 4h (5) and 18 h (6) respectively [64].

the deformed polymer even at a relatively low temperature (25°C) leads in time to a gradual restoration of the mechanical properties (and, consequently, of the structure) of the initial polymer. The latter results indicates the high lability of the structure of the deformed polymer and its susceptibility to healing of the structure formed in deformation.

Later, low-temperature deformation of this type was studied in detail by Peterlin and Olf [66]. Shortly after publication of [64] they concluded that the deformation of the polymer leads in the first instance to the increase of the specific volume of the polymer and of the fraction of the free volume. This effect is equivalent to a large extent to the reduction of the glass transition temperature to the temperature lower than the experiment temperature. An especially large increase of the specific volume and the fraction of the free volume takes place in the areas of stress concentration in the material. As a result of the high local concentration of the stress, the polymer in these ranges transformed to the rubber-like state and, consequently, becomes capable of high deformation at stresses considerably lower than the yield stress of the material, surrounding the stress raiser. Therefore, these polymers fracture in a non-brittle manner, even at very low temperatures, down to 4 K [67]. This is indicated by the analysis of the fracture surfaces of the polymers which shows that the true fracture crack is always preceded by the zone of the plastically deformed material.

It is important to note that the development of high cryogenic strains takes place non-uniformly throughout the volume of the polymer, primarily in the stress concentration areas. These areas are characterised by the formation of the zones of the plastically deformed polymer, with a fibril–porous structure (crazes). In [68] the authors describe the mechanical and thermomechanical properties of the polypropylene, deformed by the crazing mechanism in liquid nitrogen at –196°C, i.e., at temperatures considerably lower than the glass transition temperature of the polymer. The results show that the crazes retain their size in liquid nitrogen at –196°C after removing the load. At the same time, they instantaneously 'collapse' at the same temperature, if the nitrogen is removed from the volume of the crazes leading to the almost complete shrinkage of polypropylene. The curve of repeated tensile loading of the specimens of polypropylene in liquid nitrogen at –196°C has the form identical with the curve of 'dry' crazing of the amorphous polymer (polycarbonate), obtained in the previously mentioned study by Kambour and Kopp [16].

As in the study by Kambour and Kopp, the initial modulus and the yield stress in the repeated cycles of the polypropylene after shrinkage are considerably lower than the appropriate characteristics of the initial polypropylene during its deformation in liquid nitrogen. It was assumed that the stresses, causing 'collapse' of the crazes in the removal of hydrogen from them, can be calculated from the equation $\sigma = 2\gamma/r$, where γ is the surface tension of the polymer, r is the radius of the cavities in the craze. The calculations give the value of the order of 6 MPa which again is considerably lower than the yield stress of the polypropylene at this temperature. As indicated by the experimental results, it is not possible to explain satisfactorily the reversibility of the high deformation of the crazed polymer only by the effect of surface forces, assuming that the mechanical properties of the craze material are equal to those of the initial glass polymer. The crazing of the crystalline polymers can be carried out not only at cryogenic temperatures. In the following chapter, attention will be given to the crazing of crystalline polymers in liquid media at normal (including the room) temperatures. This type of crazing has a number of special features, as a result of which the structure becomes similar the structure of the hard elastic materials, so that in [11] it was referred to as delocalized crazing. Therefore, it is not unexpected that the mechanical behaviour of the crystalline polymers, deformed in the AAM, is in excellent agreement with the mechanical behaviour of the hard elastic polymers, discussed previously. Figure 5.14 shows

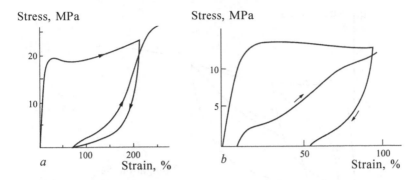

Fig. 5.14. Curves of tensile loading, shrinkage and repeated tensile loading of polypropylene in AAM (heptane) [69] (*a*) and high-density polyethylene in AAM (heptane) [70] (*b*).

the curves of tensile loading and shrinkage of isotactic polypropylene (*a*) and high-density polyethylene (HDPE) (*b*), deformed by the crazing mechanism in active liquid media. It may easily be seen that in both cases there are also features characteristic of strain softening in the previously examined cases (the initial curve has a straight initial section, the yield stress and the plateau region); the removal of stress results in high reversible strain, and the repeated loading demonstrates the mechanical behaviour, characteristic of the rubber-like polymer.

In addition, if the polymer, subjected to delocalized crazing, is released from the clamps of the tensile loading device and extracted from the liquid so that it can completely evaporate from the porous structure of the polymer, subsequent deformation of the specimen in air also demonstrates the distinctive effect of strain softening. Figure 5.15 shows the tensile loading curves and the curves of repeated tensile loading of polypropylene in air, after preliminary tensile loading in the AAM (supercritical CO_2) by the mechanism of delocalized crazing to 100% and subjected to complete shrinkage in the removal of the AAM from its porous structure [71]. In this study, the AAM was represented by supercritical CO_2, which is an efficient crazing agent and causes the same structural rearrangement in the deformed crystalline polymers (PP, HDPE) as the liquid media (hydrocarbons), used usually for the realisation of crazing in the given polymers [72]. Figure 5.15 shows that the removal of the liquid from the porous structure of the polymer, subjected to delocalized crazing, does not have any significant effect on its

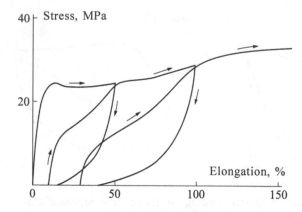

Fig. 5.15. Curves of tensile loading, shrinkage and repeated tensile loading of polypropylene in air, after preliminary tensile loading in the AAM (with CO_2) by the mechanism of delocalized crazing to 100% and undergoing complete shrinkage in the removal of the AAM from its porous structure [71].

repeated mechanical behaviour, whilst the effect of strain softening is quite strong (compare Figs. 5.14 and 5.15).

Another special feature of the hard elastic and crazed polymers is the spontaneous healing of their nanoporous structure, formed in the first tensile loading cycle. This phenomenon will be discussed in greater detail at a later stage. Here, we only mention that the hard elastic and crazed polycrystalline polymers gradually restore their initial structure with time. It is important to mention that this process takes place spontaneously at temperatures considerably lower than the appropriate melting point (solidification temperature) [73]. The restoration of this type is rapidly accelerated by increasing temperature [74]. As shown by direct microscopic studies [53], in shrinkage and restoration of the initial structure of the hard elastic polymer the 'edges' of the crazes come closer together or, in other words, the shrinkage mechanism consists of the contraction of the fibril–porous material, filling the interlamellar space. It should be mentioned that the process of healing the highly dispersed structure of the hard elastic and crazed crystalline polymers is fully identical with the appropriate process, examined previously for the crazed glassy polymers [16, 22].

It should be mentioned that the high-dispersion fibrillised material, filling the volume of the crazes, has a highly developed surface so that it is essential to take into account the contribution of the surface forces to the unusual structural–mechanical behaviour of the hard

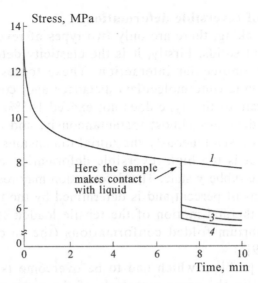

Fig. 5.16. Dependence of the stress on time in the hard elastic polypropylene, tensile loaded to 50%. At the points, indicated by the arrow, the sample was brought into contact with ethanol (2), butanol (3) and hexane (4); (1) the curve of relaxation of the polymer in air [76].

elastic and crazed polymers [75]. In experiments, the effect of the surface forces on the mechanical behaviour of these materials is reflected, for example, in the large change of the stress in the loaded specimens, if they are brought into contact with the active wetting liquid. As indicated by Fig. 5.16, this effect is manifested in the large decrease of the stress which depends on the nature of the liquid. It should be mentioned that this change of the stress is reversible and the polymer restores the initial stress during its drying or when one medium is replaced by another [77].

5.5. Mechanism of strain softening of crystalline polymers

The above-discussed experimental data indicate a large number of special features in the structural–mechanical behaviour of the rigid elastic and crazed crystalline polymers, which do not fit the generally accepted considerations of the structure and properties. One of the main problems is the need to explain the nature of the reversibility of high strains of the crystalline polymers in the temperature range below their melting point.

The nature of reversible deformation

Generally speaking, there are only two types of reversible (elastic) deformation of solids. Firstly, it is the elasticity determined by the forces of intermolecular interaction. These forces act over very small interatomic (intermolecular) distances and, consequently, the reversible strain of this type does not exceed 1–3%. Such energetic deformation develops almost instantaneously, and also the elastic solid acquires instantaneously the initial dimensions [78].

Secondly, it is the high reversible deformation of the polymer, which is in the rubbery state. The deformation may reach hundreds or even thousands of percent and is determined by the transition under the effect of thermal motion of the tensile loaded chain molecules to the equilibrium folded conformations (the so-called entropic elasticity) [79].

The main problem which had to be overcome is to explain the combination, in the same material, of the properties typical of polycrystalline materials (high modulus, presence of the yield stress and the plateau region) with the properties of the rubbery polymers (high reversible strain, low modulus, no yield stress) under repeated tensile loading. In other words, it was necessary to develop an adequate mechanism of the strain softening of crystalline polymers of this type.

In the initial studies, concerned with the determination of the nature of reversibility of high strains, the authors attempted to explain this phenomenon by means of the structures of the spring type, because it was assumed that the entropic elasticity cannot be realised in crystalline polymers. Several models were proposed for explaining the hard elastic behaviour of the laminar crystalline polymers. These models were based on the assumption of the energy nature of reversibility of their deformation [54]. Typical examples of these models are shown in Fig. 5.17.

All these models, examining the energy reversibility of deformation of the hard elastic polymers, use as the main element of the structure the hard crystalline frame with a continuous volume. This is understandable because in this case deformation can be reversible. This frame has points at which the crystalline lamellae are connected because in tensile loading of such a model porosity can form as a result of their bending and shear (Fig. 5.17 c).

However, these energetical models of the reversibility of deformation can not explain, firstly, the effect of force softening of the material and, secondly, the previously mentioned mechanism of

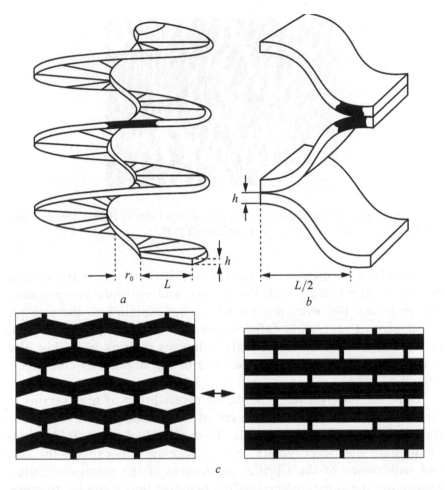

Fig. 5.17. Schematic representation of the structural models of the hard elastic material: *a*) the coil spring model; *b*) the leaf spring model [52]; *c*) the general scheme of the reversibility of the energy spring [54]

healing of the deformed hard elastic materials (Fig. 5.13), and also the strong effect of temperature on this processes (Fig. 5.12).

In addition to this, it was later shown that the deformed hard elastic polymer does not contain an integrated crystalline frame (Fig. 5.18) [53]. Figure 5.18 shows that the tensile loaded rigid elastic polymer contains crystals separated in space and connected by fibrillar aggregates of the macromolecules. In the same study it was shown that the shrinkage of the tensile drawn rigid elastic polumer is accompanied by the convergence of the crystals and the fibrilar aggregates of the macromolecules become indistinguishable.

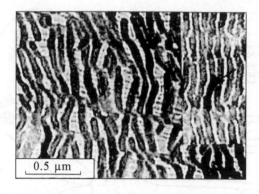

Fig. 5.18. Electron micrograph of a thin film of rigid elastic HDPE, tensile drawn to 100% (the tensile drawing axis is horizontal) [53].

Therefore, a model of the reversibility of the hard elastic materials, combining both the energy and entropic components, was proposed. However, this model can not explain how the entropic component of reversible deformation can operate in the conditions in which the material is not only below its melting point but also below the glass transition temperature of the amorphous component (Fig. 5.12) [59].

Thus, the purely energy model of reversibility of high strains in the hard elastic polymers as a result of deformation of the crystalline frame proved to be unacceptable. To explain high reversible strains of the hard elastic polymers, it was necessary to explain the role and importance of the fibrillar aggregates of the macromolecules, connecting the separated crystalline lamellae into a single structure. To achieve this, it was necessary to discuss and describe the nature of the fibril of this type. In particular, it is important to understand the phase state of such aggregates of the oriented macromolecules.

The general pattern of deformation of the hard elastic polymer and also obviously of the polymer deformed by the delocalized grazing is described in [80] as follows (Fig. 5.19). In tensile loading of the polymer with a lamellar structure the initial stages of deformation are characterised by tensile loading of mainly amorphous interlayers (Fig. 5.19 *a*, *b*). In particular, this is indicated by the direct proportionality of the large period in relation to the degree of tensile loading of the polymer [57, 81]. At a relatively high degree of tensile loading (Fig. 5.19 *c*) the amorphous regions are characterised by the start of cavitation as a result of a shortage of the material capable of compensating the Poisson compression of the deformed amorphous

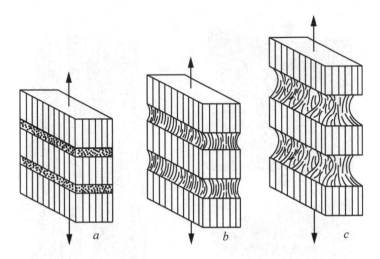

Fig. 5.19. Schematic representation of structural rearrangement taking place in the initial stages of tensile loading of the hard elastic polymer [80]. Explanation is given in the text.

regions, and this is accompanied by the nucleation of fibrillar aggregates of the oriented macromolecules. This stage of deformation of the hard elastic polymer was described earlier in Fig. 5.11.

It was important to describe the fibrils formed in the inter-crystalline space and their role in the detected high reversible deformation (shrinkage) of the tensile loaded hard elastic and crazed crystalline polymers.

In [80] the authors propose a possible fibrillization mechanism (Fig. 5.20). According to this model, the material of the fibrils is 'pulled out from the crystals'. The pulling-out of this type is evident because deformation to high degrees of elongation (hundreds of percent) is unlikely to take place only by the consumption of amorphous intercrystalline interlayers. However, Fig. 5.20 and the study [80] propose the crystallisation of the oriented macromolecules in the fibrils already in the early stages of deformation (Fig. 5.20 a). In subsequent stages, this orientation crystallisation, which is accompanied by the skewness and shift of the initial crystalline lamellae, leads to the formation of a structure characteristic of the deformed conventional crystalline polymer with the formation of a neck (Fig. 5.20 b, c). In other words, the deformed hard elastic polymer is identical to the conventional crystalline cold-drawn polymer, i.e., its structure is a conglomerate of the parallel packed crystalline fibrils with a distinctive large period.

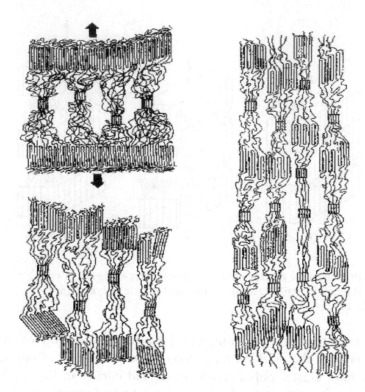

Fig. 5.20. Structural rearrangement of the rigid elastic polymer according to [80].

If we accept this model, it can not be used to explain the high reversible deformation of the tensile loaded rigid elastic and crazed crystalline polymers nor the phenomenon of spontaneous healing of their porous structure which is again accompanied by the restoration of their geometrical dimensions. The essential conditions for both mentioned processes is the high molecular mobility of the polymer molecules which is greatly restricted when the polymer is placed in the crystal lattice.

At the present time, it is not doubted that the macromolecules, restricted in the ultrathin fields with the thickness comparable with the radius of rotation of the macromolecules, can change their physical properties in comparison with the bulk because the intermediate phase plays a continuously increasing role in the general free energy of the system [81]. In the conditions of the restricted volume both the crystallisation rate and, most importantly, the degree of crystallinity of the polymer decrease with the reduction of the thickness of its layer [82–89]. For example, in a recent study [90] it was shown that the rate of crystallisation of polyethylene oxide

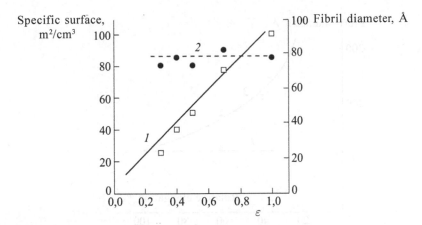

Fig. 5.21. Dependence of the specific surface of the rigid elastic polypropylene (1) and the diameter of the fibrils (2) on the degree of tensile loading of the polymer (ε) [58].

in the thin layers (13 nm–2 µm) may decrease by up to 40 times in comparison with the bulk, starting at a thickness of 150 nm. The morphology of the crystals changes at the same time.

The previously mentioned studies dealt with the crystallisation of the polymer in the thin layers. It is also assumed that the crystallisation of the polymer takes place in the fibrils which in fact, are one-dimensional formations. As mentioned previously, these formations contain in the cross-section several tens of oriented molecules. At the same time, the crystalline phase is the phase characterised by the long-range order in the distribution of the molecules which can not be organised in such a limited volume. Naturally, the crystallisation process is associated with considerable difficulties (and may not take place at all) in the volume of the fibrils formed in the first cycle of the formation of the hard elastic and crazed polymers. Analysis of the presented data indicates that the fibrillar aggregates of the macromolecules, formed during tensile loading of the hard elastic materials and crazing of glassy and crystalline polymers, are in the amorphous phase state.

Figure 5.21 shows the dependence of the specific surface of the hard elastic polypropylene fibre and the diameter of the growing fibrillar aggregates of the macromolecules on the degree of tensile loading of the polymer [58]. It may be seen clearly that the specific surface of such a material increases and reaches 100 m²/g, which indicates the rapid development of the internal interfacial surfaces. The porosity of the deformed polymer increases in the same manner, and the size of the pores, determined by the methods of mercury

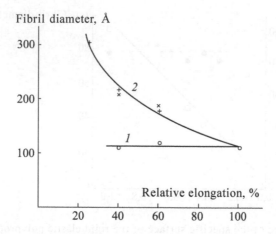

Fig. 5.22. Dependence of the diameter of the fibrils formed in tensile loading of the hard elastic polypropylene (1) and during its shrinkage after tensile loading (2) on the strain [93].

porosimetry reaches 0.2–0.015 μm [91], in complete agreement with the previously mentioned data. However, the diameter of the fibrils does not change in this case. This indicates that, as in many other known cases of crazing, the so-called surface drawing of the fibrils takes place [29]. In other words, the elongation of the fibrillar aggregates, connecting the adjacent crystals, increases as a result of the consumption of the material of these crystals [53].

In this section we will not analyse the processes of nucleation and growth of the fibrils in the intercrystalline areas of the lamellar crystalline polymers, because there are a number of detailed publications dealing with this question [53, 92]. However, it is important to investigate the 'fate' of these fibrills in the reverse process – shrinkage, which takes place in the hard elastic materials and is the primary reason for their force softening.

Thus, the fibrillar–porous structure in the hard elastic polymers forms by surface drawing of the fibrils with the conserving of their diameter. Another important question is what takes place with these fibrils in the reverse process – shrinkage. Figure 5.22 shows the dependence of the diameter of the fibrils of the hard elastic polypropylene on their diameter during shrinkage. Figure 5.22 shows that in shrinkage the diameter of the fibrils rapidly increases in complete agreement with the behaviour of the rubber-like polymer during its reversible deformation. The increase of the diameter of the fibrils is accompanied by a reduction of the specific surface of the hard elastic polymer during its shrinkage [94]. Since the volume

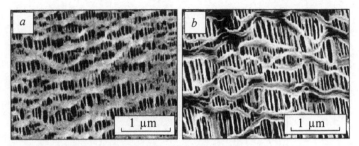

Fig. 5.23. Scanning electron micrographs of commercial membranes, produce from hard elastic polypropylene and (*a*) and high-density polyethylene (*b*).

of the rubber does not change during deformation, the shrinkage of the oriented rubber (decrease of the linear dimensions) should be accompanied by the equivalent increase of its cross section. The result indicates unambiguously the amorphous and devitrified state of the fibrillized high dispersion material, typical of the hard elastic polymer.

It would appear that this result contradicts the data published in [52] where the results indicate the negative coefficient of the elastic force in heating the hard elastic copolymer of polyoxymethylene (Celcon). However, this polymer was produced by annealing of the deformed hard elastic products in special conditions. Annealing of this type is accompanied by rapid growth of the fibrillar aggregates, formed during deformation of the lamellar polymer, growth of inter-fibrillar distances and, naturally, crystallisation of the polymer in these large aggregates of macromolecules [95, 96]. This material is fully stable in time and is a commercial product suitable for application as membranes. Examples of commercial membranes are shown in Fig. 5.23.

Naturally, such a completely crystalline product will have a negative coefficient of equilibrium force. In [97] investigations were carried out on polypropylene samples subjected to crazing in the AAM. The specimens were tensile loaded, the dimensions were fixed and the solvent was removed from their volume. Subsequently, they were annealed. If annealing was carried out in the free states, the specimens showed extensive shrinkage which depended on the annealing temperature. The stresses formed during isometric heating were recorded. The typical experimental results are presented in Fig. 5.24.

The increase of the stresses in an annealing shows unambiguously that the crazes in the polypropylene are filled with an amorphous

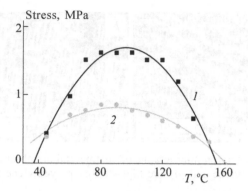

Fig. 5.24. Curves of isometric heating of polypropylene specimens, deformed in the AAM by delocalized (1) and classic (2) crazing. Preliminary deformation in AAM 100% [97].

rubber-like material. It should be mentioned that the same dependence of the force on temperature in the conditions of isometric annealing was also detected previously for low-density polyethylene, stretched in air with the formation of a neck [98].

The phenomenon of healing of crazes during ageing at room temperature (Fig. 5.13 [63]) and, even more so, in annealing also indicates the amorphous rubbery state of the material, filling the crazes. In fact, (chapter 2), the mechanism of healing interfacial surfaces in the polymers consists of mutual diffusion of the chains and (or) their fragments through the interface. This process is possible only if the macromolecules are characterised by the high large-scale mobility, i.e., they are in the rubbery state. It is important to note that the polymers show at least two types of healing of the interfacial surface. When the two identical polymers are brought into contact above their T_g temperature, the interface between them usually gradually disappears and the material becomes monolithic. This phenomenon was described for the first time in the fundamental monograph of S.S. Voyutskii which described it as follows: 'autohesion or self-adhesion is the capacity of two surfaces of the same material brought into contact to provide a strong bond preventing their separation in the contact area' [99]. This type of healing of the surfaces is associated with the mutual diffusion of the macromolecules and/or their fragments through the interface. In the general form, the variation of the area of the interfacial surface of the polymer is associated with the diffusion of macromolecules through the contact of the identical polymer surfaces. To support this assumption, S.S. Voyutskii stated that '... All factors, increasing the rate of thermal motion, support of adhesion and on the other hand, the factors reducing the diffusion rate, reduce the autohesion' [99]. In fact, this process does not differ at all from the mixing of two volumes of the same liquid. The unique feature of the polymer systems in this case is the special feature of the molecular movement of the polymer chains.

As already mentioned, the crazed and rigid elastic materials are susceptible to healing of their highly dispersed nanoporous structure. This process is spontaneous and, in the end, results in the complete the restoration of the initial structure of the polymer. In particular, this case of healing of the nanoporous structure of the rigid elastic polypropylene is shown in Fig. 5.25. The figure shows that the process of healing and restoration of the properties of the initial polymer is very long. Even after 240 days there is no complete restoration, although the mechanical properties become similar to the properties of the initial polymer. This is not surprising because diffusion is a process that lasts a very long time, especially in highly viscous media, such as polymers.

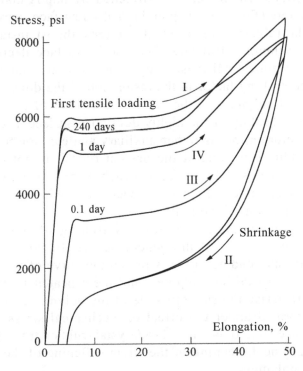

Fig. 5.25. Stress–strain behaviour of the rigid elastic polypropylene during its ageing at room temperature [52].

However, the processes of development and healing of the interfacial surface in the polymers are not exhausted by the above case. There is another aspect of this problem which so far has not been investigated or taken into account. Any deformation of the solid and of the volume, in particular, is accompanied by the change of its surface area in addition to the change of the geometrical dimensions. At the same time, whilst the volume of the deformed polymer may remain constant [79], its surface area always changes. This means that under any effect on the polymer, and especially at high mechanical strains, the material is transported from the bulk to the surface when the surface increases and, vice versa, from the surface into the bulk in cases in which the surface becomes smaller (is healed) [100].

The principal factor in this case is that the development of the interfacial surface during tensile loading of the polymer and its healing during shrinkage takes place at rates incomparable with those observed in conventional spontaneous diffusion of the chains or their fragments examined previously. Even everyday experience shows that the conventional rubber can be stretched by hands many times at almost any rate (Fig. 1.1, chapter 1), and its surface increases many times. If the rubber is released, it shortens almost instantaneously and hits our thumbs. But how, because its surface decreased (was healed) many times. All these simple observations indicate that the main and controlling factor in the restoration of the dimensions of the tensile loaded polymers is the entropic elasticity of the chains, and the process of healing of the highly developed surface, accompanying reversible deformation, is rather auxiliary and can not be its driving force. In addition to this, at the present time it is not known how the macromolecules in the deformed rubber are capable of diffusing from the surface into the bulk and vice versa with such high rates during deformation of the polymer.

Thus, the phenomenon of force softening of the glassy and crystalline polymers is of the general nature. In this phenomenon, the initial tensile loading curve is in fact the curve of tensile loading of the glassy crystalline polymer, and the repeated tensile loading curves characterise the rubbery state of the polymer to which it has been transformed under the effect of mechanical stress in the first deformation cycle. It is this glass/crystal–rubber transition which forms the physical meaning of the strain softening of the glassy and crystalline polymers.

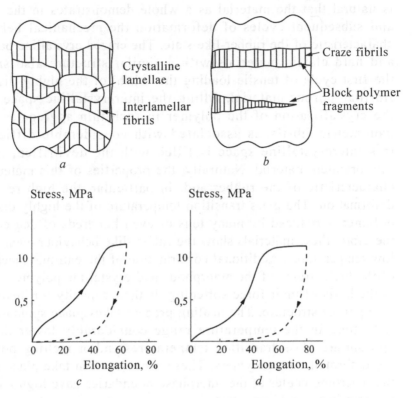

Fig. 5.26. Schematic representation of the deformed hard elastic polymer (*a*) and crazed polymer (*b*) [92] and the curves of tensile loading and shrinkage of the predeformed high-impact polystyrene (e) and the hard elastic fibres of high-density polyethylene (*d*) [53].

Conclusions

Analysis of the experimental results shows that the mechanism of strain softening of the glassy and crystalline polymers, accompanied by the development of the highly developed nanoporous structure, is of the general nature. These systems include hard elastic materials and also crazed glassy and crystalline polymers. The generality of this type was detected a relatively long time ago (Fig. 5.26). The figure shows schematically the structure of the deformed hard elastic polymer and of the crazed polymer (Fig. 5.26 *a, b*). The evident morphological agreement of both materials stresses their almost identical mechanical behaviour (Fig. 5.26 *c, d*) and is not accidental. The crazing of the glassy polymer results in the formation of the highly dispersed oriented material, filling the volume of the crazes. Since the glass transition temperature of the polymer is many tens of degrees lower than that of the bulk polymer surrounding it, it is natural that the material as a whole demonstrates in the second and subsequent cycles of deformation the mechanical behaviour characteristic of the rubber-like state. The crazed crystalline polymers and hard elastic materials with a similar structure also show in the first cycle of tensile loading the 'birth' of the highly dispersed fibrillar–porous material, filling the intercrystalline space. Since the crystallisation of the polymer in such thin aggregates as the nanometric fibrils, is associated with considerable difficulties, this intercrystalline space is filled with the devitrified, rubber-like oriented material. Naturally, the properties of this material are characteristic of the rubber and, in particular, the high reversible deformation. The glass transition temperature of the highly dispersed polymer is reduced by many tens or even hundreds of degrees and, therefore, these materials show the rubber-like behaviour even at very low temperatures. Additional confirmation of the entropic mechanism of the high strains of the amorphous and crystalline polymers, which is the basis of their force softening, is their capacity for healing the nanoporous structure. The healing process takes place spontaneously with time in the temperature range considerably lower than the appropriate glass transition temperatures and/or melting points of the deformed bulk polymers. These processes can take place only if the macromolecules at the interphase boundaries have high mobility, i.e., are in the rubbery state.

References

1.   Mullins L., Tobin N.R., Rubber Chem. Technol., 1957, V. 30, P. 555.
2.   Mullins L., Rubber Chem. Techol., 1969, V. 42, P. 339.
3.   Alfrey T., The mechanical properties of high polymers. Moscow, Publishing House of Foreign Literature, 1952.
4.   Betz G.E., Encyclopedia of polymers, Moscow, Sov. Entsiklopediya, 1972, V. 1, P. 665.
5.   Kargin V.A., Slonimski G.L., Short essay on the physical chemistry of polymers, Moscow, Khimiya, 1967.
6.   Kotaka T., Miki T., Arai K., J. Macromol. Sci. Phys., 1980, V. B17, No. 2, P. 303.
7.   Michler G.H., Adhikari R., Henning S., Jour. Mater. Sci., 2004, V. 39, P. 3281.
8.   Shibaev V.P., Encyclopedia of polymers. Moscow, Sov. entsiklopediya, 1972, V. 1. P. 270.
9.   Kambour R.P., J. Polym. Sci., Macromol. Rev., 1973, V. 7, P. 1.
10.   Passaglia E., J. Phys. Chem. Solids, 1987, V. 48, No. 11, P. 1075.
11.   Volynskii A.L., Bakeev N.F., Solvent crazing of polymers. Amsterdam, New York, Elsevier, 1996. P. 410.
12.   Lazurkin Yu.S., Dissertation, Dr. Sci. Sciences. Moscow, S.I. Vavilov Institute of Physical Problems, USSR Academy of Sciences, 1954.

13.   Narisawa I., Strength of polymeric materials, Moscow, Khimiya, 1987.
14.   Volynskii A.L., Khetsuriani T.I., Bakeev N.F., Vysokomolek. Soed. B, 1974, V. 16. P. 564.
15.   Volynskii A.L., Bakeev N.F., Vysokomolek. Soed. A, 1975, V. 17, No. 7, P. 1610.
16.   Kambour R.P., Kopp R.W., J. Polymer Sci. A-2, 1969, V. 7, P. 183.
17.   Volynskii A.L., Gerasimov V.I., Bakeev N.F., Vysokomolek. Soed. A, 1975, V. 17, P. 2461.
18.   Sinevich E.A., Prazdnichnyi A.M., Bakeev N.F., Vysokomolek. Soed. B, 1988, V. 30. P. 536.
19.   Wolfson S.A., Encyclopedia of polymers. Moscow, Sov. Entsiklopediya, 1977, V. 3, P. 541.
20.   Bucknall C.B. Toughened plastics, London, Applied Science Publishers, 1977.
21.   Polymer blends, Edited by D.R. Paul and S. Newman, Academic Press, New York.
22.   Ramsteiner F., Heckmann W., McKee G.E., Breulmann M., Polymer, 2002, V. 43, P. 5995,
23.   Michler G.H., Kunststoff-Mikromechanik: Morphologie, Deformations- und Bruch-mechanismen, Hanser-Verlag, Munchen. 1997.
24.   Wool R.P., O'Connor K.M., Polym. Engng. Sci., 1981, V. 21, P. 970.
25.   Berger L.L., Sauer B. B., Macromolecules, 1991, V. 24, P. 2096.
26.   Yang A.S.M., Kramer E.J., J. Polym. Sci. A, 1985, V. 23, P. 1353.
27.   Tan E.P.S., Lim C.T., Composites Science and Technology, 2006, V. 66, P. 1102.
28.   Hodges C.S., Advances in Colloid and Interface Science, 2002, V. 99, P. 13.
29.   Qingzheng Cheng I., Siqun Wang, Composites: Part A, 2008, V. 39, P. 1838-1843.
30.   Arnold C.. Yang M., Materials Chemistry and Physics, 1995, V. 41, P. 295-298.
31.   Kramer E.J., Adv. Polym. Sci., 1983, V. 52/53, P. 116.
32.   Forrest J.A., Dalnoki-Veress K., Adv. Colloid and Interface Sci., 2001, V. 94, P. 167.
33.   Ellison C.J., Kim S.D., Hall D.B., Torkelson J.M., Eur. Phys. J. E, 2002, V. 8, P. 155.
34.   Forrest J.A., Eur. Phys. J. E, 2002, V. 8, P. 261.
35.   Volynskii A.L., Bakeev N.F., Vysokomolek. Soed. B, 2003, V. 45, No. 7, S. 1209-1231
36.   Sharp J.S., Teichroeb J.H., Forrest J.A., Eur. Phys. J. E, 2004, V. 15, P. 473-87.
37.   Erber M., Georgi U., Müller J., Eichhorn K.-J., Voit B., European Polymer Journal, 2010, V. 46, No. 12, P. 2240.
38.   Rathfon J.M., Cohn R.W., Crosby A.J., Tew G.N., Macromolecules, 2011, V. 44, P. 134.
39.   Dinelli F., Ricci A., Sgrilli T., Baschieri P., Pingue P., Puttaswamy M., Kingshott P., Macromolecules, 2011, V. 44, P. 987.
40.   Forrest J.A., Mattsson J., Phys. Rev. E, 2000, V. 61, P. R53.
41.   Loboda-Iaikovizh J., Iaikovizh H., Hosemann R., Journal of Macromolecular Science, Part B: Physics, 1979, V. 16, No, 1. P. 578.
42.   Karpov E.A., et al., Vysokomolek. Soed. A, 1995, V. 37, P. 2035.
43.   Moonseog Jang, Sung Soo Kim, Jinho Kim, Journ. Membr. Sci., 2008, V. 318, P. 201-209.
44.   Chun-Hui Du, Bao-Ku Zhu, You-Yi Xu, Macromolecular Materials and Engineering, 2005, V. 290, No. 8, P. 786-791.
45.   Dmitriev I., Bukoek V., Lavrentyev V., Elyashevich G., Acta Chim. Slov., 2007, V. 54, P. 784-791.
46.   Garber C.A., Clark E.S., Journal of Macromolecular Science,. Part B: Physics, 1970, V. B43, P. 499.
47.   Tetsuya Tanigami, Kazuo Yamaura, Shuji Matsuzawa, Kenji Ohsawa, Keizo Miya-

saka, Journ. Appl. Polymer Sci, 1986, V. 32, No. 4, P. 4491-4502.

48.   Dosiare M., Makromolekulare Chemie, Macromolecular Symposia, 1989, V. 23, No. 1, P. 205-211.

49.   Quynn R.G., et al., Journal of Macromolecular Science, Part B: Physics, 1970, V. B44, P. 953.

50.   Noether H.D., Whitney W., Kolloid-Z. Z. Polymere, 1973, V. 251, P. 991-1005.

51.   Sprague S., Journal of Macromolecular Science, Part B: Physics, 1973, V. B8, No. 1-2, P. 157-187.

52.   Miles M., Peterman J., Gleiter H., Journal of Macromolecular Science. Part B: Physics, 1976, V. B124, P. 523-534.

53.   Cannon S.L., Mckenna G.B., Statton W.O., J. Polymer Sci.: Macromolecular Reviews, 1976, V. 11, P. 209-275.

54.   Slutsker A.I., Encyclopedia of polymers, Moscow, Sov. entsiklopediya, 1974, V. 2, P. 515.

55.   Noether H.D., Polymer Engineering Science, 1979, V. 19, No. 6, P. 427-432.

56.   Michler G.H., Adhikari R., Henning S., Journal of Material Science, 2004, V. 9, P. 3281.

57.   Efimov A.V., Lapshin V.P., Fartunin V.I., Kozlov P.V., Bakeev N.F., Vysokomolek Soed., A, 1983, V. 25, No. 3, P. 588.

58.   Efimov A.V., Bakeev N.F., Proceedings of International Conference Khimvolokna-2000, Tver', 2000, P. 101-107.

59.   Hild S., Gutmannsbauer W., Luthl R., Fuhrmann J., Gunterodt H.J., J. Polymer Sci. Part B, Polymer Physics, 1996, V. 34, P. 1953-1959.

60.   Yizhan Zhu, Norimasa Okui, Toshiaki Tanaka, Susumu Umemotoand Tetuya Sakait, Polymer, 1991, V. 32, No. 14, P. 2588.

61.   Kuleznev V.N., Shershnev V.A., Chemistry and physics of polymers, Moscow, KolosS, 2007.

62.   Kardash G.G., Andrianova G.P., Bakeev N.F., Kargin V.A., Dokl. AN SSSR, 1966 V. 166, No. 5, P. 1155.

63.   Kargin V.A., Andrianova G.P., Kardash G.G., Vysokomolek. Soed., A. V. 1967, V. 9, No. 2, P. 267.

64.   Andrianova G.P., Nguyen Vinh Chii, Kargin V.A., Dokl. AN SSSR, 1970, V. 194, No. 2, P. 345.

65.   Andrianova G.P., Kargin V.A., Vysokomolek. Soed., A, 1970, V. 12, No. 1, P. 3.

66.   Peterlin A., Olf H.G., J. Polymer Sci. Polym. Symp., 1975, No. 50, P. 243-264.

67.   Hubeny H., Dragunn H., Mater. Sci. and Eng., 1976, V. 24, P. 293.

68.   Olf H.G., Peterlin A., J. Polymer Sci., Polymer Phys., Ed. 1974, V. 12, P. 2209.

69.   Bondarev V.V., Diss. Cand. chemical. Sciences, Moscow, MSU. MV Lomonosov Moscow State University, 1983.

70.   Efimov A.V., Valiotti N.N., Bakeev N.F., Vysokomolek. Soed., A, 1991, V. 33, No. 5, P. 1042-1048.

71.   Trofimchuk E.S., et al., Vysokomolek. Soed. A, 2011, V. 53, No. 5, P. 1020-1032.

72.   Volynskii A.L. Arzhakova O.V., Yarysheva L.M., Bakeev N.F., Vysokomolek. Soed. B, 2000 V. 42, No. 3, P. 549-564.

73.   Wool R.P., Soft Matter., 2008, V. 4, P. 400-418.

74.   Wool R.P., Polymer interfaces: Structure and strength. New York, Hanser Press, 1995.

75.   Wittkop M., Kreitmeier S., Goritz D., Acta Polymer., 1995, P. 319-327.

76.   Tonguin Y.U., Qianguo D.U., Jiacong H.U., Polymer communication. 1985, No. 1.

77.   Volynskii A.L. Bakeev N.F. Highly dispersed, Highle Dispersed Superfine oriented

state of polymers, Moscow, Khimiya, 1984.
78. Kikoin A.K., Kikoin I.K., Molecular physics, Moscow, Nauka, 1976.
79. Treloar L., Physics of rubber elasticity, Moscow, Mir, 1975.
80. Treloar L. The Physics of rubber elasticity, Oxford, Oxford Univ. Press., 1949.
81. Wade W., Yang D., Edwin L. Thomas M., Journ. Mater. Sci., 1986, V. 21, P. 2239-2253.
82. Hild S., Gutmannsbauer W., Luthl R., Fuhrmann J., Gunterodt H.J., J. Polymer Sci., Part B, Polymer Physics, 1996, V. 34, P. 1953-1959.
83. Physics of polymer surfaces and interfaces, Sanchez I.C. (ed.) Boston, Butterworth-Heinemann, 1992.
84. Frank C.W., Rao V., Despotopoulou M.M., Pease R.F.W., Hinsberg W.D., Miller R.D., Rabolt J. F., Science, 1996, V. 273, P. 912.
85. Despotopoulou M.M., Frank C.W., Miller R.D., Rabolt J.F., Macromolecules, 1995, V. 28, P. 6687.
86. Despotopoulou M.M., Frank C.W., Miller R.D., Rabolt J.F., Macromolecules, 1996, V. 29, P. 5797.
87. Kim J.H., Jang J., Zin W., Macromol. Rapid Commun., 2001, V. 22, P. 386.
88. Wang Y., Ge S., Rafailovich M., Sokolov J., Zou Y., Ade H., Luning J., Lustiger A., Maron G., Macromolecules, 2004, V. 3, P. 3319.
89. Schonherr H., Bailey L.E., Frank C.W., Langmuir, 2002, V. 18, P. 490.
90. Scheonherr H., Frank C.W., Macromolecules, 2003, V. 36, P. 1199.
91. Massa M.V., Dalnoki-Veress K., Forrest J.A., The European Physical Journal E,- Soft Matter., 2003, V. 11, No. 2, P. 191-198.
92. Quynn R.G., Brody H., J. Macromol. Sci.-Phys., 1971, V. B55, P. 721.
93. Miles M.J., Baer E., J. Mater. Sci., 1979, V. 14, P. 1254.
94. Bulaev V.M., Diss. Cand. Chemical. Sciences. Moscow, MSU, 1986.
95. Ren W., Colloid. Polymer Science, 1992, V. 270, No. 8, P. 342.
96. Adams W.W., Yang D., Thomas E.L., J. Mater. Sci., 1986, V. 21, P. 2239.
97. Goritz D., Muller F.H., Colloid Polym. Sci., 1975, V. 253, P. 844.
98. Volynskii A.L., Grokhovskaya T.E., Kulebyakina A.I., Bolshakov, A.V., Bakeev N.F., Vysokomolek. Soed., A, 2009, V. 51, No. 4, S. 598-603.
99. Decandia F., Russo R., Vittoria V., Peterlin A., J. Polymer Science: Polymer Physics Edition, 1982, V. 20, P. 1175-1192.
100. Voyutsky S.S., Autohesion and adhesion of high polymers, Moscow, Khimiya, 1960.
101. Volynskii A.L., Priroda, 2011, No. 7, P. 14.

Role of surface phenomena in deformation of polymers in active liquid media

The results discussed so far confirm the important role of surface phenomena in effects of different type on the polymer systems. These effect are associated, firstly, with the fact that they cause changes in the area of the interface of the polymer and the associated transport of the material from the bulk to the surface and vice versa. It is evident that these processes should be controlled by surface-active substances capable of changing the interfacial surface energy of the polymer and, therefore, lead to structural rearrangement, taking place at the interfacial surfaces. This section is concerned with the special features of this effect on deformation of glassy and crystalline polymers.

P.A. Rehbinder was the first who paid attention at the beginning of the previous century to the fact that the deformation and failure of solids is always accompanied by changes of the interfacial surface. In other words, the work of deformation (failure) of the solid body includes the work of formation of the new surface. He attempted to influence the work of deformation (failure) of a solid body by adsorption-active substances which are capable of effective adsorption on the interfacial surface even at a low concentration in the surrounding medium and can greatly reduce the surface tension of the solids. The molecules of these substances attack the inter-molecular bonds at the tip of a growing fracture crack and, adsorbing on the freshly formed services, weaken them. Selecting the special liquids and introducing them on the surface of the deformed solid body, Rehbinder obtained a very large decrease of the work of fracture. Figure 6.1 shows this effect. It may clearly be seen that

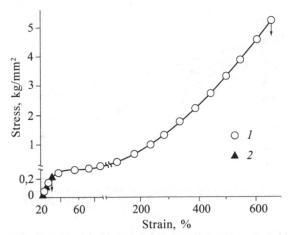

Fig. 6.1. Dependence of stress and strain of zinc single crystals at 400°C: 1) in air; 2) in a tin melt. The arrows indicate fracture strain [1].

the absorption-active medium lowers the work of fracture (the area below the tensile curve) by several orders of magnitude.

This effect is referred to as the Rehbinder effects or the adsorption reduction of the strength of solids [2]. The Rehbinder effect was discovered and described indeed when examining the effect of adsorption-active liquids mainly on the deformation-strength properties of low-molecular crystalline solids [1]. This effect is general and is observed in deformation in active liquid media of also amorphous solids, including polymers. One of the most clear confirmations of the adsorption effect of liquid media on the stress–strain behaviour of solids are the results of investigation of their deformation in solutions of homologous series of fatty alcohols (Fig. 6.2).

It was observed that in transition to each subsequent homologue, the concentration, corresponding to the given value of the effect of reduction of strength decrease 3–3.5 times, – in complete agreement with the Duclaux–Traube rule of the increase of the surface activity of the diphilic molecule with increase of the length of the hydrocarbon radical by one CH_2-group [4].

6.1. What is the adsorption-active medium?

To detect and describe the Rehbinder effect in polymers it is first of all necessary to select liquid media capable of reducing the interfacial surface energy of the polymer. In contrast to low-molecular solids,

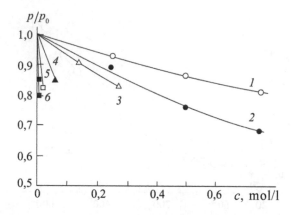

Fig. 6.2. Dependence of the relative reduction of the strength of polycrystalline naphthalene (p/p_0) on the concentration of water solutions of alcohols (c): 1) methyl, 2) butyl, 3) amyl, 4) hexyl, 5) heptyl, and 6) octyl alcohols in which deformation was carried out [3].

the polymers in interaction with the low-molecular solvents may show extensive volume swelling which sometimes reaches hundreds of wt.%. At the same time, the Rehbinder effect, in contrast to the volume swelling of the polymer, is a purely surface phenomenon. Nevertheless, the volume swelling also strongly influences the mechanical properties of the polymer. This phenomenon has been known for a relatively long time, has been studied sufficiently and is referred to as plastification [5]

Therefore, in [6] the liquids which can be characterised as surface-active or adsorption-active, were selected using the following procedure. The polymer (PET) was held in 30 different low-molecular liquids at room temperature for 1 month. This was followed by the determination of swelling of the polymer in them by the weight method. This procedure was used to select the liquids (given in the caption for Fig. 6.3) which did not cause any extensive swelling of the polymer at room temperature. In these liquids, PET was deformed and the appropriate tensile loading curves are shown in Fig. 6.3. It is clearly seen that, regardless of the almost complete absence of swelling of the non-loaded polymer, the selected liquids have the strongest effect on its mechanical response. In comparison with deformation in air, the tensile loading curves of the polymer in liquid media are displaced into the range of low stresses. This effect may be regarded as the adsorption effect because, as shown in Fig. 6.3, the initial modulus of the polymer does not change. If

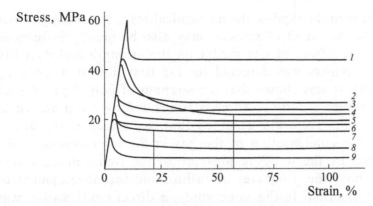

Fig. 6.3. Stress–strain curves of polyethylene terephthalate at room temperature in air (1), ethylene glycol (2), *n*-heptane (3), oleic acid (4), formamide (5), triethylamine (6), *n*-propanol (7), CCL$_4$ (8) and butyl iodide (1) [6].

swelling of the polymer would have taken place (volume interaction), we would have observed especially the reduction of the initial elasticity modulus [5]. Therefore, the reduction of the mechanical characteristics of the polyethylene terephthalate, reported in [6], in deformation in liquid media (Fig. 6.3) may be regarded without doubt as the manifestation of the Rehbinder effect in polymers, and the liquids as the adsorption-active media (AAM).

It is important to mention that in [6] the authors detected at least two liquids in which the Rehbinder effect is not detected – water and glycerin. These liquids are the most polar liquids in the group of the solvents not causing swelling of the PET at room temperature which evidently is also the reason for the fact that they do not interact by the adsorption mechanism with the PET. However, water starts to influence the mechanical response of the polymer if external hydrostatic pressure is applied in deformation. In [7, 8] it was shown that the application of external hydrostatic pressure greatly increases the efficiency of the effect of all the investigated liquid media (silicone oils, methanol and water) on the mechanical response of the polymer (PS). In addition, the water, which has no significant effect on the mechanical properties of the PS at the atmospheric pressure, starts to show such an effect when the hydrostatic pressure is applied. The authors assume that this effect is caused by the fact that at the atmospheric pressure, the water consists of large associates which cannot penetrate effectively into the areas of local deformation of the polymer. At high pressures (up to 4 kbar) these associates dissociate and the efficiency of the effect of water on the mechanical behaviour of the polymer greatly increases.

In addition to the liquids, the mechanical response of the polymer subjected to the effect of stresses may also be greatly influenced by gases. The effect of gas media on the strength and ductility of glassy polymers was detected for the first time in a study by Lazurkin [9]. It was shown that the strength of polyvinyl chloride at a temperature of $-70° \div -80°C$ is controlled by the presence of carbon dioxide. Lazurkin assumed that the effect of the gas is caused by the condensation of the gas in the microcracks of the polymer and by the wedging pressure of the liquid phase which facilitates both the processes of failure and the development of plastic deformation. In the same study, a direct confirmation was presented for the penetration of liquid hydrogen and liquid nitrogen into the microcracks, formed during the deformation of the polymers. The working part of the specimens, subjected to plastic deformation in these media, became white after heating to room temperature and bulged, increasing its volume 2–3 times. Bubbles formed and burst on the surface, releasing the gas trapped in them. Subsequently, this process was interrupted, and the sample continued to lose the gas slowly, gradually returning to the initial dimensions and normal colour. Evidently, this was the first observation of the Rehbinder effect in the polymers.

Important results were obtained when examining the effect of gas media on the mechanical response of amorphous polymers. It is well-known that the drawing of glassy and crystalline polymers in air takes place with the formation of a neck without any distinctive disruption of the integrity of the material. When the temperature is reduced the mechanical parameters of the polymer (yield stress, strength) usually increase in complete agreement with the theoretical Eyring–Lazurkin considerations [9]. With a further reduction of temperature the brittleness temperature of the polymer is reached and then the neck does not form in the polymer and fracture takes place at low strains, as in the case of brittle low-molecular solids [9, 10].

However, a further reduction of temperature to the temperatures close to the condensation temperature of the gases in which deformation is carried out (N_2, Ar, O_2, CO_2, etc) leads to the unexpected high plasticity of the polymer. In the temperature range close to the condensation temperature of each gas the fracture elongation of the polymer rapidly increases and its yield stress decreases. Consequently, the dependences of the fracture elongation and the yield stress on temperature show a maximum in the range of very low temperatures. At the same time, deformation of the same polymers at such low temperatures in vacuum or helium does not lead to any very high ductility. Plastic deformation of this type is carried out on many amorphous polymers, such as polymethyl metacrylate, polycarbonate, polyethylene terephthalate, polychlorotrifluoroethylene, polybutadiene and also on crystalline polymers – polypropylene and polytetrafluoroethylene [11–17].

The efficiency of the effect of the gas media in the range of low temperatures depends strongly on the partial pressure of the gas. It was shown [16] that when the partial pressure of nitrogen in its mixtures with inert helium is reduced the effect of low-temperature plasticity is suppressed and gradually disappears. The phenomenon of low-temperature plasticity in the presence of gas media has many common features with the previously described effect of the liquid AAM on the mechanical behaviour of the polymers. Evidently, an important role in the manifestation of the low-temperature plasticity is played by the absorption effects.

The first physical theory of cryogenic microcracking, proposed by Brown [18], attributed the controlling effect to the surface, adsorption reduction of strength, i.e., interpreted the Rehbinder effect in a unique manner. Actually, investigations of the low-temperature physical adsorption of gases (nitrogen, argon) by polymers showed that the reduction of the surface energy of the polymer varies from 11 erg/cm^2 for PET and PP to 20–40 erg/cm^2 for PMMA, PS and PVC [19]. Since the surface energy of the solid polymers usually does not exceed 50 erg/cm^2, this means that the surface energy of the polymer in low-temperature adsorption of the gases decreases by 25–75%.

The proposed theory qualitatively describes the mechanical behaviour of polymers at cryogenic temperatures in the gas environment. However, the theory, constructed exclusively on the basis of the assumptions of the adsorption effect of the medium, contains a number of contradictions. It does not take into account the processes of plastic deformation of the polymer at cryogenic temperatures.

The low-temperature plasticity of the crystalline polymer (polypropylene) was described for the first time in the studies by V.A. Kargin et al [20–30]. In these studies it was shown that the polypropylene may show extremely high strains (hundreds of percent) in the low-temperature range, down to the liquid nitrogen temperature. In addition, such cryogenic strains are reversible when the tensile loaded polymer is heated to room temperature.

Later, the low-temperature deformation of this type was studied in detail by Peterlin and Olf [13]. Following the authors of [23], they concluded that the deformation of the polymer leads in the first instance to an increase of its specific volume and the fraction of the free volume. This effect is equivalent to a large extent to the reduction of the glass transition temperature of the polymer to the temperature lower than the experiment temperature. An especially large increase of the specific volume and the fraction of the free volume takes place in the areas of stress concentration in the material. As a consequence of the high local concentration of the stresses, the polymer in this region transforms to the rubber-like state so that it becomes capable of high strains at stresses considerably lower than the yield stress of the material, surrounding the stress raiser. Therefore, the polymers fracture in a non-brittle manner even at very low temperatures, down to 4 K [24]. This is indicated by the analysis of the fracture surfaces of the polymers, which shows that the true fracture crack is always preceded by the zone of the plastically deformed material.

The presence of the gas with a high thermodynamic activity, i.e., situated in the vicinity of its condensation point, greatly facilitates the development of plastic deformation and interfaces. This is explained by the volume surface sorption of the gas. The volume plasticizing effect of the gas can operate if the gas is capable to migrate at a higher rate into the zone of active deformation of the polymer. However, at cryogenic temperatures, the rate of diffusion transport of the gas is greatly reduced. For example, the diffusion coefficient of nitrogen in, for instance, polypropylene at 77 K is equal to $6-10^{-18}$ cm^2/s [25], i.e. approximately 300 seconds are necessary for the penetration of the gas into the polymer to a distance of 10 nm. This time is long and such transport cannot facilitate the cryogenic deformation of the polymer detected in the experiments.

The situation greatly changes if the polymer contains stress raisers with a higher specific volume and a larger fraction of the free volume in comparison with the surrounding bulk polymer. In this case, because of the increase of the mobility of the polymer chains the rate of diffusion into the 'loosened' zone increases so much that the gas is sorbed, dissolves in these local zones, plasticizes the polymer and, at the same time, enables high strains in the material.

It should be mentioned that the development of high cryogenic strains of the polymer does not take place uniformly in the volume of the polymer body and takes place primarily in the areas of stress raisers characterised by the development of the zones of plastically deformed material with a fibrillar–porous structure (crazes). In [25] the authors describe the mechanical and thermomechanical properties of polypropylene deformed by the crazing mechanism in liquid nitrogen at −196°C, i.e., well below the glass transition temperature. The results show that the crazes retain their size in liquid nitrogen at −196°C after removing the load. At the same time, they immediately 'collapse' at the same temperature, if nitrogen is removed from the volume of the crazes leading to the almost complete shrinkage of polypropylene. In the repeated tests of polypropylene after shrinkage the tensile loading curves were situated at considerably lower stresses than in the case of the initial polypropylene during its deformation in liquid nitrogen. It was assumed that the stresses, causing 'collapse' of the crazes when nitrogen is removed from them, can be calculated from the equation $\sigma = 2\gamma/r$ (where γ is the surface tension of the polymer, r is the radius of the cavities in the craze). The calculations give the value of the order of 6 MPa which again is considerably lower than the yield stress of polypropylene at this temperature. As indicated by the experimental data, it is not possible to explain satisfactorily the reversibility of the high deformation of the crazed polymer only by the effect of surface forces, assuming that the material of the crazes has the mechanical properties of the initial glassy polymer.

Since both processes, volume and surface, can take place only at a relatively high thermodynamic activity of the gas, it is then clear why the effects of cryogenic plasticity are most evident in the vicinity of the condensation temperatures of the corresponding gases and disappear when the temperature is increased.

However, it is very difficult to observe the adsorption effect of the active agent in the pure form as it was carried out for low-molecular solids [26]. This is due to the chain structure of the polymer molecules so that the polymer materials show a capacity for a large increase of the fraction of the free volume which greatly facilitates the volume sorption of the low-molecular substances.

The above examined effect of cryogenic crazing in the gas media at temperatures close to their condensation temperature is the manifestation of the Rehbinder effect which does not differ at all from the Rehbinder effect in the liquid AAM. Actually, this effect is observed only when gas condensation starts in narrow pores and/or other imperfections of the surface of the real solid polymer and the presence of the liquid phase in the areas of local deformation of the

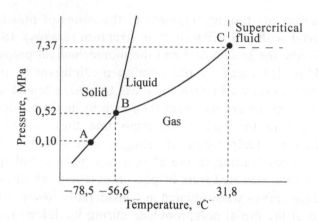

Fig. 6.4. Phase diagram of CO_2 in dependence on temperature and pressure: A) sublimation point; B) ternary point; C) critical point.

polymer results in the observed change of the mechanical response of the polymer. In particular, this circumstance also determines, for example, good agreement between the experimental theory, developed in [18, 25, 27].

Nevertheless, the Rehbinder effect can also be detected under the effect of truly gas media. In the last 20 years the interest of researchers has been concentrated on the problems of interaction of the polymers with gas media with the temperatures higher than the critical temperature, under the effect of a relatively high pressure (tens of even hundreds of times greater than the atmospheric pressure) [28–30]. These are the so-called supercritical media (or supercritical fluids) whose special feature – the absence of the gas – liquid phase boundary and the fact that the liquid phase can be produced by increasing the pressure in the system above a specific critical temperature. Figure 6.4 shows the phase diagram of carbon dioxide.

The supercritical region starts at the critical point C characterised not by one parameter as, for example, the melting point, but by two parameters – temperature and pressure. The reduction of either temperature or pressure below the critical value displaces the material from the supercritical state and, therefore, in Fig. 6.4 this region is restricted on the left and at the bottom by two straight lines – and then propagates in the direction of higher temperatures and pressure.

The supercritical media combine the properties of liquids (the relatively high density, capacity to dissolve solid substances) and gases (low viscosity, low surface tension). It is important to mention

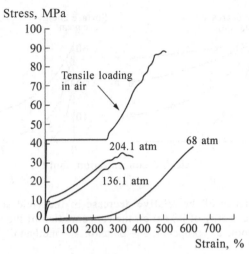

Fig. 6.5. Effect of the pressure of CO_2 on the stress–strain characteristics of polyethylene terephthalate at 35°C in the supercritical range [35].

that in comparison with the conventional gas media, in the case of the supercritical fluids it is possible to change in a wide range the density, reaching the density of the liquids [31] so that the activity of the medium can be efficiently regulated. It is not surprising that in the last couple of decades investigations were started to explain the effect of the supercritical gases on the structure and properties of the polymers. Deformation of the polymers in supercritical carbon dioxide [32, 33] and supercritical xenon [34] demonstrates the structural–mechanical special features, characteristic of the Rehbinder effect. In particular, deformation in the supercritical liquids clearly demonstrates the mechanical response of the polymer characteristic of the Rehbinder effect [35, 36]. Figure 6.5 shows the curves of tensile loading of polyethylene terephthalate in supercritical carbon dioxide.

Figure 6.5 shows that the supercritical CO_2 has the strongest effect on the stress–strain behaviour of polyethylene terephthalate. Firstly, the result indicates that not only the liquids but also the gases are capable of realising the Rehbinder effect in the polymers. Secondly, the pressure of the gases is a powerful factor which can be used to regulate the intensity of the Rehbinder effect in the polymers.

To conclude this section, we will discuss a number of data shedding light on the mechanism of the effect of the liquid AAM on the deformation of polymer glasses. The most convincing confirmation of the occurrence of the Rehbinder effect in the polymers can be obtained by discussing the effect of the adsorption

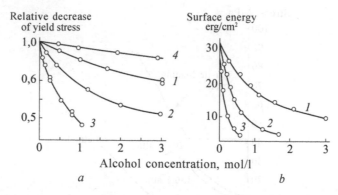

Fig. 6.6. Dependence of the relative decrease of the yield stress of PS (*a*) and interfacial surface energy of PS (*b*) on the concentration of the aqueous solution of alcohol (1 – ethanol, 2 – propanol, 3 – butanol, 4 – methanol) [37].

phenomenon on their mechanical behaviour. In Fig. 6.2 this effect is demonstrated for the processes of failure of crystalline naphthalene in water–alcohol mixtures (in the form of fulfilment of the Duclaux –Traube rule). It appears that the adsorption effect of liquid media of this type also operates in the amorphous solid bodies and is observed not only at their fracture but also in the processes of plastic, inelastic deformation. Figure 6.6 *a* shows the dependences of the relative decrease of the yield stress of the amorphous polymer of polystyrene (PS) during its deformation in the water solutions of aliphatic alcohols [37]. As indicated by Fig. 6.6 *a*, the Duclaux–Traube rule is also fulfilled in deformation of the amorphous polymer in the solutions of aliphatic alcohols. It is important to note that the direct evaluation of the surface tension of the solid phase is not possible because due to the fact that the surface forces are low the solids can not change their form and/or volume under the effect of these forces. However, this evaluation can be carried out in cases in which it is possible to measure reliably the contact wetting angle of the solid body by the liquid. For the case of the PS–water solutions of aliphatic alcohols, such an evaluation was carried out in [37] (Fig. 6.6 *b*). It may easily be seen that the interfacial surface energy at the PS–water solutions of aliphatic alcohols interface changes in the same manner as the yield stress of the PS in these media.

Comparison of Figs. 6.6 *a* and *b* indicates unambiguously the strong and non-trivial influence of surface effects on the stress–strain behaviour of the solid bodies of greatly different nature.

6.2. Structural special features of deformation of polymers in adsorption-active media

Although the Rehbinder effect is of the general nature, the chain structure of the polymer molecules brings in special features in the manifestation of this effect. The point is that the inelastic deformation of the polymer is accompanied by the unfolding and mutual orientation of the macromolecules and by this it differs greatly from the deformation of the low-molecular solids. The latter circumstance is quite evident in structural investigations.

In chapter 4 we described and investigated surface phenomena, accompanying inelastic (plastic) deformation of the amorphous polymers. It was shown that in the well-investigated process, such as plastic deformation, which forms the basis of the orientation drawing of the polymer used widely in practice, there appear extensive structural rearrangements accompanied by the formation and development of the interfacial surfaces. Nevertheless, on the macroscopic level this process is manifested in the development, at first sight, of a monolithic transparent neck. In fact, the neck of the polymer, subjected to cold drawing, is a close-packed aggregate of fibrillar formations at the nanometric level which are constructed from the oriented macromolecules characterised by distinctive interfaces which can be easily observed by electron microscopy [38, 39] and by x-ray diffraction analysis [40]. It is rational to expect that the presence of the adsorption-active media (AAM) should influence the process of fibrillization of the polymer during its orientation drawing.

Figure 6.7 shows the images of two specimens of the same amorphous polymer of polyethylene terephthalate, one of which was stretched in air (upper) and the other one in and adsorption-active liquid (lower). In the first case, the so-called neck forms in

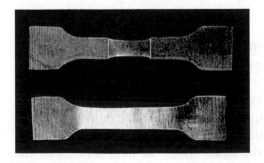

Fig. 6.7. External appearance of polyethylene terephthalate (PET) samples, stretched in air (upper) and in an adsorption-active medium (*n*-propanol) (lower).

Fig. 6.8. Scanning electron micrograph of a sample of glassy polymer (polyethylene terephthalate) deformed to 100% in the AAM.

the sample in which the macromolecules are oriented in relation to each other so that on the whole the polymer has high mechanical properties. This is the reason why the process of manufacture of polymer fibres and films always includes the stage of orientation drawing whose aim is the general molecular orientation of the polymer transformation of the polymer to the material of the neck.

In the second case, the width of the film does not decrease but the film becomes milky white and opaque. The reasons for the observed whitening become clear in microscopic studies. It appears that the formation of the monolithic transparent neck in the polymer is replaced by the formation of a unique fibrillar–porous structure (Fig. 6.8), consisting of filament-like aggregates of macromolecules (fibrils), separated by microcavities (pores). In this case, the mutual orientation of the macromolecules is reached inside the fibrils and not in the monolithic neck. Since the fibrils are separated in space, the structure contains a very large number of microcavities which intensively scatter the light and make the polymer milky white. Thus, in the case of polymers, the Rehbinder effect is manifested in the form of the development of a large number of zones of the plastically deformed polymer, having the fibrillar–porous structure. Evidently, the role of the AAM is the stabilisation of the highly developed surface of the fibrils, formed in the process of cold drawing of the polymer.

6.3. Crazing in liquid media – manifestation of the Rehbinder effect in polymers

In the previous section it was shown that the deformation of the glassy polymers develops completely non-uniformly in their volume. Homogeneous affine deformation takes place only in deformation of the rubber-like polymers. At the same time, the deformation of the

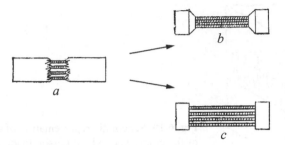

Fig. 6.9. Diagram of structural rearrangement in the deformed polymer: a) nuclei of localized deformation; b) the neck during tensile loading in air; c) the craze in tensile loading in the adsorption-active medium [44].

glassy polymers takes place by the nucleation and development of the localized zones of the inelastically (plastically) deformed polymer. These localized zones (crazes) [41] and/or shear bands [42] nucleate on the defects, imperfections of the structure, characteristic of any real solids and randomly distributed in the volume and on the surface of the polymer. As shown in the previous section, the nucleated crazes and shear bands have many common features and the only difference between them is the small difference in the orientation in relation to the direction of tensile stress [43].

It is very important to mention that the nuclei of localized deformation in the polymer glasses have a fibrillar–porous structure (Fig. 6.9 a). The further 'fate' of these nuclei depends strongly on the medium in which the polymer is deformed. If deformation is carried out in the AAM (in the conditions of reduced interfacial energy), the highly developed surface of the growing fibrils is stabilised and the process takes place by the development of crazes (Fig. 6.9 c) [44]. However, if deformation takes place in air or some other inactive medium, not capable of reducing the interfacial surface energy of the polymer, the fibrillar aggregates of the macromolecules are destabilised and merge (coalesce) in a monolithic neck (Fig. 6.9 b). Regardless of the apparent monolithic nature of the resultant neck, the latter contains coalesced interfaces which can be detected in the experiments [38–40].

6.4. Mechanism of the formation of the unique structure of crazes

Thus, the deformation of the glassy amorphous polymer by the crazing mechanism is in fact a typical example of manifestation of

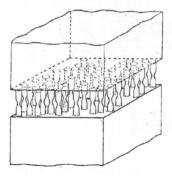

Fig. 6.10. Schematic representation of the structure of the craze. The axis of tensile loading is vertical [47].

the Rehbinder effect caused by the combined effect of mechanical stress and the adsorption-active medium (AAM). In fact [41, 45, 46] the results of a large number of investigations of the structure of greatly different crazed polymers indicate that as a result of deformation in the AAM, the non-ordered amorphous polymer transforms spontaneously into a surprisingly ordered, regular structure. This structure is shown schematically in Fig. 6.10 [47]. In the case of crazing, the orientation transformation of the polymer takes place inside the fibrillar aggregates, connecting the fragments of the initial amorphous non-oriented polymer.

In the most general form, the solid deformed in the tensile loading conditions is affected by two forces: the surface tension force, which 'aspires' to reduce the interfacial surface area, and the expanding mechanical stress, acting in the opposite direction.

In particular, the interaction of these two forces acting in opposite directions also causes the self-dispersion of the deformed polymer. The situation is identical with that observed if we increase the distance between two glass sheets separated by a layer of a working liquid (Fig. 6.11). In this case, on reaching the critical distance between the plates, the liquid layer is unavoidably dispersed into a set of digitate aggregates. If the liquid is low-molecular, the surface tension forces transform these aggregates to the spherical droplets, having the smallest specific surface. However, in the case of a polymer in tensile loaded digitate aggregates of macromolecules the latter become oriented in relation to each other and this is accompanied by the phenomenon of orientation strengthening [48] as a result of which they retain their form and individuality. Treating the crazing phenomenon as some variety of the loss of stability of the polymer system under the effect of the external factors, it is possible to explain convincingly the formation and development of the unique structure of the crazes. According to the current views,

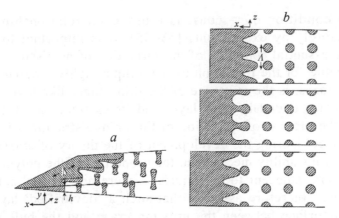

Fig. 6.11. Schematic representation of the process of fibril formation in crazing of the polymers [50]: *a*) side view, *b*) plan view.

this morphological transformation is based on the mechanism used for the first time for explaining the phenomena which take place during mixing of two incompatible liquids with greatly different densities [49]. This mechanism, referred to as the Taylor instability of the meniscus, may be described as follows. When two immiscible liquids move under the effect of the external force in a narrow channel, on reaching the critical shear stress the interface between them becomes unstable. The instability of the interface between two immiscible liquids is manifested by the fact that the liquid below viscosity starts to penetrate into the more viscous liquid and becomes dispersed in it in the form of a relatively regular system of thin 'fingers'. This mechanism is in fact the basis of the crazing phenomenon [50, 51]. In deformation of the polymer the external stress and the hydrostatic pressure, acting in the opposite direction to this stress and determined by the effect of surface forces, create critical conditions in which the polymer starts to be dispersed to a system of fibrillar aggregates.

The main relationship which has been experimentally verified many times [52], based on the theory of instability of the meniscus and used normally for describing and analysis of the crazing of the polymers, connects the external stress σ, the diameter of the resultant fibril d with the specific surface energy of the polymer Γ:

$$\sigma d = 8\Gamma(V_f)^{1/2} \tag{6.1}$$

where V_f is the volume fraction of the fibrils in the craze.

The condition σd = const is a universal relationship verified convincingly by experiments [53–55]. It is important to mention that the essential condition of the formation of meniscus instability, responsible for the formation of the unique highly ordered structure of the crazes, is the existence of the thin rubber-like layer, adjacent to the craze boundary. This layer and its existence are the areas in which the process of formation of the unique structure of the craze takes place. During the development of the theory of crazing by the meniscus instability mechanism, the properties of the polymers in the thin layers, adjacent to the interface, were not studied sufficiently and it was therefore assumed that such a 'devitrified' layer exists at the interface between the growing craze and the bulk polymer. It was assumed that the softening of the thin (several nanometres) layer is associated with the effect of the expanding mechanical stress, leading to a local increase of the free volume [56]. According to the assumptions made in [52], the transfer of the polymer to the softened rubber-like state in a narrow layer adjacent to the wall of the craze, provides the possibility for the formation of the highly dispersed regular fibrillar structure. It is assumed that the force softening of this type may be so large that it may lead to 'devitrification' of the thin layer of the polymer in the transition region and its transfer to the rubbery state. According to the assumptions made in [52], in particular the transfer of the polymer to the softened, rubbery state in a narrow layer adjacent to the craze wall, creates suitable conditions for the loss of its mechanical stability and, consequently, formation of the highly dispersed regular fibrillar structure. The authors of the above mentioned assumptions [52, 56] anticipated the discovery made at least 10 years later, i.e., the actual existence of the devitrified layers on the surface of the glassy polymers (see chapter 3).

6.5. Crazing dynamics of polymers in liquid media

Since the crazes containing a large number of microcavities, they can be easily registered using a light optical microscope, so that it is possible to use the direct method for characterising the dynamics of their nucleation and development [47]. In particular, this method was used in [58] for the determination of the multistage nature of the crazing process and its relationship with the mechanical response of the deformed polymer.

Figure 6.12 shows schematically the main special features of the development of the crazing of the polymer in the liquid medium in

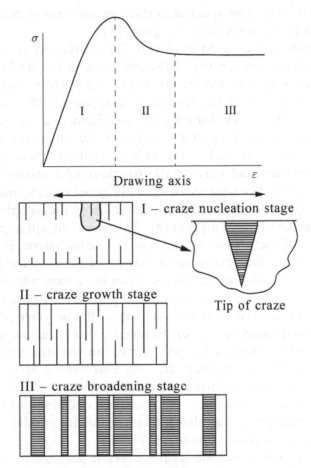

Fig. 6.12. Curve of tensile loading of the glassy polymer in the AAM and schematic representation of the individual stages of crazing of the polymer: I) the region of initiation of the crazes, II) the region of growth of the crazes, III) the region of broadening of the crazes.

the process of tensile loading of the polymer with a constant rate and compares the pattern of crazing of the polymer, obtained in direct microscopic studies, with the appropriate tensile loading curve. The resultant data demonstrate the Rehbinder effect in the polymers, because they determine the relationship of the mechanical response of the polymer deformed in the AAM with the structural rearrangement, accompanying this process. In the context of the subject of this monograph, it should be mentioned that the investigated process is accompanied by the total development of the interfacial surface of the polymer as a result of its dispersion into aggregates of the

nanometre size. The specific surface of the crazed polymer equals
hundreds of square metres per gram [41].

We examine in greater detail the structural rearrangements
accompanying the crazing of the polymers in the AAM (Fig. 6.12).
In the first stages of tensile loading of the polymer (up to the yield
stress − region I on the tensile loading curve) the surface of the
polymer shows the nucleation of a specific number of crazes. During
further tensile loading of the polymer the nucleated crazes grow
in the direction normal to the axis of applied stress, retaining the
almost constant and very small (fractions of a micron) width (the
stage of growth of the crazes). This process continues until the
growing crazes intersect the cross-section of the sample (the region
II on the tensile loading curve). This moment corresponds to the
formation of the plateau on the tensile loading curve (Fig. 6.7). This
is followed by the next stage of crazing of the polymer in the liquid
medium − widening of the crazes, when the crazes which have grown
through the entire cross-section of the polymer increase in size in the
direction of the axis of tensile loading of the polymer (the region III
on the tensile loading curve). Evidently, this is accompanied by the
main transformation of the polymer to the oriented fibrillized state.

It should be mentioned that the correlation of the individual
stages of crazing with the mechanical response of the deformed
polymer was detected not only for the conditions of the constant
tensile loading rate but also for tensile loading under the effect of
a constant load (creep conditions) [59].

The direct microscopic studies make it possible to investigate in
detail and characterise the individual stages of manifestation of the
Rehbinder effect in the polymers.

The stage of initiation of the crazes
In [60] the microphotography method was used to count directly
the number of the formed crazes under the effect of a number of
external factors. Figure 6.13 shows the dependence of the density of
the nucleated crazes on the magnitude of the constant tensile load,
applied to the polyethylene terephthalate placed in an AAM. It is
important to note that the increase of load results in a non-linear
increase of the number of nucleated crazes.

Another important factor affecting the number of nucleated crazes
is the nature of the AAM, more accurately its capacity to reduce
the interfacial surface energy of the polymer. Figure 6.14 shows the
dependence of the number of the resultant crazes on the magnitude

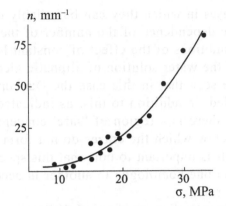

Fig. 6.13. Dependence of the density of the resultant crazes in tensile loading of polyethylene terephthalate in *n*-propanol on the magnitude of constant load.

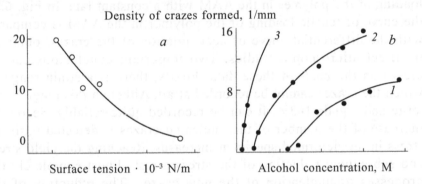

Fig. 6.14. Dependence of the density of the crazes, formed in polyethylene terephthalate under the effect of constant load (23.3 MPa), on the surface tension at the polymer – liquid medium interface (a) and on the concentration of aliphatic alcohols (1 – ethanol, 2 – propanol, 3 – butanol) in water (b).

of the reduction of the interfacial surface energy of polyethylene terephthalate in the conditions of the effect of constant load in different AAMs. The value of the interfacial surface energy in this case was calculated by the Owens–Wendt equation [61] because the direct determination for the liquids, efficiently wetting the polymer, is at present an impossible task. Nevertheless, it should be mentioned that the value of the interfacial surface energy, with other conditions being equal, has the strongest effect on the number of nucleated crazes: with the decrease of the interfacial surface energy of the polymer the number of the nucleated crazes increases.

It should be mentioned that there is a method for a more accurate evaluation of the interfacial surface energy of the solids. This possibility is associated with the determination of the contact wetting angles [62] in cases in which they can be reliably measured. Figure 6.14 *b* shows the dependence of the number of the resultant crazes formed in the conditions of the effect of constant load on the molar concentration of the water solution of aliphatic alcohol, used as the AAM. It may be seen that in this case the Duclaux–Traube rule is efficiently fulfilled. In addition to this, as indicated by Fig. 6.14 *b*, for each alcohol there is a region of 'safe' concentrations, i.e., the concentrations below which the crazes do not form in the sample at the given stress. It is important to note that this special concentration also changes from one homologue to another in accordance with the Duclaux–Traube rule.

All the previously discussed special features of the stage of initiation of the crazes can also be studied in the conditions of tensile loading of the polymer in the AAM with a constant rate. In Fig. 6.15 the curve of tensile loading of the polymer in the AAM is compared with the differential curve of accumulation of the crazes, obtained in direct microscopic studies. Two important conclusions can be drawn on the basis of these data. Firstly, there is a strain range in which the crazes cannot be recorded at all. Although this range is not large and equals 1–3%, it can be recorded quite reliably. Secondly, increase of the number of the nucleated crazes is detected until the stress in the deformed specimen increases. Reaching the yield stress and subsequent reduction of the stress arrest almost completely the processes of nucleation of the new crazes. The reduction of the tensile loading rate (stress) leads regularly to a reduction of the number of nucleated crazes, and it is possible to reach such a low level of the stress at which only one craze is formed in the sample

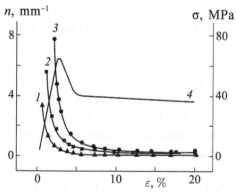

Fig. 6.15. Dependence of the density of crazes, formed in polyethylene terephthalate during its tensile loading at a constant rate of $8.3 \cdot 10^{-6}$ (1), $3.3 \cdot 10^{-5}$ (2) and $1.7 \cdot 10^{-4}$ m/s (3), the curve of tensile loading of polyethylene terephthalate in ethanol at a rate of $1.7 \cdot 10^{-4}$ m/s (4).

[63]. In this case, the entire orientation transformation of the polymer will take place only in one local zone (craze), similar to what takes place in tensile loading of the polymer in air with the formation of a neck.

The study of the experimental material, presented in this section, shows that the stage of initiation of the crazes has a number of special features, irrespective of the methods of loading of the polymer in the liquid medium. The most important special feature of the Rehbinder effect, manifested in the deformation of the polymer in the AAM, is the fact that the number of nucleated crazes depends on the level of the stress applied to the polymer and on the nature of the AAM.

The stage of growth of crazes

We now examine the processes of development of the ensemble of the crazes, nucleated in the polymer. Direct microscopic studies of the crazing of the polymer in the liquid medium, carried out for the first time in [64], proved to be highly effective. The typical results of these investigations are presented in Fig. 6.16. These data were obtained in taking microphotographs of the polymer directly in the process of tensile loading in the AAM with a constant rate. The figure shows the data obtained in the measurement of the length of the growing crazes in relation to the duration of tensile loading of the specimen with a constant rate. It can be seen immediately that, as shown in [57], different loading conditions of the polymer leads to identical consequences as regards the second stage of crazing of the polymer in the liquid medium – the stage of growth of the crazes.

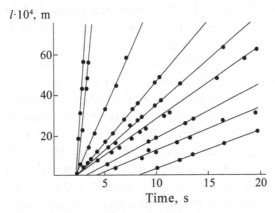

Fig. 6.16. Typical dependences of the length of the crazes, growing in polyethylene terephthalate during its deformation in ethanol at a rate of $1.67 \cdot 10^{-4}$ m/s on the tensile loading time [64].

Regardless of whether the polymer is stretched at a constant rate or under the effect of constant load, deformation develops by the growth of the family of the crazes in the manner shown in Fig. 6.16.

Figure 6.16 shows the extremely important fundamental special features of the Rehbinder effect in the polymers. Firstly, the dependence of the length of each craze on the deformation time is linear, irrespective of the type of loading. This dependence of the length of the craze on time indicates that the growth rate of the crazes is constant. Secondly, the nature of development of the individual crazes in the case that first to form are the crazes with the highest rates and then the crazes growing at lower rates. In particular, this is indicated by the 'fan-like' set of the dependences of the length of the individual crazes on the duration of loading (Fig. 6.16). Otherwise, at least some of these relationships should intersect at specific times of development of deformation. Finally, although each individual craze grows at a constant rate, these rates are not identical. As indicated by Fig. 6.16, the difference in the growth rates is sufficiently large and, therefore, in [57, 64] this stage of crazing was studied by analysis of the appropriate curves of distribution of the growth rates of the individual crazes. It must be mentioned that the growth rate distributions of the crazes, detected for the first time in [64], can be accurately reproduced if the number of measurements is sufficiently large (no less than 200). It is interesting to note that in [65] such a large difference in the growth rate of the individual crazes was erroneously attributed to the scatter of the experimental data.

The typical results of statistical processing of the data for the growth rates of the individual crazes are presented in Fig. 6.17. It can be seen that the growth rate distributions have a distinctive maximum corresponding to the most probable growth rate. The increase of the tensile loading rate (deformation stress) displaces the maximum on the distribution curves into the region of high growth rates and greatly increases their width. Direct comparison of the tensile loading curves with the data for the buildup of the crazes in the loaded polymer confirms the conclusion according to which the crazes growing at the highest rate are the first to form (at low strains) in the range of the yield stress on the tensile loading curve. Reaching the yield stress not only greatly reduces the number of nucleated crazes but also catastrophically reduces their growth rate in this stage of tensile loading [64]. The latter data demonstrate the

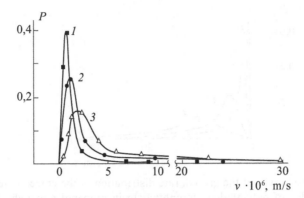

Fig. 6.17. Curves of the distribution of the growth rates of the crazes in tensile loading of samples of polyethylene terephthalate in ethanol at a rate of $8.3 \cdot 10^{-6}$ (1), $3.3 \cdot 10^{-5}$ (5) and $1.7 \cdot 10^{-4}$ m/s (3) [64].

distinctive relationship between the first two stages of crazing – nucleation and growth of the crazes.

As mentioned previously, the loading conditions do not have any appreciable effect on the main relationships of the crazing of the polymer in liquid media. Nevertheless, the tensile loading conditions at a constant load (creep conditions) appear to be more suitable for analysing this phenomenon. The point is that in tensile loading at a constant rate the stress in the deformed specimen changes in a complicated fashion and this may make the interpretation of the experimental data difficult.

Figure 6.18 shows the growth rate distributions of the crazes produced in tensile loading of polyethylene terephthalate in the AAM at a constant load. It may be seen that the distributions with a distinctive maximum and asymmetric form are also obtained in this case. Increase of the load displaces the distribution curves in the direction of higher growth rates of the crazes. This is in full agreement with the data obtained at a constant tensile loading rate [57, 64].

The stage of growth of the crazes has a distinctive statistical form, and the maximum on the distribution curve corresponds to the most probable growth rate of the crazes for the given polymer in the given conditions. Analysis of this dependence in relation to the different external conditions provides important information on the crazing mechanism of the polymer in liquid media. In particular, these data enabled us to use the traditional kinetic Eyring–Lazurkin approach

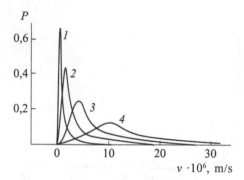

Fig. 6.18. The curves of the growth rate distribution of the crazes in tensile loading of the samples of polyethylene terephthalate in propanol under the conditions of constant load 14.9 (1), 16.4 (2), 17.9 (3) and 19.3 MPa (4).

Logarithm of the craze growth rate, m/s

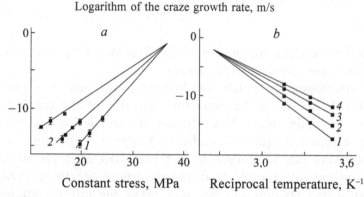

Constant stress, MPa   Reciprocal temperature, K^{-1}

Fig. 6.19. Dependence of the most probable growth rate of the crazes in polyethylene terephthalate during tensile loading in *n*-propanol on the magnitude of the constant load at temperatures of 10 (1), 20 (2) and 30 (3) °C (*a*) and on reciprocal temperature at stresses of 15 (1), 18 (2), 21 (3) and 24 (4) MPa (*b*).

[66, 67] for determining the structural and energy parameters of the crazing of the polymer in the liquid medium.

Figure 6.19 shows the dependences of the most probable growth phase of the crazes (the rates correspond to the maximum under curves of distribution of growth rates of the crazes) on stress (*a*) and temperature (*b*). It may be seen clearly that these dependences are linear, and the sets of the produced curves converge in both cases in a fixed centre. The presence of the fixed centre was detected several times previously in the analysis of different processes of fracture and deformation of solids. Without discussing the reasons for the existence of the fixed centre, it should nevertheless be mentioned that dependences of this type are characteristic of the typical activation

Table 6.1. Parameters of the Eyring–Lazurkin equation for amorphous polyethylene terephthalate

Liquid medium	ΔF, kJ/mol	$\gamma \times 10^4$, m³/mol
Air	56	
Ethylene glycol	63	
Hexadecane	68	
Hexane	56	
Propanol	30	
Ethanol	28	
Solutions of n-propanol in water with concentration, mol/l		
0.5	59	19
1	57	20
1.5	42	20
2	35	19

processes. The data in Fig. 6.19 were obtained in [52] using the Eyring–Lazurkin equation. This processing gives the value of the activation energy of this process of 53.2 kcal/mol when using propanol as the AAM. This value is close to the values obtained in investigating the rates of growth of the shear bands in PS [52]. Thus, the transfer of the polymer to the oriented state in crazing in the liquid medium is a typical thermally activated process of plastic deformation. In [68] the authors carried out standard analysis of the deformation kinetics of polyethylene terephthalate in different AAM using the Eyring–Lazurkin equation: $\sigma_c/T = U_0/(\gamma T) + R/\gamma \ln (v/A)$, where σ_c is the stress sustaining the inelastic deformation of the polymer (usually the yield stress), U_0 is the activation energy of the process, γ is the activation volume, A is the empirical constant, R is the universal gas constant, and T is absolute temperature.

The typical results of this investigation are presented in Table 6.1. It may be seen that within the experimental error range the activation volume remains almost completely constant, whereas the activation energy is controlled by the nature of the AAM used and, in fact, determines the effect of the AAM on the mechanical response of the deformed polymer. The role of the AAM in crazing is reduced to decreasing the activation barrier of plastic deformation.

The stage of widening of the crazes

The stage of widening of the crazes starts when the crazes intersects cross-section of the polymer sample. This stage has much in common with the stage of growth of the crazes, investigated in the previous section. The process of widening of the crazes is a typical activation process of plastic, inelastic deformation of the polymer [68]. If we draw analogy with the tensile loading of the polymer in air, it may be concluded that the stage of growth of the crazes corresponds to the stage of formation of the neck (the strain range in the vicinity of the yield stress). However, the stage of widening of the crazes fully corresponds to the stage of transformation of the initial polymer to the 'matter' of the neck (the plateau region on the tensile loading curves). This stage in particular is characterised by the main transition of the polymer to the oriented state in the process of crazing of the polymer in the liquid medium (transformation of the initial polymer to the 'matter' of the neck) (see Fig. 6.12).

The identical characteristic of the stages of growth and widening of the crazes is confirmed by the experimental data obtained in [69, 70]. In these studies it was shown that, firstly, as in the growth stage, each individual craze widens at a constant rate. Secondly, each individual craze has its own rate of widening. The process of widening of the crazes, like the process of their growth, can be analysed efficiently by means of the appropriate curves of distribution of their widening rates (Fig. 6.20). This graph shows that both the rate of widening and the growth rate depend on the level of external loading. In [70] it was shown that the rate of widening of the crazes and, naturally, the corresponding distribution curves are influenced most markedly (like the process of growth of the crazes)

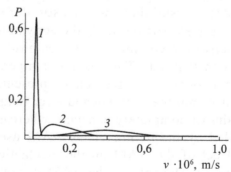

Fig. 6.20. The curves of distribution of widening rates of the crazes in tensile loading of the specimens of polyethylene terephthalate in methanol in the conditions of the effect of constant load of 16.1 (1), 18.0 (2) and 19.6 (3) MPa at 20°C [69].

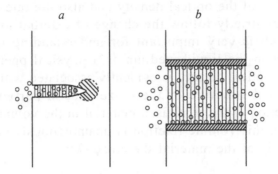

Fig. 6.21. Diagram illustrating the process of tensile loading of the glassy polymer in the AAM: a) the initial stage of stretching of the polymer (until the tensile loading curves reaches the plateau region); b) the stage of development of deformation of the polymer at constant load (the plateau region on the tensile loading curve). The crosshatched zone is the delocalized transition of the polymer to the oriented state ('matter' of the craze) [44].

by the nature of the liquid medium and temperature. In other words, both crazing stages are very similar to each other and as regards their nature are typical activation processes of plastic deformation. Evidently, the main difference between them is the size and geometry of the zone of transition of the polymer to the oriented state (the sharp tip of the craze in the first case and the flat wall of the craze in the second case) and, evidently, the nature of the stress state in the zone of this transition. The special features of the stages of growth and widening of the crazes are illustrated schematically in Fig. 6.21.

Nevertheless, detailed microscopic studies of the stage of widening of the crazing of the polymer in the liquid medium yielded important data without which the physical pattern of this phenomenon would not be complete. In particular, this concerns the establishment of the relationship between the rate of widening of the craze and its internal structure. This conclusion was drawn in [70, 71] on the basis of direct microscopic experiments. These investigations make it possible to examine the displacement of the wall of the craze into the bulk of the non-oriented polymer under the effect of applied load. At the same time, it is possible to evaluate the changes in the structure of the craze because the optical microscope can be used to record the optical density of the object studied in transmission. The experimental results show that the variation of the load sustaining the deformation of the polymer changes not only the rate of widening of each case but also the optical density of the polymer inside the craze.

This variation of the optical density and also the rate of broadening of the crazes strictly follow the change of external load

This result is very important for understanding the crazing of the polymer in the liquid medium as a physical phenomenon. The variation of optical density is evidently associated with the variation of the internal structure of the craze and, in particular, with the volume fraction of the fibrillised material in the volume of the craze (V_f). In turn, this volume fraction is unambiguously connected with the draw ratio of the material the craze (λ):

$$\lambda = 1/V_f \qquad (6.2)$$

In microscopic studies, with the results presented in [70], the value λ is determined by the direct independent method using special reference grids, deposited on the surface of the polymer prior to the start of the formation. This approach can be used for the direct determination of the draw ratio in each individual craze in dependence on the deformation conditions.

Evidently, if the rate of widening of each craze is unambiguously connected with the draw ratio of the fibrillised material inside the craze, there should be not only the distribution of the rate of widening but also the distribution of the draw ratio in the polymer, deformed by the mechanism of classic crazing. In [72] this type of distribution was in fact detected in direct microscopic experiments (Fig. 6.22). The results show that the draw ratio is not identical for different crazes, developed in the same material during its deformation in the AAM. As indicated by Fig. 6.22, there is a very wide distribution of the draw ratio in the crazes (in the investigated cases from 130 to 420%). In the same temperature –rate conditions the deformation of the polymer in air results in the value of the so-called natural draw ratio in the neck of 310–320%. Consequently, in the crazing of the polymer in the liquid medium the fundamental

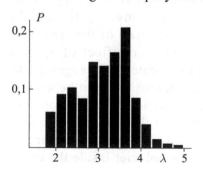

Fig. 6.22. Histogram of the distribution of the draw ration of polyethylene terephthalate in crazes produced in tensile loading in ethanol under the effect of a constant load of 16.4 MPa [72].

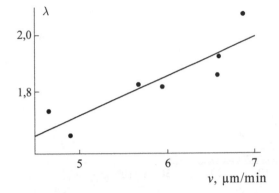

Fig. 6.23. Dependence of the draw ratio of polyethylene terephthalate in the crazes on the rate of widening in ethanol under the effect of a constant load of 12.6 MPa [70].

characteristics such as the natural draw ratio loses its meaning. It is important to note that there is an evident relationship between the draw ratio of the polymer in the volume of the craze and the rate of widening. Figure 6.23 shows the dependence of the draw ratio in the individual craze on the rate of widening of the craze [70]. It may be seen that as the rate of widening of the craze increases, the orientation of the polymer in their volume becomes more precise.

Taking into account the data relating to the stage of widening of the crazes, it may be concluded that the statistical nature of the dynamics of crazing of the polymer in liquid media leads to the statistical special features in the structure of the produce polymer. It is important to note that in all stages of crazing in the AAM the polymer transforms to the highly dispersed state and this is evidently accompanied by the total growth of its interfacial surface and complicated structural rearrangement determined by the transport of material from the bulk of the polymer to its surface.

6.6. Main factors determining the dynamics of crazing of the polymer in the AAM

Thus, the crazing of the polymer in the AAM is a complicated statistical process taking place simultaneously in a large number of local areas (crazes). We determine which properties of the AAM determine the dynamics of the transfer of the polymer to the oriented state during its crazing in the liquid medium. Figure 6.24 shows the growth rate distribution of the crazes, produced in polyethylene terephthalate during its deformation in the conditions of constant load in the medium for aliphatic alcohols which are members of the same homologous series. It can be seen that, regardless of the constant load, the distribution curves regularly change. Evidently, in

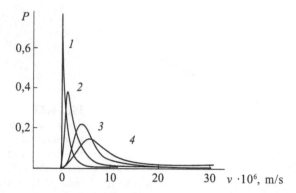

Fig. 6.24. The growth rate distribution curves of the crazes in tensile loading of polyethylene terephthalate samples under the effect of a constant load of 17.9 MPa in ethanol (1), *n*-propanol (2), *n*-butanol (3) and *n*-hexanol (4).

the investigated case, all the changes in the growth rate distribution are determined exclusively by the nature of the AAM because all other conditions of deformation were the same. The evolution of the distributions in transition from one member of the homologous series of alcohols to another shows that with the reduction of the molecular mass of the alcohols growth rate distribution curves rapidly widen, and the value of the most probable growth rate of the crazes increases.

It may be assumed that the changes in the growth rate distribution of the crazes are associated in particular with the transport properties of the liquid media. The point is that the development of the crazes can take place only if the AAM is supplied at the right time and in a sufficient amount in the areas of the orientation transformation of the polymer (the tip of the craze).

Figure 6.25 shows the dependence of the most probable growth rate of the crazes on their viscosity in deformation in the homologous series of the aliphatic alcohols. In may be seen that these values are inter-connected. As the viscosity of the medium in which it is deformed increases, the average growth rate of the crazes decreases and vice versa. This assumption is probably accurate because it may be seen that the increase of the deformation rate of the polymer in the AAM may lead to complete suppression of crazing [73, 74]. Evidently, this effect is detected in cases in which the deformation rate of the polymer becomes so high that the active liquid, as a result of the reduce transport rate, does not 'manage' to penetrate efficiently into the areas of active deformation of the polymer (the tips of the crazes). As a result, the crazing process in the AAM develops under

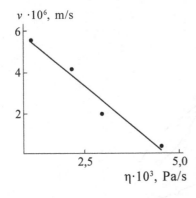

$v \cdot 10^6$, m/s

2,5   5,0
$\eta \cdot 10^3$, Pa/s

Fig. 6.25. Dependence of the most probable growth rate of the crazes in polyethylene terephthalate, growing under the effect of the constant load of 17.9 MPa in the same liquid media as in Fig. 6.16, on their viscosity.

high tensile loading rates by the same mechanism as in the area, i.e., by the development of shear bands and a monolithic neck.

The following question should be answered: are there any other properties of liquid media capable of influencing the growth rate of the crazes? To answer this question, in [75] the AAM was represented by aqueous solutions of aliphatic alcohols. The point is that these solutions in the range of low alcohol concentration have small and similar viscosity values, i.e., the transport properties of the liquid in these media are approximately the same. Figure 6.26 shows the dependences of the most probable rates of growth of the crazes in polyethylene terephthalate during tensile loading in the aqueous solutions of the aliphatic alcohols. It may be seen that the effect of the AAM is very strong also in this case. This effect is expressed in fulfilling the Duclaux–Traube adsorption rule in the range of low alcohol concentrations. The special features of this rule were discussed in greater detail previously, when examining the stage of initiation of the crazes. Here, it should only be mentioned that the fulfilment of this rule indicates that the interfacial surface energy of the investigated system decreases. In turn, the decrease of the interfacial surface energy in this case is evidently associated with

$v \cdot 10^6$, m/s

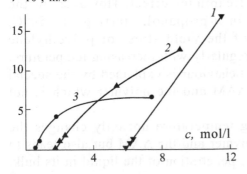

c, mol/l

Fig. 6.26. Dependence of the most probable growth rate of the crazes in polyethylene terephthalate in tensile loading under the effect of a constant load of 20 MPa in the aqueous solutions of aliphatic alcohols on their concentration.

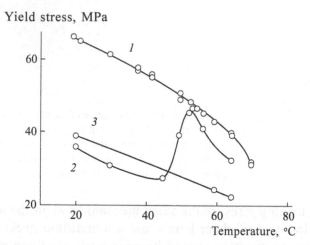

Fig. 6.27. Temperature dependences of the yield stress of polyethylene terephthalate in its deformation in air (1), in *n*-propanol (2) and *n*-octane (3) [76].

the adsorption of the surface-active substance at the phase boundary. At first sight, we are concerned here with some contradiction: the thermodynamic equilibrium factor (reduction of the interfacial surface energy) has an obvious effect on the purely kinetic activation process (the growth rate of the crazes).

The special features of the Rehbinder effect were detected in investigating the effect of temperature on the Rehbinder effect in polymers [76]. Figure 6.27 shows the temperature dependence of the yield stress of polyethylene terephthalate in its deformation in air (curve 1) and liquid AAM (curves 2 and 3). In the range of relatively low temperatures (to ~50°C) all the three dependences decrease regularly with increasing temperature in complete agreement with the Lazurkin–Eyring relationship [66, 67]. This temperature range is characterised by the effect of the AAM manifested in the displacement of these relationships into the range of low stresses and is a consequence of the Rehbinder effect. However, in the deformation of the polymer in *n*-propanol, starting at ~50°C, the temperature dependence of the yield stress of polyethylene terephthalate starts to increase regularly with increasing temperature. Such an anomalous mechanical behaviour is explained by the special features of interaction of the AAM and the polymer which is not detected in other solids.

The point is that increasing temperature not only changes the adsorption interaction of the polymer and the AAM but also leads to the swelling of the polymer, i.e., penetration of the liquid in its bulk.

In the case of a liquid with a low affinity for the polymer (*n*-octane), the increase of temperature has almost no effect on the nature of its interaction with the polymer. As a result, the reduction of the yield stress is detected in the entire investigated temperature range. At a relatively high affinity of the polymer and the AAS (*n*-propanol), the increase of temperature results in extensive volume swelling of the polymer. Consequently, the surface layer softens and transforms to the rubber-like state. This transformation prevents the formation and development of crazes because the crazing is possible only in the glassy polymer. The consequence of this rearrangement of the polymer is the increase of the yield stress of the polymer which was detected in the experiments (Fig. 6.27).

Summing up, it may be concluded that the stage of growth of the crazes and, consequently, the mechanical response of the polymer, deformed in the AAM, are controlled by at least two factors: the kinetic factor, associated with the access of the active liquid to the zone of inelastic deformation of the polymer (the tip of the craze) and the thermodynamic factor, determined by the polymer–liquid medium inter-molecular interaction.

6.7. The multiplicity factor of the number of areas of localized plastic deformation

Thus, we have examined the main factors influencing the mechanical response of the polymer, deformed in the medium capable of decreasing its surface tension. The main factors of such type can be divided into two components: polymer–AAM adsorption interaction, and the kinetic factor determined by the timely supply of the active liquid to the areas of local deformation of the polymer. In fact, these factors also play the controlling role in the case of manifestation of the Rehbinder effect in low molecular solids [2].

There is another factor, capable of exerting a strong effect on the mechanical response typical only of the polymers. The point is that the deformed polymer may show the initiation of different numbers of the crazes, depending on the deformation conditions. As shown previously, the number of the nucleated crazes is influenced by the activity of the liquid medium [66, 77] and the level of the stress which in turn is determined by the effect of tensile loading of the polymer, and by a number of other factors [78].

The examined mechanism of the effect of the tensile loading rate of the polymer and a number of nucleated crazes (Fig. 6.28).

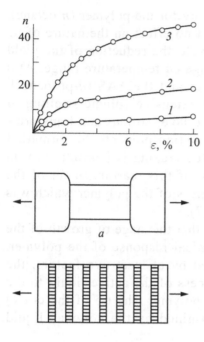

Fig. 6.28. Dependence of the number of crazes per 1 mm of the length of a polyethylene terephthalate sample during its tensile loading in ethanol with a rate of $7.67 \cdot 10^{-6}$ (1), $3.1 \cdot 10^{-5}$ (2) and $1.58 \cdot 10^{-4}$ m/s (3) [79].

Fig. 6.29. Diagram of the process of structural rearrangement of the polymer deformed in air with the formation of a neck (*a*) and in the AAM by the crazing mechanism (*b*) [79].

Figure 6.28 shows that changing, in particular, the tensile loading rate of the polymer in the AAM it is possible to regulate in a wide range the number of nucleated crazes. In turn, this means that in the case of crazing it is possible to regulate the number of the areas of local transition of the polymer in the oriented state in a wide range. Figure 6.29 shows the schematic representation of the transition of the polymer to the oriented state in conventional transition of the polymer to the neck (*a*) and in crazing (*b*). It may also be seen that in the first case the transition of the polymer to the oriented state takes place in no more than two areas and the interface between the neck and the undeformed part of the polymer. In the second case, this transition takes place in a large number of zones (at the interface between each individual craze and the non-oriented part of the polymer), and the number of these zones can be easily regulated. This circumstance is of considerable importance for the orientation process as a whole.

Figure 6.30 shows the dependence of the stress of stationary develop and of deformation (the plateau region in Fig. 6.12) on the rate of preliminary tensile loading of polyethylene terephthalate in the AAM by 10%. As indicated by the figure, the different numbers of crazes nucleated in the polymer in this case. This means that the

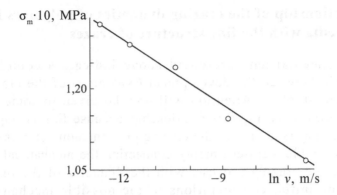

Fig. 6.30. Dependence of the stress of stationary development of deformation of polyethylene terephthalate on its tensile loading in ethanol at a rate of $3 \cdot 10^{-6}$ m/s on the rate of preliminary tensile loading in the same medium by 10%.

transition of the polymer to the oriented state takes place in each individual case in different numbers of the local zones. It may also be seen that as the number of the zones of transition (crazes) to the oriented state in the polymer increases, the level of the stress at which this process takes place decreases.

The point is that, as shown previously, crazing is a thermally activated processes of plastic deformation, governed by the Lazurkin equation:

$$v = k e^{\alpha \sigma / RT} \tag{6.3}$$

where v is the localized rate of transition of the polymer to the oriented state through the interface between the neck and the non-oriented part of the polymer, σ is the stress maintaining the rate of transition of the polymer to the oriented state, k is a constant, R is the universal gas constant, and T is absolute temperature. Evidently, with other conditions being equal, the rate of localized transition through the phase boundary between the neck and the non-oriented part of the deformed polymer will be m times smaller in the case in which the number of the zones of this transition increases n times. In other words, the rate of transition of the polymer to the oriented state during crazing will be n times smaller in each of these zones in comparison with tensile loading to the neck with the same strain rate.

6.8. Relationship of the crazing dynamics of polymers in liquid media with the fine structure of crazes

The experimental data presented above indicate a complicated relationship between the development (widening) of the craze and its internal structure. Attention will now be given to some of the current views regarding this relationship because this is important for general understanding of the crazing phenomenon. At present, we have the unstable meniscus theory connecting the mechanical stress, sustaining the crazing process, with the diameter of the resultant fibrils. The initial considerations of the possible mechanism of formation of the fibrils in the structure of the growing craze were published in the studies by Argon and Salama [50, 51]. According to their considerations, the formation of fibrillar aggregates of the deformed polymer takes place as a result of the instability of the meniscus [49]. This phenomenon is detected during expansion of the layer of the liquid, for example, in a narrow gap between the two glass plates moving away from each other. Similar concepts were used as a basis of the model of widening of the crazes by Kramer [52]. When examining the deformation of the polymer in a narrow zone between the bulk and deformation-softened polymers, the latter breaks up into a system of finger-shaped menisci (fibrils).

An important advantage of this theory is the fact that it determines the quantitative relationship between a number of quantities, determining the experiments (see equation (6.1)).

Equation (6.1) can be easily verified because both its right and left parts contain quantities independently measured in the experiment. Equation (6.1) shows that the product σd = const, i.e., the product of the stress by the fibril diameter is a constant value. This relationship was subjected to a large number of experimental verifications because reliable methods of determining the mean diameter of the fibrils by small angle x-ray scattering [80, 81], small angle electron diffraction [82, 83], and penetration of the liquid under the effect of a pressure gradient [84], have been developed.

The fulfilment of the rule σd = const was experimentally detected for the first time in [85] where the method of small angle x-ray scattering was used to determine the diameter of fibrils, formed during dry crazing of PC and PMMA. The product σd was constant for a wide range of temperatures and tensile loading rate. Later, this was stated for greatly different polymers during their crazing both in air [86, 87] and in the AAM [53–55].

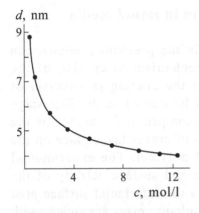

Fig. 6.31. Dependence of the average diameter of the fibrils in the crazes in polyethylene terephthalate, deformed in the aqueous solutions of *n*-propanol, on its concentration [55].

In the context of the discussed topics, it is important to note that the crazing of the polymer in the AAM makes it possible to evaluate an important characteristic such as the interfacial surface energy of the polymer (see equation (6.1)). Evaluation for polyethylene terephthalate and polycarbonate was carried out in [54] and a rational value of the interfacial surface energy of the polymer was obtained. The effect of the AAM on this value was also evaluated. In [55] the authors determined the important relationship between the fibril diameter in the structure of the craze and the surface activity of the liquid media.

Figure 6.31 shows the dependence of the average diameter of the fibrils in the crazes in polyethylene terephthalate, determined by the method of small angle x-ray scattering, on the concentration of the water solution of propyl alcohol in which the polymer was deformed [55]. Figure 6.31 shows that the nature of the AAM has the strongest effect on the diameter of the fibrils in the craze structure. It can easily be seen that the average diameter of the fibrils in the crazes in polyethylene terephthalate during its tensile loading in the water–alcohol mixture changes in the same manner as the most probable growth rate of the craze in the same conditions (compare Figs. 6.26 and 6.31). In other words, the average fibril diameter changes with the change of the composition of the alcohol–water mixture in accordance with the Duclaux–Traube rule. It should be mentioned that with increasing alcohol concentration in water the average fibril diameter is more than halved.

Thus, in this section it is shown that the parameters of the internal structure of the crazes depend on the stress, sustaining the process of this type of inelastic deformation of the polymers and, which is important, on the nature of the AAM.

6.9. Crazing mechanism of polymers in liquid media

The experimental material, presented in the previous sections, can be used for proposing the general mechanism of crazing of the polymer in the AAM. Examination of the crazing mechanism of the polymer in the liquid medium will be started in the first stage – initiation of crazes. One general assumptions be made for the adequate study of this stage: nucleation of crazes takes place on the surface microheterogeneities of a real polymer. The experimental data show that the effect of the stress and surface activity of the liquid medium or its capacity to reduce the interfacial surface area of the polymer on the number of the resultant crazes are superposed. All this confirms that the processes of nucleation of the crazes are regulated by the well-known Griffith thermodynamic criterion [88].

This criterion was developed for describing the process of failure of elastic–brittle solids. It is assumed that the work of fracture is transformed in this case to the surface energy of the resultant fracture surfaces. After all, the application of this criterion for the analysis of crazing of the polymer in the liquid medium does not mean that this type of deformation of the polymer is identical with elastic-brittle fracture. As mentioned several times previously, crazing of the polymer in the liquid medium is one of the types of inelastic plastic deformation of the solid polymers. The fulfilment of the Griffith criterion on the indicates that the surface microheterogeneities, present in the actual polymer, lose their stability and become the centres of localized plastic deformation (crazes) in accordance with the Griffith criterion. It may be seen that the loss of stability as a phenomenon is important for the stage of initiation of the centres of plastic deformation of the polymer (crazes). In the analytical form, this criterion can be written in the following form

$$\sigma^2 \sim (E\gamma_{1,2})/r \qquad\qquad (6.4)$$

here σ is the stress; E is the modulus of the polymer, $\gamma_{1,2}$ is the surface energy of the polymer–liquid medium interface, and r is the radius of surface microheterogeneity (defect) which becomes unstable at stress σ and surface energy $\gamma_{1,2}$.

This equation does not include in the explicit form the number of the nucleated crazes. Nevertheless, this number is present here in the implicit form because the radius of the critical defect, which becomes unstable, is a variable quantity. Actually, as the radius of the defect decreases, the degree of stress concentration by the defect decreases and vice versa. This means that at the given external stress and the given surface energy of the polymer the process of its inelastic deformation includes the number of the defects (nucleated crazes) for which the value σ satisfies the Griffith conditions. In other words, as the stress applied to the polymer increases, the number of defects (less dangerous defects) included in the process of initiation of new crazes decreases.

The considerations regarding the local–critical nature of the process of initiation of the crazes have been used to develop a new quantitative theory of this phenomenon [89, 90]. In these studies, using this approach, the authors determined the dependence of the number of nucleated crazes on the loading rate of the polymer in the AAM and the geometrical special features of the deformed polymer sample and obtained satisfactory agreement between theory and experiment. Having the experimental dependence of the number of nucleated crazes under loading rate and geometrical characteristics of the specimen, the local–critical approach was used to solve the inverse problem: determination of the size distribution of surface microdefects [89].

Figure 6.32 shows the size distribution curve of surface defects for a polyethylene terephthalate film deformed in the AAM. It can be seen that the given AAM (n-propanol) transforms to the crazes in the given temperature–force conditions the relatively narrow size distribution of the surface defects. The most 'popular' defects have the size of hundreds or thousands of angströms. The resultant values correspond to the available experimental data [91] obtained in the direct evaluation of the dimensions of the surface defects in actual polymers.

Thus, in the loading of the polymer in the AAM the set of the surface microdefects, differing in their capacity to concentrate stresses (the degree of 'danger'), initiates a set of nucleation crazes. In analysis of the process of growth of the crazes it is necessary to make another assumption which, nevertheless, is completely evident: the driving force for growth of each craze is the *local* stress in front of the tip of the craze which is proportional but not equal to the applied external stresses. The formation of this local stress will be investigated. As mentioned previously, each surface defects in the actual polymer is capable of concentrating the external stress around it to different degrees. This means that a set of local stresses forms at the given applied stress on the microdefects. The nucleation of

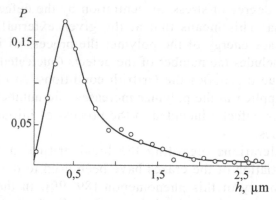

Fig. 6.32. The size distribution curve of surface microdefects, initiating crazes during tensile loading of the sample in ethanol at room temperature.

the crazes in this set of the microdefects takes place evidently of different local stresses. As shown previously, the different levels of the stress form different structures of the nucleation crazes. In this case, the structure means primarily the volume fraction of the fibrils in the craze V_f, which is inversely proportional to the draw ratio of the craze material, according to equation (6.2).

This is the end of the initiation stage and we return to the study of the second crazing stage – the growth of crazes. The main topics for understanding the physical mechanisms of this stage are: why does each craze has a constant growth rate ? why does the family of crazes form some distribution of these rates ? what is the mechanism of the effect of external factors on these distributions and why, in particular, is the Duclaux–Traube rule fulfilled in crazing of the polymer in the solutions of aliphatic alcohols ?.

To answer these questions, it is necessary to return to some special features of the initiation stage. It is evident that irrespective of the type of loading, the loading of the polymer does not develop instantaneously and this means that with increase of the stress the most dangerous defects are the first to 'operate' and this is followed by other less dangerous defects. This is the reason why the dependences of the length of the individual crazes on the loading time form a relatively well-developed 'fan' i.e., do not intersect each other (Fig. 6.16). Naturally, the highest growth rate is shown by the crazes which nucleated first at the most dangerous defects. It is evident that the only driving force for growth (and widening) of the craze is the mechanical stress.

Here, it is important to stress that the nucleation crazes have a different structure because they nucleated under the effect of different stresses. The latter is evident because it is not doubted that the actual polymer contains a set of surface defects differing in the degree of danger, i.e., the capacity to concentrate the stress. It is clear that when the nucleated craze starts to grow it is this nucleation structure that forms the local stress in front of the tip of each craze. The other question is not clear. How can the nucleated craze 'remember' the nucleation structure which the craze 'inherited' from its defect? As indicated by the experimental data [57], the growing craze maintains its 'intrinsic' growth rate even in the case in which its tip has moved away from the initial defect to a very large distance. As mentioned previously, the craze propagates with an almost constant rate over many millimetres from the nucleation defect with the size of tens or hundreds of a fraction of the micron.

We believe that the reasons for this excellent memory can be described as follows. The local stress at the stress raiser is responsible for the nucleation of the craze and not its growth. The growth of the craze is determined by the completely different local stress. The stress forms around the tip of the already nucleated craze and the stress determines its growth. As mentioned previously, this stress is unambiguously connected with its internal structure which also determines the growth rate of the craze and maintains this rate constant. The strong relationship between the internal structure of the craze and the growth rate enabled Kambour to explain the constant value of this rate as follows [45]: 'the movement of the tip of the craze from one position to another is accompanied by a similar displacement of the field of stresses surrounding the tip of the craze, without any change of the stress concentration'.

Since the structure of each craze differs from all other structures because of the statistical nature of the defects which generate the structures, the growth rates of the crazes, dictated by the local stresses, are completely different. This is also the reason for the existence of the growth rate distribution of the crazes. It should be mentioned that in the stage of widening of the crazes the previously mentioned relationships between the growth rate of the craze and its internal structure was determined in direct microscopic studies [70]. This cannot be realised in the growth stage of the crazes. The point is that in this stage of crazing the width of the growing craze, especially in the region of its tip, is so small (fractions of a micron) that the change of the orientation of the polymer by the microscopic method becomes an almost impossible task. However, this relationship can also be determined by calculations [92]. An example of these calculations is described in [70]; the distribution of the stress along the boundaries of the crazes with different volume fractions of the fibrils is shown in Fig. 6.33. It may be clearly seen that the local stress in the tip of the craze increases with the reduction of the volume fraction of the fibrils. As mentioned previously, this local stress is the only driving force of the growth of each craze. This result is fully confirmed by the experimental data obtained in direct microscopic studies of the stage of widening of the crazes, and also the previously mentioned considerations regarding the mechanism of the observed phenomena.

These results can also be used to explain the dependence of the growth rate of the crazes on the adsorption activity of the liquid used as the AAM. As shown previously (Fig. 6.26), the crazing of PET in aqueous solutions of aliphatic alcohols is governed by the Duclaux–Traube adsorption rule. It would appear that in this case we have a contradiction because it is not clear why the thermodynamically equilibrium factor (reduction of the interfacial surface energy) has an evident effect on the purely kinetic activation process (the growth rate of crazes). The structural approach to the crazing of the polymer in the liquid medium explains this effect. The effect of the aqueous solutions on the growth rate of the crazes does not operate directly and operates through the formation of the internal structure of the crazes. As mentioned previously, the adsorption activity of the liquid

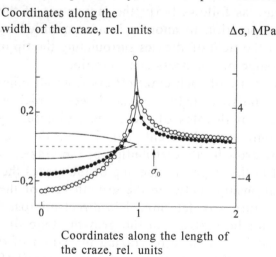

Coordinates along the length of
the craze, rel. units

Fig. 6.33. Stress distribution along the boundaries of the craze with different volume fractions of the polymer $V_f = 0.2$ (open points) and 0.5 (solid points).

forms the corresponding internal structure of the craze which, in turn, forms the local stress ahead of the craze tip which also determines the growth rate.

Thus, the crazing in the liquid media is one of the mechanisms of manifestation of the Rehbinder effect in polymers. This is a complicated multistage process. Its first stage – initiation of the crazes – is associated with the microscopic surface defectiveness of the actual polymer and is controlled by the Griffith criterion. The growth and widening of the crazes are the thermally activated processes of plastic deformation of the polymer. Their statistical nature is determined by the close relationship with the initiation stage whose statistical nature in turn is determined by the surface micro-defectiveness of the actual polymer. It is important to note that all the main stages of crazing in the liquid medium are in fact a critical phenomenon in which the transfer of the amorphous polymer to the highly ordered state is associated with reaching the critical conditions by the polymer transferring it to an unstable state.

6.10. Delocalized crazing of the polymers in liquid media

Thus, the crazing of the glassy polymer in the AAM is in fact a surface phenomenon characterised by the rapid development of the interfacial surface of the polymer. The main distinguishing feature of the deformation of this type is the accompanying development of the porosity. The second special feature of crazing is the fact that the porosity is localized in special discrete zones – crazes which form and develop in polymers under the effect of the mechanical stress. This type of crazing, referred to as classic and was discussed in the previous section, has been sufficiently studied and described in a large number of publications, many of which have been generalised in the excellent reviews [43, 45, 52, 93, 94].

At the same time, it is well-known that the classic crazing is not the only type of deformation of the polymers which is accompanied by the development of porosity. In the literature, there are a large number of data on the so-called hard elastic materials whose deformation is also accompanied by the development of porosity and specific surface [95–97]. This type of deformation is typical of crystalline polymers, characterised by the unique lamellar strictly asymmetric structure. The development of porosity in these materials is observed only if their tensile loading takes place in the direction perpendicular to the major axis of the lamellae. An important special

feature of the hard elastic materials is the development of porosity
not only in individual areas but throughout the entire volume
uniformly. This type of deformation of the polymers was discussed
in greater detail in chapter 5.

In addition to this, in recent years it has been shown that the
structural–mechanical behaviour, similar to the hard elastic materials,
can also be shown by many crystallising polymers (polyolefins,
polyethylene terephthalate, polyamides, polyvinyl chloride, etc),
which do not have any special asymmetric crystalline structure [98,
99]. This type of crazing, detected in deformation of the crystalline
polymers in active liquid media, is referred to as delocalized [41]
or intercrystalline [100] crazing. We examine in greater detail the
main special features of delocalized crazing and its identical features
and similarity as regards the classic crazing, discussed previously.

As mentioned previously, the delocalized crazing is detected in
the deformation of crystalline polymers in the active liquid media.
A special feature of the crystalline polymers is their capacity for
limited swelling in solvents. In fact, the swollen crystalline polymers
are identical with the cross-linked polymers in which the cross-
linked members are crystallites. Because of this special feature the
swollen samples of the crystalline polymers retain high strength
and elasticity so that they can be processed by cold drawing in
the presence of plastification (swelling) media to high degrees of
elongation. Thus, if the amorphous glassy polymers can be deformed
by the classic crazing mechanism only in the adsorption-active liquid
media (AAM), i.e., the liquid is not capable of causing any extensive
swelling of the non-stress polymer, the crystalline polymers can also
be deformed in the liquid media causing extensive swelling. The
latter circumstance results in large differences in the manifestation
of the Rehbinder effect in the crystalline polymers.

It is well-known that the liquids, causing volume swelling of the
polymer, are dissolved in its amorphous regions without affecting the
crystallites. Consequently, even prior to the start of deformation the
volume of the crystalline polymer contains a low-molecular liquid
and this is the main difference between this case and the deformation
of the glassy polymer in the AAM by the classic crazing mechanism.
In particular, the investigation of the processes of deformation of
the crystalline polymers in the liquids, capable of causing their
true volume swelling, made it possible to determine a number of
interesting relationships governing this process [98, 99, 101–103].
Firstly, it should be mentioned that the deformation of the crystalline

polymers in partially compatible liquids is accompanied by the rapid development of porosity. This process may be characterised by measuring the amount of the liquid penetrating into the volume of the deformed polymer. Figure 6.34 shows the relative increase of the volume of high-density polyethylene (HDPE) (1) and polypropylene (2) in the process of drawing in the swelling liquid (n-heptane) at room temperature. For comparison, the same figure shows the corresponding dependence of the porosity of polyethylene terephthalate, deformed by the mechanism of classic crazing in the AAM (n-propanol).

It may be seen that the cold drawing of the polymers in the liquid compatible with these polymers is accompanied by the efficient penetration of the liquid into the porous structure of the polymer. Porosity rapidly grows in the initial stages of tensile loading and the growth is then gradually interrupted, starting at approximately 200% elongation. The same figure shows the dependence of porosity on the draw ratio for the polymer, deformed by the classic crazing mechanism (polyethylene terephthalate in propanol). The nature of this dependence is completely different. The dependence of porosity on the degree of drawing in this case passes through a maximum. This is associated with the process of collapse of the porous structure of the polymer. The mechanism of this process will be examined in greater detail later [104, 105] and here we only mention that the process of collapse of the porous structure is not characteristic of delocalized crazing or, in any case, is far less intensive.

Detailed microscopic study of the polymer samples, deformed by the delocalized crazing mechanism, shows that their deformation is not accompanied by the nucleation of individual crazes with their stage by stage development, characteristic of classic crazing (see previous sections). Figure 6.35 shows the scanning electron micrographs of the low-temperature cleaves of the crazed specimens of HDPE and polypropylene, prepared by cold drawing in n-heptane to a strain of 200%. It may be seen that the deformation of the crystalline polymers in the swelling liquid is not accompanied by the formation of classic crazes, and the developing porosity is distributed uniformly in the entire volume of the samples. In addition, the data cannot be used to identify the individual pores in the structure of the polymer and investigate their morphology. It appears that Fig. 6.35 shows the micrographs of the non-porous materials. However, as mentioned previously, the porosity in these materials developed quite rapidly (Fig. 6.34, curves 1 and 2). In particular, this is also indicated

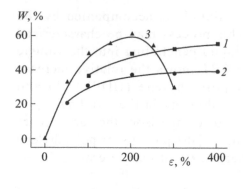

Fig. 6.34. Porosity W of the polymer in dependence on the draw ratio ε in deformation of the films of HDPE (1) and polypropylene and (2) in the swelling liquid (n-heptane) and polyethylene terephthalate (3) in the AAM (n-propanol).

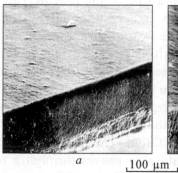

Fig. 6.35. Scanning electron micrographs of low-temperature cleaves of the crazes specimens of HDPE (a) and polypropylene (b), obtained by cold drawing to 200% in n-heptane.

by the results of small angle x-ray scattering (SAXS). The typical patterns of SAXS are presented in Fig. 6.36. It may be seen that in both classic crazing (Fig. 6.36 a) and in delocalized crazing (Fig. 6.36 b) porosity forms in the polymers. In classic crazing the x-ray diffraction diagrams show two mutually perpendicular reflections, whose origin is determined by the fibrillar–porous structure of the polymer and has been studied several times in the literature [41, 44]. In delocalized crazing, the diffusion scattering takes place elongated in the meditional direction with a stroke at the equator or without it (Fig. 6.36 b). Unfortunately, these data are evidently insufficient for the complete description of the morphology of the porous structure of the craze polymers, produced by the delocalized crazing mechanism. Nevertheless, the experimental results make it possible to assume that the formation and development of porosity in the crystalline polymers during their deformation in the presence of compatible (swelling) liquid media differs greatly from the corresponding process in the

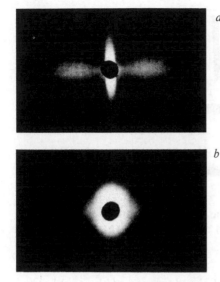

a

b

Fig. 6.36. Typical small angle x-ray diffraction diagrams of the crazed specimens, produced by the classic crazing mechanism (deformation of the films of polyethylene terephthalate in ethyl alcohol) (*a*) and delocalized crazing (deformation of the polypropylene films in *n*-heptane) (*b*).

case of classic crazing. Thus, another difference of the delocalized crazing in comparison with classic crazing is the delocalization of porosity throughout the entire volume of the specimen which has been reflected in the name of this phenomenon [99].

The data obtained in optical microscopy can be used for a more detailed description and definition of the previously mentioned differences [102]. Figure 6.37 shows the thin sections of the crazed specimens of PA-6, produced by classic (Fig. 6.37 *a*) and delocalized crazing (Fig. 6.37 *b*) in the presence of liquid media, containing a contrasting dye. It may be seen that in the case of classic crazing, the distribution of the contrasting dye is localized in the crazes in the form of open and continuous channels. Evidently, in this case, the transfer of the liquid is of the distinctive phase nature and the transport of the liquid together with the dissolved dye to the volume of the polymer takes place by the viscous flow of the dye through the porous structure of the growing crazes. In delocalized crazing, the penetration of the liquid component of the polymer takes place by a completely different mechanism. The thin sections of the crazed specimens show clearly the front of the contrasting dye, uniformly distributed in the deformed polymer.

The evident difference in the mechanism of penetration of the liquid into the bulk of the polymer during its deformation by the mechanism of classic and delocalized crazing is reflected in the mechanical response of the polymer during its uniaxial loading.

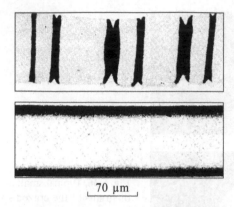

_70 µm_

Fig. 6.37. Typical optical micrographs of the thin sections of the craze polymers, produced by the mechanisms of classic (_a_) and delocalized (_b_) crazing in PA-6 samples [102].

Figure 6.38 shows the dependences of the decrease of the yield stress in drawing the samples of PA-6 in the vapours of active liquids on the relative vapour pressure [102]. In the case of delocalized crazing (Fig. 6.38, curve 2), the molecular diffusion of the active component takes place in the entire range of the values of the relative vapour pressure and this has a strong effect on the mechanical response of the polymer. In classic crazing, there is a certain 'safe' range of the values of the relative vapour pressure at which there are no changes in the limit of plasticity. This safe range forms due to the fact that the mechanism of classic crazing operates only as a result of the presence of the liquid phase of the active medium and the start of the condensation process is determined by the critical dimension of the surface defects and by the relative vapour pressure. The condensation of the liquid on the surface microdefects results in the nucleation of classic crazes, and the transport of the liquid to the areas of active deformation of the polymer takes place in this case by viscous flow of the liquid in the porous structure of the crazes. In other words, the occurrence of the mechanism of classic crazing indicates the active transport of the liquid by its viscous flow, whereas the delocalized crazing is accompanied by the volume penetration of the liquid component into the areas of active deformation of the polymer. It should be mentioned that the special features of the mechanism of transport of the active liquid in the case of classic and delocalized crazing are determined only by the type of crazing and do not depend on the nature of the polymer [41].

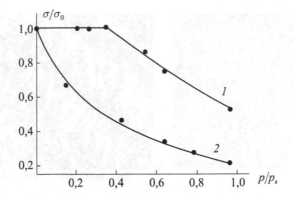

Fig. 6.38. Dependence of the relative decrease of the yield stress of the polymer on the relative vapour pressure p/p_s for PA-6 samples, deformed by classic (1) and delocalized (2) crazing [102].

The morphology of the polymers with delocalized crazing was not clear until recently and was based on a number of assumptions made using the results of x-ray diffraction analysis and the permeability of liquids under the effect of a pressure gradient [106–108]. These data were used to assume that the deformation of crystalline polymers in the AAM takes place mainly in the amorphous regions between the lamellae with the transfer of the polymer to the fibrillised state and the formation of cavities between the fibrils. Nevertheless, these experimental methods can be interpreted on the basis of some modelling speculative considerations.

Detailed information on the structure of the polymer subjected to delocalized crazing can be obtained by the method of atomic force microscopy in which the objects can be studied directly in the liquid media. Figure 6.39 *a* shows the micrograph of the surface of the initial material, produced by the method of scanning electron microscopy in this investigation. The extruded HDPE, produced by extrusion with blowing, is a close-packed aggregate of crystalline lamellae, oriented normal to the direction of the extrusion axis [109].

Deformation of this polymer directly in the active liquid was investigated by atomic force microscopy [110]. The results of these investigations are presented in Fig. 6.39 *b*. The surface of the deformed HDPE shows clearly the lamellae (1), oriented mostly in the direction normal to the drawing axis. The thickness of the lamellae, determined on the profiles of the cross-section, is approximately 50 nm. Fibrils (2) oriented strictly along the drawing axis can be seen between the lamellae. The length of the fibrils or

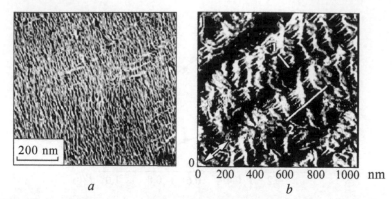

a   *b*

Fig. 6.39. Scanning electron micrographs of an extruded sample of HDPE [109] (*a*) and the atomic force microscope image of the same sample produced directly in the AAM after tensile loading to 200% (*b*) [110].

the distance between the adjacent lamellae along the drawing axis varies in the range 50–150 nm. The image also shows clearly cavities between the fibrils. On the basis of these data it can be concluded that in tensile loading of the extruded HDPE in the active liquid the crystalline lamellae move away from each other and fibrillar aggregates of the macromolecules form between them. It can be seen that the structure of this type resembles the structure of the materials of the hard elastic type, examined in detail in the previous section [95–97]. It was also shown their that the mechanical behaviour of the polymer, subjected to delocalized crazing, is also very similar to the corresponding behaviour of the hard elastic polymers which evidently is not accidental. It should be mentioned that the development of porosity in HDPE, as a result of the mechanism of delocalized crazing in [110], took place using the liquid which did not cause volume swelling of the unstressed polymer (water–alcohol mixture).

The possibility of occurrence of delocalized crazing in the crystalline polymer in the non-swelling active media will be examined later, and here we analyse the nature of interaction of the polymer and the swelling liquid medium. As already mentioned previously, of principle importance is the fact that in the case of delocalized crazing the plasticizing liquid medium is situated in the amorphous regions of the crystalline polymer prior to the start of deformation. Subsequent drawing of this polymer is accompanied by the penetration of an additional amount of the low-molecular component into the bulk of the polymer (Fig. 6.34). In this case, it is not important which plasticising component is present in the polymer prior to the start of deformation. This may be the same plasticising liquid in which subsequent deformation is carried out, or any other plasticising medium. In any case, if the amorphous regions are subjected to preliminary plastification, the drawing process will take place by the mechanism of delocalized crazing. In addition to this, if the subsequent drawing of the crystalline polymer subjected to preliminary plastification takes place in the non-swelling liquid, compatible with the liquid, already present in the amorphous regions of the polymer, the deformation process developed by the mechanism of delocalized crazing [111, 112]. Consequently, the presence of the plastification component in the amorphous regions of the crystallising polymer results in the deformation of the polymer by the mechanism of delocalized crazing.

There is another factor supporting the formation of the polymer by the mechanism of delocalized crazing. It is the crystalline structure of the initial polymer. It appears that as the degree of crystallinity of the initial polymer increases, the efficiency of delocalized crazing also increases. A suitable example illustrating this factor may be the difference in the behaviour of high and low density polyethylene. Evidently, the equilibrium degree of swelling, with other conditions being equal, is higher in low-density polyethylene (LDPE) than in HDPE. It appears that this circumstance should result in more extensive crazing in LDPE than in HDPE. However, regardless of this, in drawing in the plasticising liquids, in contrast to HDPE, this polymer does not show any extensive development of porosity. Evidently, to retain the porosity, characteristic of deformation by the mechanism of delocalized crazing, it is essential to have a very strong stable crystalline frame capable of resisting the effect of surface forces 'trying' to reduce the interfacial surface. Such a frame can form only at a relatively high degree of crystallinity of the initial polymer.

The effect of the crystalline structure of the initial polymer on the nature of its deformation in active liquid media can be strong and complicated. Such a complicated effect was also detected for a polymer – isotactic polypropylene (PP). It was shown [111–114] that the quenched polypropylene, having a low-ordered smectic crystalline structure, is deformed in the non-swelling adsorption active liquids by the classic crazing mechanism. At the same time, drawing of the quenched polymer in the swelling liquid media takes place by delocalized crazing in complete agreement with the previously described behaviour of other crystallising polymers.

It is well-known [115, 116] that the annealing of the specimens of quenched polypropylene results, depending on the temperature and annealing time, in the growth of the crystallinity of the polymer and is gradual transition to the monoclinic crystalline modification. It was shown [111–114] that the annual specimens of the polypropylene, characterised by a relatively high degree of crystallinity, show delocalized crazing not only during drawing in the swelling liquids but also in deformation in the non-swelling adsorption active media. This type of delocalized crazing is evidently associated with the special features of the structure of the annealed polypropylene. In [117] it was shown that annealing of the quenched specimens of the isotactic polypropylene results in the extensive 'loosening' of its amorphous regions up to the appearance of recorded porosity. It is possible that this special feature of the polypropylene is the reason for its deformation by the mechanism of delocalized crazing in the liquids not causing its volume swelling. Evidently, the wetting active medium penetrates by surface diffusion to the interfacial boundaries of the crystals in the volume of the polymer under mechanical loading. In subsequent stages, the penetrated AAM promotes crazing of the boundaries of the crystallised in the entire volume of the polymer at the same time and this is also the main distinguishing feature of delocalized crazing.

As mentioned previously, the transition from classic to delocalized crazing is determined by two factors: plastification of the crystalline polymer prior to the start of its deformation and/or increase of the degree of crystallinity in annealing of the quenched polymer. In the case of isotactic polypropylene annealing can gradually change is crystalline structure increasing the degree of crystallinity. In turn, this circumstance makes it possible to ensure the gradual transition of the deformation mechanism from classic crazing to delocalized crazing. Figure 6.40 shows the electron micrographs of the polypropylene samples, illustrating this transition. It can be clearly seen that the classic crazes gradually lose their characteristic features and there is a gradual transition to the structure shown in Fig. 6.35.

Summing up, it may be concluded that delocalized crazing in the liquid media is a special type of inelastic deformation of the crystallising polymers having a number of principal differences from the well examined classic crazing. These differences can be formulated as follows.

The initiation of classic crazing is determined by the defectiveness of the surface of the initial actual polymer, containing a large number of microdefects differing in the degree of danger. The nucleation of each individual craze is of the local–critical nature, and the density of the crazes depends on the deformation conditions and the perfection of the surface of the polymer. As soon as a craze is nucleated on one of the structural defects of the polymer surface, the further growth of the craze takes place in the direction normal to the direction of uniaxial drawing in a relatively homogeneous elastic medium. In the case of the crystalline polymer the elastic properties of the amorphous and crystalline components are approximately the same when the amorphous component of the polymer is below the glass transition temperature. In this case, in growth in such a medium with the craze 'does not notice' the crystalline phase of the polymer and deformation takes place by the classic crazing mechanism.

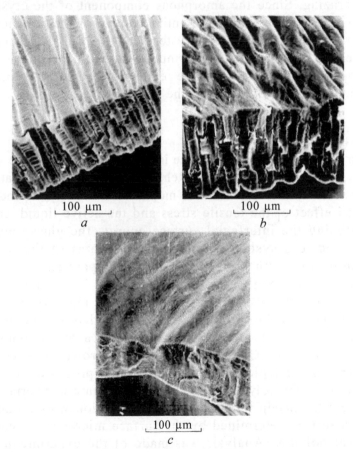

Fig. 6.40. Scanning electron micrographs of polypropylene specimens deformed in *n*-propanol by 100%: initial non-annealed polypropylene sample (*a*); polypropylene annealed at 60 (*b*) and 100°C (*c*).

The principle of increase of the differences in the properties of the amorphous and crystalline phases is the basis of the occurrence of delocalized crazing in the polypropylene during its deformation in non-swelling (absorption-active) liquids. Evidently, in the final analysis, this principle leads to a situation in which the active liquid is not required for the initiation of delocalized crazing. In this case, because of the large differences in the properties of the crystalline and amorphous phases, inelastic deformation takes place primarily by the movement of the crystals away from each other and by the growth of the micro-cavities in the loosened amorphous regions. Evidently, in this deformation there is only the stage of widening of these regions throughout the entire volume simultaneously and there no stages of nucleation and transverse growth, characteristic of the classic crazing. Since the amorphous component of the crystalline polymer is delocalized over the entire volume of the polymer (this polymer can be regarded as a system consisting of fine crystals, dispersed in the amorphous continuous matrix), the development of the porosity, nucleated in these regions, will also be delocalized throughout the entire volume of the polymer.

Conclusions

Thus, in this chapter we analyse the information regarding the special features of manifestation of the Rehbinder effect in deformation of the polymers in adsorption-active media. It is shown that under the combined effect of the tensile stress and the active liquid, capable of decreasing the interfacial surface energy, the glassy polymer is dispersed to a system of fibrillar aggregates of the oriented macromolecules with a size of 1–20 nm, separated in space by the microcavities of the same size (crazing). The crazing of the polymer in the liquid media is a complicated multistage process. The first stage is the initiation of crazes and is associated with the microscopic surface defectiveness of the actual polymer and is controlled by the Griffith criterion. The growth and widening of the crazes are typical thermally activated processes of plastic deformation of the polymer. Their statistical nature is determined by the close relationship with the stage of initiation whose statistical nature in turn is determined by the surface microdefectiveness of the actual polymer. Analysis was made of the experimental data relating to a special type of inelastic deformation of the crystallising polymers in liquid media – delocalized crazing. Characteristic special features of this type of deformation of the polymers and its principal difference from classic crazing have been detected. In contrast to classic crazing, in delocalized crazing, the porosity does not develop in the local zones (crazes) and develops throughout the entire volume of the deformed polymer simultaneously. In this case, the porosity development mechanism has many common features with the appropriate processes taking place in the hard elastic materials. As with any other manifestations of the Rehbinder effect in the crazing of the polymers in the liquid media, to understand its mechanism it is important to take into account the surface phenomena, and the deformation of the polymer in the active liquid media is in fact a complicated multistage surface phenomenon.

References

1. Rehbinder P.A., Selected Works. Surface phenomena in disperse systems. Physicochemical mechanics, Moscow, Nauka, 1979. 203 pp.
2. Goryunov Yu.V., Pertsov N.V., Summ B.D., The Rehbinder effect, Moscow, 1966.
3. Sinevich EA, Pertsov N.V., Shchukin E.D., DAN SSSR, 1971, V. 197, P. 1376.
4. Voyutsky S.S., Lectures in colloid chemistry. Moscow, Khimiya, 1976. 126 pp.
5. Kozlov P.V., Papkov S.P., Physical and chemical fundamentals of plasticizing polymers, Moscow, Khimiya, 1982.
6. Aleskerov A.G., Dissertation, Cand. Chemical Sciences, Moscow, MGU, 1979.
7. Matsushige K., Radcliffe S.V., Baer E., J. Macromol. Sci., 1975, V. B11, No. 4. P. 565–592.
8. Matsushige K., Radcliffe S.V., Baer E., J. Appl. Pol. Sci., 1976, V. 20, P. 1853.
9. Lazurkin S., Disseration, Dr. Sci. Sciences, Moscow, S.I. Vavilov Institute of Physical Problems, USSR Academy of Sciences, 1954.
10. Kuleznev V.N., Shershnev V.A., Chemistry and physics of polymers, Moscow, KolosS, 2007.
11. Brown N., Phil. Mag., 1975, V. 32, No. 5, P. 1041-1050.
12. Brown N., Fisher S., J. Polymer Sci. Polym. Phys., Ed. 1975, V. 13, No. 7, P. 1315-1331.
13. Peterlin A., Olf H.G., J. Polymer Sci. Polym. Symp., 1975, No. 50, P. 243-264.
14. Schirrer R., Polymer, 1988, V. 29, No. 9, P. 1615-1618.
15. Brown N., Metzger B., J. Polymer Sci. Polym. Phys., Ed. 1980, V. 18, No. 9, P. 1979–1992.
16. Brown N., Imai Y., J. Polymer Sci. Polym. Lett. Ed., 1975, V. 13, No. 9, P. 511-516.
17. Imai Y., Brown N., J. Polymer Sci. Polym. Phys. Ed., 1976, V 14, No. 4, P. 723-739.
18. Brown N., J. Polymer Sci. Polym. Phys., Ed. 1973, V. 11, P. 2099
19. Braught N., Brunning B.B., Scholz J.J., J. Colloid. Interface Sci., 1969, V. 13, P. 263.
20. Kardash G.G., Andrianova G.P., Bakeev N.F., Kargin V.A., DAN SSSR, 1966, V. 166, No. 5, P. 1155.
21. Kargin V.A., Andrianova G.P., Kardash G.G., Vysokomolek. Soed. A, 1967, V. 9, No. 2, S. 267.
22. Andrianova G.P., Nguyen Vinh Chii, Kargin V.A., DAN SSSR, 1970, V. 194, No. 2, P. 345.

23.   Andrianova G.P., Kargin VA, Vysokomolek. Comm. A. 1970. T. 12, No. 1. C. 3.
24.   Hubeny H., Dragunn H., Mater. Sci. and Eng., 1976, V. 24, P. 293.
25.   Olf H.G., Peterlin A., J. Polymer Sci. Polym. Phys., Ed. 1974, V. 12, P. 2209.
26.   Sakodynskii K.I., Panina L.I., Polymer sorbents for molecular chromatography, Moscow, Nauka, 1977, 165 pp.
27.   Volynskii A.L., Yarysheva L.M., Ukolova E.M., Kozlova O.V., Vagina T.M., Kechek'yan A., Kozlov P.V., Bakeev N.F., Vysokomolek. Soed., A. 1987, V. 24, No. 12, P. 2614.
28.   Cooper A.I., J. Mater. Chem., 2000, V. 10, P. 207.
29.   Kazarian S.G., Polymer Science, C, 2000, V. 42, No. 1, P. 78-101.
30.   Supercritical carbon dioxide, edited by M.F. Kemmere, T. Meyer. Wiley-VCH, 2005, P. 358.
31.   McHugh M.A., Krukonis V.J., Supercritical fluid extraction: Principles and practice, Stoneham, Butterworth-Heineman, 1993.
32.   Trofimchuk E.S., et al., Dokl. RAN, 2009, V. 428, No. 4, P. 480-483.
33.   Trofimchuk E.S., et al., Vysokomolek. Soed. A, 2011, V. 53, No. 5, P. 1820-1832.
34.   Trofimchuk E.S., et al., Dokl. RAN, 2012, V. 443, No. 3, P.326-329.
35.   Hobbs T., Lesser A. J., J. Polym. Sci. B: Polym. Phys., 1999, V. 37, P. 1881-1891.
36.   Leiner M.G., Song J., Lesser A.J., J. Polym. Sci. Part B: Polym. Phys., 2003, V. 41, P. 1375-1383.
37.   Sinevich E.A., Ogorodov R.P., Bakeev N.F., DAN SSSR, 1973, V. 212, P. 138.
38.   Sikorski J., In: Fibre structure, Ed. J.W.S. Hearle and R.H. Peters, Trans. from English, ed. N.V. Mikhailov, Moscow, Khimiya, 1969, 205 pp.
39.   Zhurkov S.N., Marikhin V.A., Romankova L.P., Slutsker A.I., Vysokomolek. Soed., 1962, V. 4, P.282.
40.   Kuksenko V.S., Orlova O.D., Slutsker A.I., Vysokomolek. Soed. A, 1973, V. 15, P. 2517.
41.   Volynskii A.L., Bakeev N.F., Solvent crazing of polymers, Amsterdam, NY, Tokyo, Elsevier, 1995, P. 410.
42.   Li J.C.M., Polym. Eng. Sci., 1984, V. 24, No. 10, P. 750.
43.   Friedrich K., Adv. Polym. Sci., 1983, V. 52/53, P. 266.
44.   Volynskii A.L., Bakeev N.F., Superfine oriented state of polymers, Moscow, Khimiya, 1985.
45.   Kambour R.P., J. Polym. Sci., Macromol. Rev., 1973, V. 7, P. 1.
46.   The Physics of glassy polymers, ed. R.N. Haward, B.Y. Young. London–New York, Chapman and Hall, 1997, P. 508.
47.   Passaglia E., J. Phys. Chem. Solids, 1987, V. 48, No. 11, P. 1075.
48.   Slutsker A.I., Oriented state. Encyclopedia of polymers, Moscow, Sov. Entsyklopediya, 1974. V. 2, P. 523.
49.   Taylor G.I., Proc. Roy. Soc. London, A, 1950, V. 201, P. 192.
50.   Argon A.S., Salama M., Mater. Sci. Eng., 1976, V. 223. P. 219.
51.   Argon A.S., Salama M., Phil. Mag., 1977, V. 36, P. 1217.
52.   Kramer E.J., Adv. Polym. Sci., V. 52/53, Berlin; Heidelberg, Springer-Verlag, 1983, P. 1.
53.   Efimov A.V., Szczerba V.Y., Bakeev N.F., Vysokomolek. Soed. B, 1989, V. 31, No. 11, S. 715.
54.   Efimov A.V., Szczerba V.Y., Rebrov A.V., Ozerin A.N., Bakeev N.F., Vysokomolek. Soed. A, 1990, V. 32, No. 4, P. 828.
55.   Efimov A.V., Szczerba V.Y., Rebrov A.V., Ozerin A.N., Bakeev N.F., Vysokomolek. Soed. A, 1990, V. 32, No. 2, P. 456.

56.   Gent A.N., J. Macromol. Sci. B, 1973, V. 8, No. 3-4, P. 597.
57.   Pazukhina L.Yu., Dissertation, Cand. Chemical. Sciences. Moscow, M.V. Lomono-
      sov University, 1983.
58.   Volynskii A.L., Aleskerov A.G., Bakeev N.F., Vysokomolek. Soed. B, 1977, V. 19,
      P. 218.
59.   Yarysheva L.M., Pazukhina L.Yu., Kabanov N.M., et al., Vysokomolek. Soed. A,
      1984, V. 26, No. 2, P. 388.
60.   Yarysheva L.M., Chernov I.V., Kabalnova L.Y., et al., Vysokomolek. Soed. A, 1989,
      V. 31, No. 7, P. 1544.
61.   Van Krevelen D.V., Properties and chemical structure of the polymers, Moscow,
      Khimiya, 1976.
62.   Shchukin E.D., Pertsov A.V., Amelina E.A., Colloid chemistry, Moscow, Vysshaya
      shkola, 2004. P. 445.
63.   Sinevich E.A., Bakeev N.F., Vysokomolek. Soed. B, 1980, V. 22, No. 7, P. 485.
64.   Yarysheva L.M., Pazukhina L.Yu., Lukovkin G.M., et al., Vysokomolek. Soed. A,
      1982, V. 24, No. 10, P. 2149.
65.   Jakus K., Ritter Y.E., Latsen C.A., Polym. Eng. Sci., 1981, V. 21, P. 854.
66.   Glasstone S. Leydler K., Eyring H., Theory of absolute reaction rates, Moscow, IL,
      1948.
67.   Lazurkin S., Fogelson RL, Journal Tech. Physics, 1951, V. 21, No. 3, P. 267.
68.   Volynskii A.L., Efimov A.V., Bakeev N.F., Vysokomolek. Soed. A, 2001, V. 43, No.
      8, P. 1361–1369.
69.   Volynskii A.L., Chernov I.V., Yarysheva L.M., et al., Vysokomolek. Soed. A, 1992,
      V. 34, No. 2, P. 119.
70.   Chernov I.V., Dissertationm , Cand. chemical sciences, Moscow, Lomonosov Mos-
      cow State University, 1997.
71.   Lukovkin G.M., Chernov I.V., Volynskii A.L., et al., Vysokomolek. Soed. B, 2000,
      V. 42, No. 8, P. 1446.
72.   Chernov I.V., Yarysheva L.M., Lukovkin G.M., et al., Vysokomolek. Soed. B, 1989,
      V. 31, No. 6, P. 1049.
73.   Volynskii A.L., Aleskerov A.G., Zavarova T.B., et al., Vysokomolek. Soed. A, 1977,
      V. 19, No. 4, P. 845.
74.   Yarysheva L.M., Pazukhina L.Yu., Lukovkin G.M., et al., Vysokomolek. Soed. A,
      1981, V. 23, No. 4, P. 859.
75.   Yarysheva L.M., Lukovkin G.M., Volynskii A.L., et al., Advances in colloidal chem-
      istry and physico-chemical mechanics, Ed. E.D. Shchukin, Moscow, Nauka, 1992.
76.   Sinevich E.A., Bakeev N.F., Vysokomolek. Soed. A, 1982, V. 24, P. 1912.
77.   Lukovkin G.M., et al., Vysokomolek. Soed. A, 1984, V. 26, No. 10, P. 2192.
78.   Volynskii A.L., et al., Ros. Khim. Zh., 2005, Vol 69, No. 6, P. 118.
79.   Volynskii A.L. et al., Vysokomolek. Soed. A, 1982, V. 24, No. 11, P. 2357.
80.   Brown H.R., Kramer E.J., J. Macromol. Sci. Phys. B, 1981, V. 19, No. 3, P. 487.
81.   Moskvina M.A., Volkov A.V., Efimov V., et al., Vysokomolek. Soed. B, 1988, V. 30,
      No. 10, P. 737.
82.   Yang A.C.M., Kramer E.J., J. Polymer Sci., Polym. Phys. Ed., 1985, V. 23, No. 7, P.
      1353.
83.   Berger L.L., Kramer E.J., Macromolecules, 1987, V. 20, No. 8, P. 1980.
84.   Yarysheva L.M., Gal'perina N.B., Arzhakova O.V., et al., Vysokomolek. Soed. B,
      1989, V. 31, No. 3, P. 211.
85.   Paredes E., Fischer E.W., Makromol. Chem., 1979, V. 180, P. 2707.

86. Lukovkin G.M., Pazukhina L.Yu., et al., Vysokomolek. Soed. A, 1986, V. 28, No. 1, P. 189.

87. Dettenmaier M., Adv. Polym. Sci., 1983, V. 52/53, P. 58.

88. Berger L.L., Macromolecules, 1989, V. 22, No. 10, P. 3162.

89. Griffith A.A., Phil. Trans. Roy. Soc., 1921, V. A221, P. 163.

90. Lukovkin G.M., et al., Vysokomolek. Soed. A, 1987, V. 29, No. 1, P. 198.

91. Lauterwasser B.D., Kramer E.J., Phil. Mag. A, 1979, V. 39, No. 4, P. 469.

92. Narisawa J., Jee A.F., Material Science and Technology, V. 12. Structure and properties of polymers. 1993, P. 701.

93. Donald A.M., Physics of glassy Polymers, ed. by R.N. Haward, B.Y. Young, London–New York, Chapman and Hall, 1997, P. 508.

94. Cannon S.J., McKenna G.B., Statton W.O., J. Polym. Sci. Macromol. Rev.. 1976, V. 11, P. 209.

95. Tunigami T., Yamaura K., Matsuzawa S., Ohsawa K., Mijasaka K., J. Appl. Polym. Sci., 1986, V. 32, P. 4491.

96. Volynskii A.L., Ukolova E.M., Shmatok E.A., Arzhakova O.V., Yarysheva L.M., Lukovkin G.M., Bakeev N.F., DAN SSSR, 1990, V. 310, No. 26, P. 380.

97. Volynskii A.L., Shmatok E.A., Ukolova E.M., Arzhakova O.V., Yarysheva L.M., Lukovkin G.M., Bakeev N.F., Vysokomolek. Soed. A, 1991, V. 33, No. 5, P. 1004.

98. Bykova I.V., Dissertation, Cand. chemical. Sciences. Moscow, Karpov Institute, 1997.

99. Efimov A.V., Bondarev V.V., Kozlov P.V., Bakeev N.F., Vysokomolek. Soed. A, 1984, V. 26, No. 10, P. 1640.

100. Volynskii A.L., Yarysheva L.M., Ukolova E.M., Kozlova O.V., Vagina T.M., Kechek'yan A.. Kozlov P.V., Bakeev N.F., Vysokomolek. Soed. A, 1987, V. 29, No. 12, P. 2614.

101. Yarysheva L.M., Shmatok E.A., Ukolova E.M., Lukovkin G.M., Volynskii A.L., Bakeev N.F., Vysokomolek. Soed. B, 1990, V. 32, No. 4, P. 529.

102. Volynskii A.L., Aleskerov A.G., Grokhovskaya T.E., Bakeev N.F., Vysokomolek. Soed. A, 1978, V. 28, No. 11, P. 2114.

103. Volynskii A.L., Loginov V.S., Bakeev N.F., Vysokomolek. Soed. B, 1981, V. 23, No. 4, P. 314.

104. Volynskii A.L., Shtanchaev A.Sh., Bakeev N.F., Vysokomolek. Soed. A, 1984, V. 39, No. 11, P. 2445.

105. Efimov A.V., Valiotti N.N., Dakin V.I., Ozerin A.N., Bakeev N.F., Vysokomolek. Soed. A, 1988, V. 30, No. 5, P. 963-968.

106. Volynskii A.L., Arzhakova O.V., Yarysheva L.M., Bakeev N.F., Vysokomolek. Soed. B, 2000, V. 42, No. 3, P. 549-564.

107. Godshalla D., Wilkesa G., Krishnaswamy R.K., Sukhadia A., Polymer, 2003, V. 44, P. 5397-5406.

108. Adams W.W., Yang D., Thomas E.L., J. Mater. Sci., 1986, V. 21, P. 2239.

109. Argon A.S., Pure Appl. Chem., 1975, V. 43, No. 1-2, P. 247.

110. Yarysheva A.Yu., Bagrov D.V., Rukhlya E.G., Yarysheva L.M., Volynskii A.L., Bakeev N.F., Dokl. RAN, 2011, V. 440, No. 5, P. 655-657.

111. Shmatok E.A., Kozlova O.V., Yarysheva L.M., Volynskii A.L., Bakeev N.F., Dokl. AN SSSR, 1988, V. 302, No. 6, P. 1428.

112. Shmatok E.A., Yarysheva L.M., Volynskii A.L., Bakeev N.F., Vysokomolek. Soed. A, 1989, V. 31, No. 8, P. 1752.

113. Shmatok E.A., Arzhakova O.V., Yarysheva L.M., Volynskii A.L., Bakeev N.F., Vysokomolek. Soed. A, 1990, V. 32, No. 3, P. 577.

114.   Shmatok E.A., Kalinina S.V., Arzhakova O.V., Yarysheva L.M., Volynskii A.L., Ba-
        keev N.F., Vysokomolek. Soed. A, 1990, V. 32, No. 6, P. 1282.
115.   McAlister P.B., Carter T.J., Hinde R.M., J. Polymer Sci. Polymer Phys. Ed., 1978, V.
        16, No. 1, P. 49.
116.   Grembowicz J., Zan J.F., Wunderlich B., J. Polymer Sci., Polymer Symp., 1984, No.
        71, P. 19.
117.   Shchipacheva N.A., Dissertation, Cand. Chemical. Sciences, Moscow, M.V. Lo-
        monosov MSU, 1974.

The structure and properties of crazed polymers

Thus, the deformation of the glassy polymer in the adsorption-active medium (AAM) leads to its self-dispersion into the finest (nanosized) aggregates of the oriented macromolecules which is evidently accompanied by the development of a high level of the interfacial surface. The excess of the interfacial surface has a strong effect on the entire set of the properties of the polymer subjected to crazing and, consequently, it shows the structural–mechanical behaviour not typical at all of the glassy or crystalline state of solid polymers. This is accompanied by the transition of the polymer to the new structural–physical state [1]. This section describes and analyses these unusual properties. These problems were also partially discussed in chapter 5 in the analysis of the data obtained for the force softening of the glassy and crystalline polymers.

7.1. Structural–mechanical aspects of deformation of crazed polymers

As shown previously, the deformation of the polymer in the AAM is accompanied by intensive crazing of the polymer so that the latter acquires a large interfacial surface reaching several hundreds of square metres per gram [2]. The latter circumstance makes these polymers thermodynamically unstable and susceptible to structural rearrangement, leading to the spontaneous reduction of the excess area of the interfacial surface. Figure 7.1 shows the dependence of reversible deformation of a number of polymers on the degree of their tensile loading in the AAM by the crazing mechanism. The same figure compares these data with the reversible strain of the

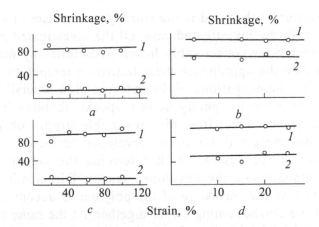

Fig. 7.1. Dependence of reversible deformation on the draw ratio for polyethylene terephthalate (*a*), polycarbonate (*b*), plasticised PMMA (*c*) and polystyrene (*d*): 1) in all graphs the shrinkage after drawing in the AAM (*n*-propanol); 2*a* and 2*b* – shrinkage of polyethylene terephthalate and polycarbonate after drawing in air; 2*c* and 2*d* – shrinkage of polystyrene and PMMA after drawing in the AAM in the moist condition. In all cases, the deformation was carried out at room temperature [3].

same polymers, deformed in air (polyethylene terephthalate and polycarbonate) and in the moist condition after drawing in the AAM (for PMMA and PS [3]).

Figure 7.1 shows that the glassy polymers, deformed in the AAM, show high reversible (close to 90%) deformation, which is not typical at all of the glassy state. This deformation is detected when the AAM is removed (evaporated) from the volume of the crazes. In addition, the polymers showing brittle behaviour during their deformation in air at room temperature (PMMA and PS) are capable of relatively high strains in the AAM (up to 30% or more). All the investigated polymers also show high strains directly in the AAM after release from the clamps of the tensile loading device [1]. Deformation of this type is slightly lower than that the detected in the complete removal of the AAM (Fig. 7.1), but it still equals several tens of percent which is also not characteristic of the glassy state of the polymer.

It is now regarded as generally accepted that the capacity for high reversible strains is the unique property of the polymers which becomes evident only in the case in which the deformed polymer is at the temperature higher than the glass transition temperature (in the rubbery state). The conditions for segmental mobility of the macromolecules, which determine the nature of the elasticity of the polymers, exist only at the temperatures higher than the glass transition temperature. The only currently known mechanism of high reversible strains of the solid is the entropy mechanism of elasticity of rubber [4]. In the investigated case, all the investigated polymers were at temperatures considerably lower (sometimes by many tens of degrees) than the appropriate glass transition temperature.

The detailed investigations of the detected high reversible strain has made it possible to specify certain special features by which these strains greatly differ from high reversible strains, observed in rubber-like polymers with no highly developed surface.

Firstly, all the processes which determine the high reversible strain take place inside the developed crazes. Direct microscopic studies show that the shrinkage of the polymer is accompanied by the walls of the crazes coming closer together. At the same time, the fragments of the initial bulk polymer, localized between the crazes, do not undergo any changes during shrinkage.

Thus, the high reversible strain, shown in Fig. 7.1, is not characteristic of the glassy polymers as a whole but only of the highly dispersed material, localized inside the crazes. This is not the only difference of the structural–mechanical behaviour of the crazed glassy polymers from the appropriate behaviour of the monolithic bulk specimens. Another important special feature of the mechanical behaviour of the craze polymers is observed if the AAM is removed from the polymer with fixed dimensions after tensile loading [1]. The results of direct microscopic studies have been used to visualise this process [3]. Figure 7.2 a shows the light micrograph of a sample of polyethylene terephthalate deformed in the AAM to 25%. It can be seen that the micrographs, presented in Fig. 7.2, were made at the fixed dimensions of the deformed specimen. It may also be seen that as a result of this deformation crazes, filled with the oriented high dispersion polymer, form in the polymer (Fig. 7.2 a). The

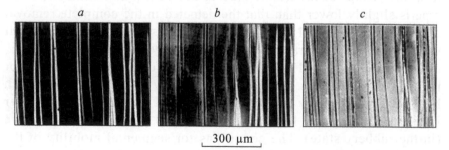

a   b   c

300 μm

Fig. 7.2. Optical micrographs of polyethylene terephthalate, deformed in the AAM by 25%: *a*) a sample photographed immediately after tensile loading; *c*) the dried sample; *b*) the intermediate state [3].

molecular orientation of the material, filling the crazes, is indicated in particular by the strong birefringence, because this photograph was made in the crossed polaroids. The evaporation of the AAM from the volume of the crazes results in a large change of the structure (Fig. 7.2 b). With evaporation of the AAM from the volume of the crazes, the birefringence is extinguished in every individual craze from one of its ends to the other end, and this process resembles the movement of mercury in the capillary of a thermometer. After completion of the process, strong birefringence is detected in the non-oriented fragments of the polymer, situated between the crazes. This is determined by the development of high stresses (up to 15 MPa) in the polymer which can be recorded in the experiments [5]. The process of removal of the AAM from the volume of the crazes is accompanied by extensive rearrangement of the structure. This is indicated by the light microscopy results (Fig. 7.3). The image compares two micrographs, obtained from the same region of the polyethylene terephthalate sample tensile loaded in the AAM to 25%. The photographs were obtained at fixed dimensions of the specimen prior to (*a*) and after (*b*) evaporation of the AAM from the volume of the crazes. Figure 7.3 shows that the crazes, containing the AAM (Fig. 7.3 *a*), do not scatter the light and the photograph shows only the contours indicating their walls. The removal of the AAM from the structure of the crazes (Fig. 7.3 b) makes their content visible because of the intensive scattering of light. The detected effect is due to the fact that in the AAM the highly dispersed material of the crazes is relatively stable so that the dimensions of the fibrillar aggregates of the macromolecules are small (units of nanometres) are

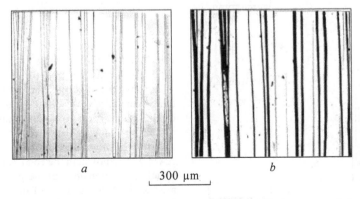

a   *b*
⌊___ 300 μm ___⌋

Fig. 7.3. Light micrographs of the sample of polyethylene terephthalate tensile loaded in the AAM (*n*-propanol) by 25%: a) moist sample; b) the same sample after removing the AAM by drying [1].

not capable of scattering the visible light. For this reason, they are not visible in the optical microscope. The removal of the AAM from the volume of the crazes makes the material of the crazes unstable and leads to its coalescence. As a result of coalescence, the fibrillar aggregates form extended phases which greatly scattered light.

Thus, the investigations of the mechanical behaviour of the crazed specimens [1] revealed a large set of the properties, completely untypical of the bulk glassy and crystalline polymers. The mechanism of the phenomenon, observed in [1–3], will be discussed later and here we present the scheme of structural rearrangement, accompanying the removal of the AAM from the volume of the crazes, based on direct structural studies. Figure 7.4 shows the diagram of structural rearrangement taking place in the structure of the crazes as a result of the removal of the AAM from the volume. The deformation of the polymer in the AAM is accompanied by the formation and growth of the crazes. The structure of this crazes is shown in Fig. 7.4 a. The removal of the AAM from the volume of the crazes results in the destabilisation of their highly dispersed structure and, consequently, the material of the crazes coalesces and this is accompanied by the reduction of the interfacial surface area of the craze. When the polymer relaxes in the free state, the process develops along the path indicated in Fig. 7.8 a→c by the contraction of the fibrillised materials crazes. It is evident that the contraction of the fibrils in the structure of the crazes not only results in the reduction of the surface area of the crazes but is also accompanied

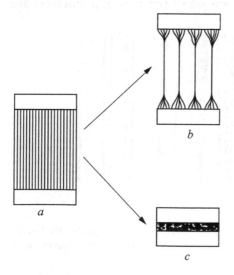

Fig. 7.4. The diagram illustrating the structural rearrangement in the removal of the absorption-active medium from the volume of the craze: *a*) the structure of the craze formed in deformation of the polymer in the AAM; *b*) the structure of the craze, from which the AAM was removed in the conditions in which the dimensions of the polymer are fixed; *c*) the removal of the AAM from the volume of the craze takes place when the polymer is in the free state [1].

by the walls of the crazes coming together and this explains the very phenomenon of reversibility of deformation.

When the removal of the AAM takes place from the crazed specimen of the polymer with fixed dimensions, the process of reduction of the interfacial surface area takes place by a different mechanism (Fig. 7.4 a b). In this case, the fibrillar aggregates of the macromolecules also merge but with the side surfaces adjacent to the fibrils (coalesce) and this is accompanied by the reduction of the total surface area. As regards the structural aspect, the process takes place by a different mechanism and, consequently, the polymer retains a high level of porosity. However, the diameters of the pores and fibrils greatly increase in this case. It should be mentioned that, in this case, the removal of the AAM from the volume of the crazes results in the formation of a structure not capable of reversible deformation at the experiment temperature.

Further investigations [1–3] revealed many special features of the mechanical behaviour of the crazed polymers as a result of which their reversible deformation greatly differs from the well-studied high reversible deformation of the rubber-like polymers. It is well-known [4] that the cross-linked rubbers retain the capacity for reversibility of deformation up to fracture elongation, i.e., at elongations of many hundreds or even thousands of percent. A different pattern is observed in the cold drawing of the polymers in the AAM. It appears that the reversibility of deformation of the polymer, deformed in the AAM by the crazing mechanism, depends greatly on the extent of tensile loading of the polymer. At relatively high values of the relative elongation of the polymer in the AAM the magnitude of its reversible deformation greatly changes, irrespective of its nature and the nature of the AAM in which deformation takes place.

Figure 7.5 shows the dependence of relative deformation on the draw ratio in the AAM for a number of polymers. These data are obtained by measuring the reversible strain of the polymer after its tensile loading in the AAM, released from the clamps of the tensile loading device and the removal of the AAM from the porous structure (drying of the specimens). It may be seen that the high reversible strains (70–95%) are characteristic of the polymers tensile loaded in the AAM only in the region of relatively low values of elongation. For each investigated polymer there is some limiting value of elongation above which the reversible strains greatly decrease and approach the shrinkage values, observed in air and in fact never reach them. The nature of the liquid medium in which deformation

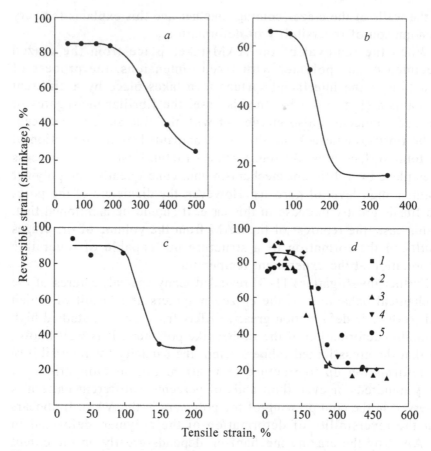

Fig. 7.5. Dependence of reversible strain on tensile strain in tensile loading of polyvinyl chloride in *n*-propanol (*a*), PA-6 in methyl ethyl ketone (*b*), polyvinyl chloride in methanol (*c*) and polyethylene terephthalate in *n*-propanol (Δ), CCl₄ (o), formamid (×), triethylamine (+) and butyl iodide (*d*) [2].

is carried out does not have any significant effect on the shrinkage in the case in which the liquid is fully removed from the polymer. For example, in the case of polyethylene terephthalate, the results obtained in five different media, are satisfactorily described by a single curve (Fig. 7.5 *d*).

Attention will be given to the structural rearrangement, accompanying deformation of the polymer by the crazing mechanism in a wide strain range. Morphological special features of the polymer, subjected to cold drawing in the AAM, will be studied on the example of polyethylene terephthalate. Figure 7.6 shows the photographs of polyethylene terephthalate samples, tensile loaded in the AAM (*n*-propanol) with different draw ratios [6]. It may be seen

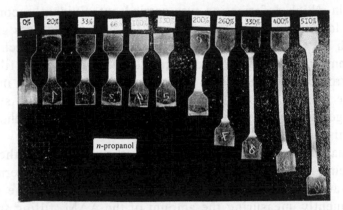

Fig. 7.6. Photographs of the polyethylene terephthalate samples, tensile loaded to various degrees of elongation in the AAM (*n*-propanol) [6].

clearly that at low elongations the cold drawing of the polymer in the AAM takes place without lateral contraction and is accompanied by the almost complete restoration of the dimensions. The characteristic behaviour of the working part of the specimen is determined by the fact that the deformation takes place by the development of crazes with a porous structure. It is important to note that in this tensile loading range the deformation of the polymer is almost completely reversible (the dimensions of the specimens are almost identical with the dimensions of the initial non-deformed polymer). However, starting at approximately 150–180% elongation, the polymer shows extensive lateral contraction so it starts to resemble externally the sample deformed in the with the formation of a neck. In contrast to the glassy polymers, deformed in air, in this case, there is no distinctive necking in any part of the specimen or its subsequent propagation throughout the entire working section. As indicated by Fig. 7.6, the reduction of the width of the working part of the specimen during deformation of the specimen in the AAM takes place quite uniformly. The external appearance of the samples, shown in Fig. 7.6, indicates that the process of cold drawing of the polymer in the AAM is accompanied by some structural transition is accompanied by the rearrangement of the porous structure visible even by the naked eye. It may also be seen that starting at approximately 150% elongation the degree of reversible strain greatly decreases and, consequently, the dimensions of the deformed polymer start to increase rapidly.

The samples of the polymer, shown in Fig. 7.6, were photographed after interrupting tensile loading and extracting them from the liquid in which deformation was carried out. Evidently, continuous tensile loading of the polymer causes the polymer to pass gradually through all stages of structural rearrangement and, which is very important, its developing porous structure is continuously filled with the surrounding liquid. The latter circumstance is very important because there is another possibility of investigating the structural rearrangement of the polymers, deformed in the AAM by the crazing mechanism.

Actually, some methods of studying the structure of the porous adsorbents, for example porosimetry, are based on the determination of the amount of the substance filling the pores in the sample. Consequently, measuring the amount of the AAM, filling the pores of the deformed polymer, we may obtain additional information on its structure because the amount of the trapped liquid can be used to characterise the most important parameter of the porous structure – the total volume of the pores.

Figure 7.7 shows the dependence of the mass of the AAM (n-propanol), included in the porous structure of the polymer (polyethylene terephthalate) on the magnitude of deformation of the polymer [7]. It may be seen that the mass of the liquid medium, trapped by the polymer, initially increases with increasing draw ratio and then, passing through the maximum, decreases. Comparison of the data presented in Figs. 7.6 and 7.7 indicates that the variation of the external appearance of the samples, deformed in the AAM by different draw ratios, and the processes associated with the transport of the liquid in the porous structure of the polymer are in an obvious correlation. Actually, the increase of the amount of the liquid takes place in the same strain range as the range in which the polymer does not undergo lateral contraction, and its main elongation is obtained as a result of growth of the microcavities, filled with the liquid.

It should be mentioned that all the stages of crazing, examined in the previous sections (see chapter 6), relate to the strains lower than

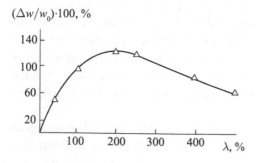

$(\Delta w / w_0) \cdot 100, \%$

Fig. 7.7. Dependence of the relative mass of the AAM ($\Delta W / W_0$), trapped in the porous structure of the glassy polymer (polyethylene terephthalate) on the draw ratio of the polymer (λ) in the liquid medium.

the beginning of the stage in which the width of the polymer sample decreases during tensile loading. As indicated by the data presented in Figs. 7.6 and 7.7, the contraction of the polymer in the range of relatively high elongations is accompanied by the release of part of the liquid into the surrounding space (Fig. 7.7). These data indicate that tensile loading of the polymer in the AAM is accompanied by a unique structural transition in which the direction of mass transfer of the surrounding liquid changes. The mechanism of this phenomenon was determined by the detailed x-ray diffraction and electron microscopic studies, carried out in [6]. Analysis of these experimental results has been used to propose the following mechanism of the process of deformation of the polymer in the AAM (Fig. 7.8). At low tensile stresses (Fig. 7.8 *a*) crazes form in the polymer with the ends connected by thin fibrils. These fibrils are separated in space and, consequently, the system stores the surplus interfacial surfaces. A unique colloidal system forms, with its special feature being the presence in it of long and very thin fibrils, connecting the edges of the crazes. The transverse dimensions of these fibrils vary from 2 to 20 nm so that they can be regarded as colloidal particles, capable of taking part in Brownian motion. However, the ends of these strands are secured on the opposite sides of the crazes so that the colloidal particles of this type are capable of carrying out Brownian motion only in the direction normal to the main axes of the fibril, similar to the movement of a string with two secured ends. As any colloidal system, this system is thermodynamically unstable but is stabilised by the surface of the active medium, decreasing the surface energy of the polymer.

Increasing tensile stress is accompanied by the continuous transition of the initial bulk polymer to the oriented and highly dispersed state inside the crazes as a result of the consumption of non-oriented parts of the polymer. This is accompanied by an increase of the length of the individual fibrils inside the crazes which,

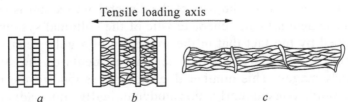

Fig. 7.8. The diagram illustrating the collapse of the structure of the polymer, taking place at high strains of the polymer in the AAM: *a–c*) different stages of tensile loading.

together with the increase of the interfacial surface area, should increase their flexibility, compliance and mobility. The restriction of the mobility of the fibrils inside the crazes, caused by the non-oriented part of the polymer, greatly decreases with increasing strain. The fibrils of the polymer, with sufficiently high mobility, can effectively contact with each other (Fig. 7.8 *b*) under the effect of thermal motion. It should be mentioned that the normal stresses, formed in tensile loading, also cause the individual fibrils to come together. It is evident that and the effect of the mechanical stress and thermal collisions, as detected in other colloidal systems, the adsorption layer on the surface of the individual fibrils is disrupted. This unstable areas of the fibrils can converge to the distances of the effect of the intermolecular forces and consequently merge (coalesce) with each other. This process reduces the interfacial surface area, is thermodynamically advantageous and, therefore, can spontaneously. It is evident that together with the reduction of the interfacial surface area of the process of merger of the fibrils results in a large increase of the length of the oriented polymer phase.

With the development of deformation, the possibility of this type of reduction of the interfacial surface area (coalescence) will continuously increase. Since in the tensile loading conditions this is possible only in the coalescence of the adjacent fibrils with each other by their lateral surface is, the experimentally detected decrease of the cross-section of the specimens, deformed to high degrees of tensile loading, takes place (Fig. 7.6).

Thus, the process of tensile loading of the polymer by the crazing mechanism is accompanied by the formation of a unique colloidal solution, filling the volume of the crazes, in which the disperse phase is represented by long and very thin fibrillised aggregates of the oriented macromolecules, and the dispersion medium are the AAM molecules. From this viewpoint, it is easy to explain the previously discussed phenomena of the loss of the part of the liquid trapped by the polymer during its transition from the porous to a more compact structure. The loss of the part of the dispersion medium is synaeresis, the phenomenon highly characteristic of the colloidal systems. This is explained by the fact that in the formation of a gel a relatively small number of contacts forms between the elements of the structure in the early stages. This number does not correspond to the maximum dense (and, consequently, thermodynamically most advantageous) state of the system. Subsequently, as a result of the regrouping of the structural elements, caused by the thermal motion, the number of the contact increases and this results unavoidable in the compression of the gel and injection from it of the dispersed media. Evidently, unique syneresis also takes place in the investigated case which, according to the Pankov classification [8], can be referred to as forced syneresis because it takes place when the external force field is applied. Consequently, the examined structural transition is the manifestation of the coalescence of the high dispersion structure of the polymer and since this coalescence takes place in the presence of the dispersed medium, it is accompanied by the syneresis phenomenon. Evidently, to enable the release of the trapped liquid into the surrounding space during synaeresis, it is necessary to ensure is viscous flow through the porous structure of the polymer. It may be seen that the crazing of the polymer in the AAM is a relatively complicated surface phenomenon which is accompanied both by the development of the interfacial surface (the first stages of deformation of the polymer in the AAM) and its healing (collapse of the structure in the final stages of the process). In other words, in tensile loading in the AAM the polymer undergoes a unique structural transition from the porous to a more compact structure. The result of this transition, the polymer acquires a characteristic structure consisting of closely packed aggregates of the fibrils, oriented in the direction of the tensile loading axis, indicating its similarity with the amorphous and polycrystalline polymers, deformed in air with the formation of a neck. However, the adhesion of the fibrils, taking place in the presence of the AAM, cannot naturally be complete as in air. Therefore, the sample deformed to high elongation values may show a decrease in its transverse dimensions but nevertheless contain a large number of micro-cavities [1, 9].

Thus, the crazed polymers show a highly unusual structural–mechanical behaviour. Firstly, it is a high reversible strain which is detected in the craze polymers after their tensile loading in the AAM and relaxation in the free state. Secondly, the complete absence of reversibility of deformation in the same craze polymers in the cases in which the relaxation after tensile loading in the AAM takes place at fixed dimensions. Finally, the dependence of the magnitude of reversible strain on the draw ratio of the polymer in the AAM. At a relatively high degree of tensile loading of the polymer there is a sufficiently distinctive transition at which the high reversible strain of the crazed polymer rapidly decreases and becomes similar to the reversible strain of the glassy polymer, deformed in air with the formation of a neck.

It has been shown many times in investigations [10–13] that the material of the crazes in the glassy polymers is characterised by large differences in the mechanical properties in comparison with the polymer in the glassy state. Even after removing the AAM from the volume of the crazes, the repeated deformation of the polymer greatly differs from the deformation of the glassy polymer. These differences were examine in detail for the first time in a study by Kambur and Kopp [10], analysed in chapter 5. In this study, the authors investigated the mechanical behaviour of a 'dry' single craze, formed in the polycarbonate, and showed that its tensile loading takes place with a distinctive effect of deformation softening, and the repeated curve of tensile loading forms at a considerably lower stresses and has two yield limits. In the same study, it was found that the 'dry' crazes of the polycarbonate, deformed to 50–60%, are capable of slowly restoring their dimensions after removing the load. It should be mentioned that this phenomenon is not characteristic at all of the glassy polymer, oriented in the cold drawing mode. To describe the mechanical properties of the craze materials, the authors of [10] used the model constructed by Gent and Thomas [14] develops for the analysis of the mechanical behaviour of foamed rubber.

Later, the mechanical properties of the crazed samples of polyethylene terephthalate were studied in [9]. In contrast to [10], the authors investigated the macroscopic properties of the crazed polymer and not the properties of the individual crazes. For this purpose, the specimens of amorphous polyethylene terephthalate were tensile loaded in the AAM by the classic crazing mechanism, subsequently released from the clamps of the tensile loading device and the liquid medium was removed from the porous structure of the polymer. Removal of the active liquid resulted in the almost complete shrinkage of the polymer, similar to that discussed previously (Figs. 7.1 and 7.4). The samples, obtained by this procedure, were subjected to repeated deformation in air and their mechanical properties were investigated.

Figure 7.9 shows the repeated tensile loading curves of the specimens produced in these conditions [9]. It may be seen that the curves for the specimens, containing the crazes, greatly differ from the well-known tensile loading curves of the glassy polymers. It is important to note that the obvious agreement between the tensile loading curves of the crazed polyethylene terephthalate [9] and the tensile loading curve of the individual craze (compare Figs. 5.5 and 7.9). The characteristic feature of these curves is the presence of the two yield stresses so that prior to reaching the plateau zone, the curves become S-shaped. It should be mentioned that after reaching the elongations, corresponding to the plateau region of the curve, a neck develops in the deformed specimens, i.e., the width of the working section greatly decreases, as in the case of the thin the glassy polymers subjected to cold drawing. However, the structure of such a neck is slightly unusual. In contrast to the usually observed necks in the polyethylene terephthalate, it is opaque and contains, according to the results of small angle x-ray scattering, a large number of asymmetric cavities, oriented in the direction of the tensile loading axis [9].

Figure 7.9 also shows that the strain at which the tensile curve reaches the plateau region depends on the degree of preliminary tensile loading of the polymer in the AAM. The elongation, at which necking starts, is indicated by the dotted line in Fig. 7.9, is in good agreement with the degree of preliminary tensile loading of the polymer in the AAM. The sample appears to 'remember' the extent of its deformation by the crazing mechanism.

The process of restoration of the dimensions of the polymer, deformed by the crazing mechanism, has its own special features [15]. It appears that the stabilising effect of the AAM does not prevent the occurrence of spontaneous structural rearrangement of the material of the crazes taking place in time and leading to a reduction of its interfacial surface area. Figure 7.10 shows the tensile loading curve of the glassy polymer (polyethylene terephthalate) in the AAM (*n*-propanol). After 100% tensile loading, the stress was removed from the sample and the polymer was let to relax for different periods of time. During this recovery, shrinkage of the polymer took place with time. This process is characterised in Fig. 7.10 by the displacement of the start of the repeated curve of tensile

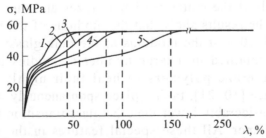

Fig. 7.9. Tensile loading curves of PET samples, subjected to preliminary tensile loading in *n*-propanol by the classic crazing mechanism, by 25 (1), 50 (2), 85 (3), 100 (4) and 150% (5) [9]

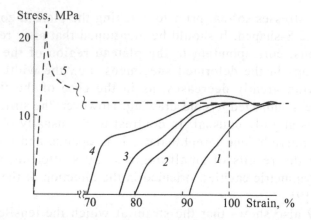

Fig. 7.10. Repeated tensile loading curves of samples of polyethylene terephthalate in *n*-propanol, obtained after 2 (1), 15 min (2), 1 (3) and 7 days (4). 5 – the first tensile loading curve of polyethylene terephthalate in the AAM [15].

loading along the deformation axis. In the repeated deformation, examination showed interesting special features of the reversible deformation of this type.

The restoration of the initial dimensions of the polymer is spread over time. The repeated curves of tensile loading of the polymer in the AAM are very similar to the repeated curves of tensile loading of the crazed glassy polymer (polycarbonate) [10] and the crazed crystalline polymer (polypropylene) [16]. As already mentioned, the process of healing of the porous structure of the polymer (in the process of restoration of the dimensions is a healing process) can take place only at a sufficiently high molecular mobility of the contacting surfaces. The previously observed effect of deformation softening of the material of the crazes [10–16] has been confirmed in direct experiments [17–19]. For example, in [20] the atomic force microscopy method was used for independent measurements of the elasticity moduli of the material of the crazes and the surrounding bulk polymer. The results show that the modulus of the material the crazes is only 3–10% of the modulus of the bulk glassy polymer.

The results obtained in a large number of studies indicate the capacity of the craze polymers to heal their highly developed interfacial surface [10, 21], taking place spontaneously with time at temperatures considerably lower than the glass transition temperature of the bulk polymer. All these special features in the mechanical behaviour indicate the special state of the polymer transferred to the craze material. Naturally, a number of attempts have been made to explain these experimental results. In [10, 16], the authors calculated the possible role of surface forces in the reversible deformation of the crazed polymers. It was assumed that the reduction of the length of the fibrils, resulting in the convergence of the walls of the crazes, and therefore the very possibility of the reversible deformation, takes place under the effect of surface forces. However, the calculations show [10, 16] that the surface forces are not sufficient to cause this phenomenon on their own. Nevertheless, the experimental material accumulated over decades and discussed previously indicates that the material of the crazes has unique properties, the main of which is the high compliance and susceptibility to healing the interfacial surface in the temperature range below the glass transition temperature of the bulk polymer.

The apparent contradiction between the high molecular mobility of the material of the crazes below the glass transition temperature of the bulk polymer can be solved by the results of investigations started in the middle of the 90s of the previous century. These investigations, discussed in detail in chapter 5 [22–27], show that the surface layers of the glassy polymers are characterised by a greatly reduced glass transition temperature. Since the crazing is a process of dispersion of the polymer into aggregates with the dimensions of tens or hundreds of angstroms, it is not surprising that the material the crazes has a considerably lower glass transition temperature. This circumstance makes it possible and without contradiction to explain the entire previously discussed of the anomalous physical–mechanical properties of the crazed polymers.

The high reversible strains are associated with the entropic elasticity of the crazed polymer with a reduced glass transition temperature. The large reduction of the degree of the high reversible deformation with the removal of the liquid medium from the volume of the crazes in the specimens with fixed dimensions, observed in [3], and also the large reduction of this parameter at high draw ratios of the polymer in the AAM [6], can also be easily explained using the data discussed previously. In fact, the removal of the liquid from the volume of the crazed polymer in the conditions in which shrinkage of the polymer is prevented, and the collapse of the porous structure of the crazes at high degrees of drawing of the polymer are accompanied by identical structural changes. In both cases, the fibrillar aggregates of the macromolecules stick together (coalescence) (Figs. 7.4 and 7.8). This is evidently associated with the increase of the thickness of the fibrillar aggregates of the macromolecules in the structure of the craze. As shown in [24–27], the glass transition temperature of the polymer depends on the dimension of the polymer phase and has a glass transition temperature gradient with increase of the distance from the free surface of the polymer. The increase of the dimension of the polymer phase (the diameter of the fibrils) results in a corresponding increase of its glass transition temperature. Consequently, the glass transition temperature is higher than the experiment temperature, the polymer transforms to the glassy state and its reversible deformation is suppressed.

7.2. Thermomechanical properties of crazed polymers

Thus, the deformation of the glassy and crystalline polymers in the AAM transfers these materials to some new structural state. This state is characterised by the entire complex of the structural and mechanical properties greatly different from the well examined and known properties of the polymers in the glassy state [28]. All the anomalies in the structural–mechanical behaviour of the craze polymers can be explained without contradiction taking into account a number of surface phenomena characteristic of the high dispersed structure of the crazes [1, 2]. The controlling factor for understanding the structural–mechanical behaviour is the recently discovered fundamental property of the amorphous glassy polymers – reduction of the glass transition temperature in their surface layers in thin films in comparison with the appropriate characteristics of the bulk polymers. This property, regardless of its fundamental nature, was unknown for a long period of time because the thin (nanometric) surface layers had no significant effect on the macroscopic properties of the bulk polymers. However, the structure of the craze polymers includes the highly dispersed material of the crazes and the structural elements of these materials have the nanometric dimensions. In particular, this circumstance is the reason why the crazed polymers have a new, unusual set of properties.

Amorphous glassy polymers
As shown previously, the crazed polymer can be 'prepared' by at least two different methods. After tensile loading in the AAM the liquid can be removed from the volume of the crazes, the dimensions of the polymer sample can be fixed, and this procedure can also be carried out with samples in the free state. As shown schematically in Fig. 7.4, in the case of free specimens the shrinkage of the

polymer is almost complete. The mechanism of this type of shrinkage was discussed previously. When the AAM is removed from the polymer with the fixed dimensions, the polymer completely loses its capacity for reversible deformation and retains its dimensions as long as required. Since the crazing of the polymer is accompanied by its molecular orientation inside the fibrillar aggregates of the macromolecules, in the removal of the AAM from the volume of the crazes in the sample with the fixed dimensions the result is the oriented glassy polymer. The thermomechanical properties of the oriented glassy polymers are well-known [28–30]. Their important special feature is the thermally stimulated shrinkage which takes place as soon as the oriented polymer is heated to its glass transition temperature.

As regards the crazed polymers, this problem has been studied far less extensively. In particular, it is well-known that in an annealing above the glass transition temperature the interphase boundaries, characteristic of the structure of the crazes in polymers, are 'healed' [31, 32]. The problems of healing of the interfacial surface in the polymers (including in the crazed polymers) have been discussed in greater details in chapters 2, 4 and 5. Here, we only mention that the processes of healing can take place only at a sufficiently high molecular mobility on the surfaces of the polymer which are brought into contact. Evidently, the examination of the thermomechanical properties of the oriented polymers, in particular, of the fundamental property such as the restoration of their dimensions during annealing, may provide valuable information for the mechanism of deformation of the polymer glasses.

Thus, the thermomechanical properties of the crazed polymers can be investigated most efficiently using samples whose structure is shown schematically in Fig. 7.4 *b*. The structure forms in the deformation of the polymer in the adsorption-active medium with subsequent removal from the sample, with the shrinkage of the sample prevented. Evidently, in these conditions, the result is the structure consisting of a system of mutually oriented fibrillar aggregates of the macromolecules. Since the polymer inside each fibril is oriented along the tensile loading axis, the sample as a whole acquires a certain molecular orientation easily recorded by x-ray diffraction [9]. Thus, in the conditions of tensile loading of the polymer in the AAM by the crazing mechanism it is possible to produce the oriented glassy polymer suitable for thermomechanical studies.

In [33] investigations were carried out to compare the amorphous glassy polymers (polycarbonate and atactic polymethyl methacrylate), deformed in the conditions of uniaxial tensile loading in an adsorption active medium (*n*-propanol) and in air. The polycarbonate specimens were deformed in air at room temperature to complete transfer of their working section into the neck. The PMMA specimens were oriented at 100°C because at room temperature this polymer fails by brittle fracture. The PC and PMMA were also deformed in the AAM (*n*-propanol) at room temperature and then dried with the fixed dimensions.

Figure 7.11 shows the dilatometric curves, produced in annealing of the samples of the PC and PMMA with preliminary orientation in the air (curves 4 and 2) and in the AAM (*n*-propanol) (curves 3 and 1). As expected, the polymers, oriented in air, fully restored their dimensions in the glass transition temperature range indicated by the dotted line in the figure. The PC is also characterised by some shrinkage (approximately 15%) in annealing at a temperature below the glass transition temperature. The mechanism of this contribution to the thermally stimulated shrinkage of the glassy polymer was investigated in chapter 4. The low-temperature section of the shrinkage in the case of the PMMA, oriented in air at 100°C, is not observed at all.

The restoration of the dimensions of the PMMA and PC, oriented in air, in the glass transition temperature range is the well-known fact associated with the entropic effect of the restoration of the most probable conformation set of the molecules.

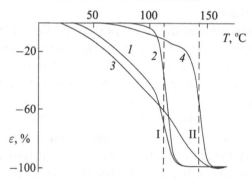

Fig. 7.11. Temperature dependence of the restoration of the dimensions of the samples (ε) of PMMA (1, 2) and PC (3, 4), deformed in the uniaxial tensile loading conditions: 1, 3) the samples of PMMA and PC, tensile loaded in *n*-propanol by the crazing mechanism; 2, 4) the samples of PMMA and PC, tensile loaded in air (PMMA at 100°C, PC at room temperature). The vertical dashed lines indicate the glass transition temperatures of PC (II) and PMMA (I).

At the same time, it was observed unexpectedly that the polymers, deformed in the AAM (*n*-propanol) by the crazing mechanism, show almost complete shrinkage (90%) below the appropriate glass transition temperatures (curves 3 and 1). Such large low-temperature shrinkage of the amorphous polymers oriented to a high degree are completely unusual and it was therefore necessary to examine in detail the manifestation of these properties, in particular, it was necessary to explain whether the shrinkage manages to reach its equilibrium value in annealing in the process of recording the dilatometric curves.

For this purpose, a polymer sample heated to a specific temperature, was held for 30 min. The results show that when heating was interrupted, the shrinkage of the specimen was also interrupted and restarted only if the temperature was increased. The dependence of the linear dimensions of the sample on the temperature was close to the corresponding dependence shown in Fig. 7.11. Consequently, the given annealing temperature corresponds to the specific dimensions of the sample, and the dilatometric curves, shown in Fig. 7.11, obtained in heating at a rate of 2°C/min, reflect the almost equilibrium state of the system produced in the conditions of stepped increase of temperature.

In [33], attention was also paid to the effect of the draw ratio and temperature in tensile loading the polycarbonate in the adsorption-active medium on the nature of its subsequent shrinkage during annealing. The results show that the contribution of the low-temperature component to the shrinkage of the polymer is almost completely independent of the degree of tensile loading and in all cases equals ~90%. It was also found that the low-temperature shrinkage is not dependent on the deformation temperature.

As shown previously, the main difference between the polymers deformed in air and in the AAM is the development of special zones of the plastically deformed polymer, filled with the high dispersion fibrillised material. It is rational to assume that the anomalous thermomechanical behaviour of the crazed polymers is determined by the processes taking place in the volume of the crazes. This assumption has been confirmed in [33] in which light microscopy was used to measure the distances between the walls of the crazes in the PC during heating. The results of these experiments are shown in Fig. 7.12. It may be seen that long before T_g (practically starting at room temperature) the width of the crazes greatly decreases, whereas the non-oriented areas between them may even slightly

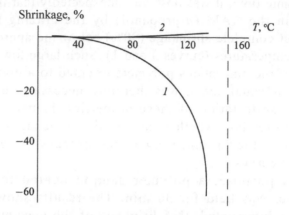

Fig. 7.12. Temperature dependence of their relative variation of the linear dimensions of the crazes (2) and non-oriented areas between them (1) in the direction of the axis of the alloying for the polycarbonate specimens, tensile loaded in the AAM. The vertical dashed line is the glass transition temperature of the polycarbonate [33].

increase their size as a result of thermal expansion. Consequently, the observed low-temperature shrinkage is caused by the processes taking place inside the crazes or, in other words, by the properties of the high dispersion oriented polymer, filling the crazes. When the T_g temperature of the bulk polymer is reached, the crazes, with the width already produced by 90–95%, are 'healed' and become invisible in the microscope.

Thus, the nature of the reversibility of the deformation in annealing of the amorphous glassy polymers, subjected to cold drawing in the liquid medium, has its special features. The amorphous glassy polymers, deformed in the adsorption active media by the crazing mechanism, almost completely restore their dimensions during annealing below the glass transition temperature. Evidently, the detected low-temperature shrinkage is associated with the properties of the highly dispersed material of the crazes, formed in the process of deformation of the polymers in the adsorption active media.

It is important to mention another special feature of the thermomechanical behaviour of the crazed polymers. Figure 7.11 shows that the low-temperature shrinkage of the crazed polymers does not have the cooperative, threshold nature. Whilst the polymer, oriented by the development of the neck, rapidly shrinks in a narrow range in the vicinity of the glass transition temperature, the shrinkage of the crazed polymer is widely 'spread' over the temperature scale (up to 100° or greater). Inside this range it is possible to interrupt annealing in any area and cool the polymer down [34]. As a result of this procedure, it is quite easy to record any value of shrinkage.

Thus, on the one side, the thin layers of the polymers are characterised by a large expansion on the temperature scale of α-relaxation and, on the other side, as already mentioned, the shrinkage of the glassy polymer, deformed by the crazing mechanism, is not simply shifted to the temperature range below the glass transition temperature. For the polymer, deformed by the crazing mechanism, this shrinkage is 'spread' (up to 100°C or greater) on the temperature scale.

It may be assumed that the mechanism of low-temperature thermally stimulated shrinkage of the crazed amorphous polymers is associated with dispersion feature of the structure of the polymer or, more accurately, with the special features of the structure which the polymer acquires during its deformation subsequent removal of the AAM from the volume of the crazes. It should be remembered that the structure of the crazes is a system of parallel fibrillar aggregates of the oriented macromolecules (Fig. 6.10). The typical transverse dimension of the fibrils equal units or tens of nanometres, i.e., fit in the range of the dimensions characterised by the previously discussed anomalies in the properties and, what is important, a large reduction of the glass transition temperature of the polymer [22–27]. In addition, the diameter of the fibrils in the structure of the craze differs. In most cases, the central part of each craze contains the so-called midrib, i.e., reduction of the width of the fibril by a factor of approximately 1.5–2. The reduction of width of this type is usually recorded in the electron microscope (Fig. 7.13). Figure 7.13 shows the transmission electron micrograph of a single craze (a) and the appropriate electron diffraction diagram [35]. It may be seen that the optical density in the central part of the craze is considerably lower than at the periphery of the craze. This density reduction reflects the reduction of the diameter of the fibrils in the central part of the craze (midrib). The small-angle electron diffraction pattern shows that the equatorial diffusion reflection, reflecting the diameter of the fibrils in the structure of the craze, is 'blurred' and stretched in the horizontal direction. This form of the reflection indicates the considerable inhomogeneity of the fibrils as regards the diameter in the structure of the single craze. The non-uniformity of the dimensions in the structure of the craze indicates the inhomogeneity of the local glass transition temperatures, associated with the size of the polymer

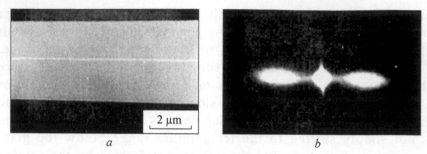

2 µm

a   b

Fig. 7.13. Transmission electron micrograph of the craze in PS (*a*) and the corresponding electron diffraction pattern (*b*) [35]. The tensile loading axis is vertical.

phase according to the assumptions made in [22–27] (see chapter 3). This fact indicates that the glass transition temperature of each individual fibril is not constant along the length of the fibril because, as shown in the previous sections, there is a strong dependence of the glass transition temperature on the geometrical dimensions of the polymer phase.

To understand the physical reasons for the existence of the stretched range of thermally stimulated shrinkage in the craze polymers, it is important to take into account another experimental facts. The point is that as a result of the statistical nature of the crazing as a physical phenomenon [36], the draw ratio differs for the individual crazes formed in the same specimen during its deformation in the AAM. The statistical nature of the crazing phenomenon is clearly indicated by the data Fig. 7.14.

Figure 7.14 shows the dependence of the mean draw ratio of the polymer in the individual crazes, determined in direct microscopic measurements, on the width of the craze [37]. It may clearly be seen that as the width of the craze increases, i.e., as the time during which it is loaded in the expansion stage increases, the draw ratio of the polymer inside the craze also increases.

This circumstance is of considerable importance for the structure of the specimen of the polymer subjected to crazing. As shown previously [38] (see Fig. 6.22, chapter 6), in the sample subjected to drawing in the AAM, the distribution of the draw ratio in the individual crazes, developed simultaneously in the deformed specimen, is very wide. In turn, the draw ratio λ is linked unambiguously with the volume fraction of the fibrils V_f in the craze:

$$\lambda = 1/V_f \qquad (7.1)$$

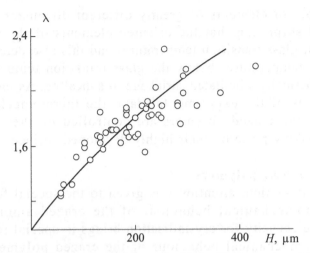

Fig. 7.14. Dependence of the draw ratio (λ) in fibrils on the width of the crazes (H) of polyvinyl chloride, deformed in ethanol [37].

The relationship (7.1) shows that the volume fraction of the high dispersion fibrillised polymer in the individual crazes varies in a wide range. In turn, this variation determines the wide range of the values of the specific interfacial surface in the individual crazes and this also determines the wide distribution of the glass transition temperature in the crazed specimens.

In the presence of a wide temperature range of low-temperature restoration of the oriented polymer, a significant contribution may also provided by another phenomenon. According to Di Marzio's theory, the draw ratio of the polymer also influences its glass transition temperature [39]:

$$\ln\left(T_g / T_g^0\right) = -\left(R / 2\Delta C_p\right) M_u / M_c \left(\lambda^2 + 2/\lambda - 3\right) \qquad (7.2)$$

where T_g^0 is the glass transition temperature of the non-oriented polymer, M_c is the mean numerical value of the molecular section of the chain between cross-links, M_u is the molecular mass of the elementary chain loop, λ is the draw ratio, ΔC_p is the variation of heat capacity during orientation, and R is the universal gas constant. In [40], this reduction of the glass transition temperature ($\sim 10°C$) was demonstrated by the DSC method for polyethylene terephthalate.

Thus, the polymer deformed in the liquid medium by the crazing mechanism is characterised by the highly non-uniform structure and represents a complicated highly dispersed oriented system with a large number of elements of greatly different dimensions. It is therefore not surprising that the oriented elements of the system show different glass transition temperatures and this also determines the wide low-temperature (below the glass transition temperature) range of its thermally stimulated shrinkage in annealing. As indicated by the analysis of the experimental data, the thermomechanical properties of the crazed polymers are controlled by the surface phenomena, taking place on their highly developed surface.

Crystallising glassy polymers
In the previous section, attention was given to the special features of the thermomechanical behaviour of the crazed amorphous polymers. The capacity for crystallisation brings in special features in the thermomechanical behaviour of the crazed polymer. The first chapter describes the phenomenon of self-elongation of the oriented crystallising polymer (polyethylene terephthalate) during its annealing. It was found that the deformation of this polymer in the AAM by the crazing mechanism also results in the highly unusual thermomechanical behaviour of the polymer.

It is important to mention that the phenomenon of spontaneous elongation in annealing of the crazed polyethylene terephthalate was found a long time ago [41, 42]. In this case, self-spontaneous elongation takes place at such a higher rate that the final dimensions of the specimens after annealing are considerably greater than prior to annealing. The crazed polymer has a relatively complicated structure in which the regions of the crazed material, i.e., the regions with the fibrillised porous structure alternate with the regions of the non-deformed polymer. The latter circumstance greatly complicates the interpretation of the thermomechanical behaviour of such objects and, therefore, the detailed mechanism of thermally stimulated change of the dimension of the crazed polyethylene terephthalate still requires explanation.

Figure 7.15 shows the external appearance of two polyethylene terephthalate samples stretched in the ethanol medium by 50% and dried with the fixed dimensions. In these conditions, the polymer is deformed by the formation and development of microscopic zones, containing the oriented fibrillised material (crazes). After drying the produced samples in the clamps of the tensile loading device and removing the mechanical stress, the dimensions of the samples do not change at all. This procedure was used to produce a sample shown in Fig. 7.15 (left). Annealing of the sample at 170°C resulted

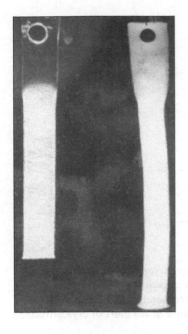

Fig. 7.15. External appearance of the crazed samples of polyethylene terephthalate prior to annealing (left) and after annealing at 170°C (right) [43].

in large changes also visible with the naked eye (Fig. 7.15, right). Firstly, the non-crazed part of the polymer becomes milky white and opaque as a result of the so-called cold crystallisation. Secondly, it may be clearly seen that the sample was greatly elongated as a result of annealing in the direction of the tensile loading axis and its size in the perpendicular direction decreased. The data in Fig. 7.15 clearly indicate the phenomenon of self-spontaneous elongation observed in the annealing of crazed polyethylene terephthalate.

It should be mentioned that the thermomechanical behaviour of the crazed polyethylene terephthalate in annealing is relatively complicated. Figure 7.16 shows the thermomechanical curves of the specimens of polyethylene terephthalate oriented in different conditions. One of the specimens (curve 1) was tensile loaded in air at room temperature with the formation of a neck. As expected, this specimen shows shrinkage during annealing. The shrinkage is incomplete because of the process of cold crystallisation. At the same time, the sample, deformed in ethanol by the crazing mechanism shows a very unusual thermomechanical behaviour (Fig. 7.16, curve 2). After some low-temperature shrinkage, starting at approximately at the glass transition temperature of polyethylene terephthalate, the polymer shows the increase of the dimensions in the direction of the axis of tensile loading (self-spontaneous elongation phenomenon).

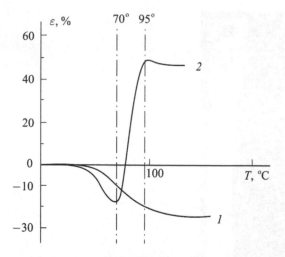

Fig. 7.16. Thermomechanical curves of polyethylene terephthalate deformed in air with the formation of a neck (1) and stretched in ethanol by 50% by the crazing mechanism (2). The vertical dot and dash lines correspond to the T_g temperature of polyethylene terephthalate (70°C) and the temperature of the start of crystallisation of amorphous polyethylene terephthalate (95°C).

It should be mentioned that this phenomenon occurs in a relatively narrow temperature range from 65–70 to 95°C. It is therefore necessary to examine the mechanism of phenomena accompanying the thermally stimulated rearrangement of the crazed polyethylene terephthalate. Firstly, what is the reason for the shrinkage of the polymer at temperatures lower than its glass transition temperature? Secondly, why is the shrinkage in the range of the glass transition temperature of the polymer replaced by spontaneous self-elongation of the polymer and what determines the temperature conditions of these phenomena? Finally, thirdly, what are the driving forces of low-temperature shrinkage and self-elongation?

Since the crazed polymer has a complicated structure (non-oriented areas alternate with the crazes with a fibrillar–porous structure), it is important to determine how the complicated structure of the crazed polymer determines the variation of its geometrical dimensions as a whole. The first chapter describes and justify is the method of visualisation of the structural rearrangement taking place in the polymers during deformation or shrinkage [44–48]. This method is based on the deposition of a thin (nanometric) coating on the surface of the polymer prior to deformation or shrinkage. The subsequent variation of the geometrical dimensions of the polymer during deformation shrinkage results in a special type of surface structure formation in the coating which contains information on the mechanism of structural rearrangement in the polymer [44–48]. Detailed investigation of the structural rearrangement, accompanying the annealing of the polyethylene terephthalate crazed in the AAM, was carried out in [44].

Using these results, it will be attempted to describe a pattern of structural rearrangement accompanying annealing of the crazed polyethylene terephthalate. The microscopic data, obtained in [44], indicate that the shrinkage of the crazed polyethylene terephthalate below its glass transition temperature is determined by the processes taking place in the volume of the crazes. It is important to mention that this shrinkage takes place in the temperature range considerably lower than the glass transition temperature of bulk polyethylene terephthalate (Fig. 7.16). The unusual structural–mechanical and, in particular, thermomechanical behaviour of the craze polymers has been detected many times previously. In the previous section, detailed attention must be paid to the thermomechanical behaviour of the amorphous glassy polymers subjected to crazing in the AAM. It was shown that these polymers show thermally stimulated shrinkage in the temperature range below the T_g of the bulk polymer. This thermomechanical behaviour has been explained without contradiction by the relatively recently discovered phenomenon of the decrease of T_g of the amorphous polymer in thin (nanometric) films and surface layers [22–27]. Since the low-temperature shrinkage of the crazed polyethylene terephthalate is determined by the processes taking place in the volume of the crazes, it appears evident that the low-temperature shrinkage of this polymer has the same mechanism as the shrinkage of the amorphous polymers investigated in the preceding section. Actually, the estimates show [49] that at a typical diameter of 10 nm the fibril contains no more than 10 chains. It follows from here that all the mutually oriented macromolecules the fibrils of the crazes are located in the surface layer and their glass transition temperature is lower than that of the bulk polymer. The decrease of the mechanical properties in the material of the crazes in comparison with the bulk polymer was confirmed directly in [15], where the studies of the dynamic mechanical properties were carried out to determine T_g in the fibrils of the crazes in the PS. The authors showed a large reduction of this characteristic in comparison with the T_g temperature of the bulk polymer and, in addition to this, they developed a method of determining the diameter of the fibrils in the structure of the polymer on the basis of the decrease of T_g.

Since the glass transition temperature depends on the distance of
the macromolecules from the surface [22–27], the fibrillised material
of the crazes has a reduced T_g temperature 'spread' in the range of
low temperatures. Thus, the low-temperature shrinkage of the crazed
glassy polymers is determined by the entropic contraction of the
fibrils, connecting the edges of the crazes, on reaching their local
glass transition temperature.

Attention will now be given to the possible reasons for the self-
spontaneous elongation of the crazed polyethylene terephthalate. It
is important to note (Fig. 7.16) that the low-temperature shrinkage
of the crazed polyethylene terephthalate is interrupted approximately
in the range of its glass transition temperature and is replaced by
self-spontaneous elongation. Termination of the shrinkage during
annealing is evidently explained by the start of crystallisation
of the oriented fibrillised material in the crazes of polyethylene
terephthalate. The same behaviour is observed for the polyethylene
terephthalate, oriented in air with the formation of a neck (Fig. 7.16,
curve 1). These phenomena are determined by the fact that the 'cold'
crystallisation of polyethylene terephthalate is greatly facilitated by
orientation of the polymer [51, 52].

The electron microscopic data, obtained in [44], suggest that the
self-spontaneous elongation of the crazed polyethylene terephthalate
is determined by the fact that the areas of the polymer, localized
between the crazes, take part in the process of the general variation
of the geometrical dimensions of the polymer. It is important to note
that the very phenomenon of self-spontaneous elongation has been
known for a relatively long time and was detected when examining
the structural and mechanical behaviour of a number of crystallising
polymers [53–55]. In particular, the phenomenon of self-spontaneous
elongation has been detected many times also for the polyethylene
terephthalate, subjected to different temperature–force effect [56–59].
In the majority of the cited investigations, the phenomenon of self-
spontaneous elongation is described without assumptions regarding
its mechanism. The most justified viewpoint is that of Bosley [60],
who assumes that the mechanism of self-spontaneous elongation
is determined by the orientation crystallisation of the polymer.
According to this model, prior to heat treatment the polymer should
be slightly oriented so that its volume contains a certain amount
of the oriented crystallisation nuclei. In annealing of this system,
crystallisation starts on these nuclei, leading to the state in which the

segments of the amorphous chains are subjected to axial compression and, consequently, the sample stretches.

A similar interpretation of the phenomenon of self-spontaneous elongation of the crazed polymers was proposed in [61, 62]. In these studies, it is shown that the self-spontaneous elongation can be induced not only by heating of the crazed polymer but also by other types of influence, initiating its crystallisation, for example, the effect of a swelling solvent or radiation heating in irradiation of the polymer with high energy electrons.

The microscopic theory, proposed in [44], was used to determine more accurately the pattern of structural rearrangement, accompanying self-spontaneous elongation. Important is the conclusion of the study according to which the orientation crystallisation, starting at the oriented nuclei, localized at the boundary of the craze and the non-oriented part of the polymer, spreads or its entire volume. In fact, the data presented in Fig. 7.15, indicate that the general contraction of the non-oriented part of the polymer takes place. In particular, the non-oriented part of the polymer or, more accurately, the processes of orientation crystallisation, taking place in this part of the crazed specimen, are responsible for the observed phenomenon of self-spontaneous elongation.

This can easily be confirmed by varying the mechanical properties of the non-oriented fragments of polyethylene terephthalate, localized between the crazes. The point is that up to now we have examined the crazing and thermomechanical behaviour of crazed polyethylene terephthalate which was in the amorphous glassy state. However, this polymer easily crystallises in annealing above its T_g. As a result of this 'cold' crystallisation, it is possible to change the phase state and mechanical properties of the polymer in a wide range.

Figure 7.17 *a* shows the dilatometric curves of the crazed samples of polyethylene terephthalate, which were annealed at different temperatures prior to tensile loading in the AAM. It may be clearly seen that the thermomechanical behaviour of the crazed polyethylene terephthalate drastically changes. The crystallisation of the polymer results in the replacement of the self-elongation process by the shrinkage process. This is understandable because the crazing of the polymer, regardless of its phase state (amorphous or crystalline), is accompanied by the dispersion of the material in the volume of the crazes to the nanosized level, and the refining of the polymer of this type results in a large decrease of its T_g which is also the driving force for thermally stimulated shrinkage (see chapters 4–7).

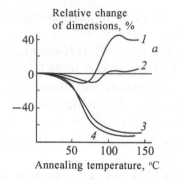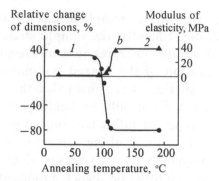

Fig. 7.17. Temperature dependence of the characteristics of the sample of polyethylene terephthalate, deformed in the AAM by 50%: *a*) the relative change of the linear dimensions (1), the annealing temperature of the initial polymer 100 (2), 110 (3) and 170°C (4); *b*) the dependence of the linear dimensions (1) and the elasticity modulus of polyethylene terephthalate at 76°C (2) on the preliminary annealing temperature of polyethylene terephthalate [2].

Figure 7.17 *b* shows that the change of the self-elongation of the crazed polyethylene terephthalate by its shrinkage takes place in the temperature range of annealing characterised by a large increase of the modulus of elasticity of polyethylene terephthalate at a temperature of 76°C, i.e., at its glass transition temperature. It should be mentioned that this temperature is the start of the process of self-elongation of polyethylene terephthalate during its heating.

The important role of the processes, taking place in the non-oriented sections of the crazed polyethylene terephthalate, is confirmed also by the data of optical microscopy of the specimens containing a small number of the crazes. For this purpose, the polyethylene terephthalate film was stretched in the AAM by a small value (3–5%) and dried with fixed dimensions. In these conditions, the polymer shows a relatively small number of crazes so that it is possible to study their evolution during annealing. Figure 7.18 shows the light micrographs obtained from a section of the crazed sample of polyethylene terephthalate during annealing. It may be seen that in these conditions the polymer contains a small number of crazes positioned at large distances from each other (tens or hundreds of microns). As a result of their porous structure, the crazes are clearly visible in the transparent film of polyethylene terephthalate (Fig. 7.18 *a*). When the specimen is heated to 70°C it may be seen that the crazes starts to widen (Fig. 7.18 *b*). The thickness of each craze increases 2–3 times. Analysis of the microscopic data makes it possible to assume that the observed widening is associated with the

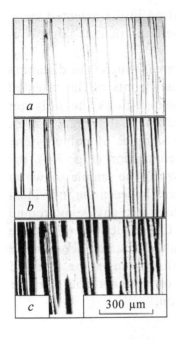

Fig. 7.18. Light micrographs of the crazed samples of polyethylene terephthalate prior to annealing (*a*) and after annealing at 70°C (*b*) and 95°C (*c*). The tensile loading axis is horizontal.

process of orientation crystallisation. The crystallised polyethylene terephthalate has an opaque milky white colour and this creates the impression of the widening of each craze. In fact, this optical effect is associated with the fact that crystallisation takes place by the movement of the front of the crystallised (opaque) polymer into the depth of the specimen from the wall of each craze. The boundary between this bulk polymer is in contact with the non-oriented craze. It may clearly be seen that this growth takes place in the direction of the tensile loading axis of the polymer (normal to the axis of each craze). This process is completed at ~95°C. At this moment, the front of oriented crystallisation propagate into the bulk polymer over a distance of many tens of microns. However, the crystallisation process is interrupted because evidently gradual disorientation of the polymer takes place in the found at the boundary between the amorphous bulk polymer in the polymer crystallised in the oriented state. The polymer, which has travelled such a large distance from the nuclei of orientation crystallisation, 'forgets' about them and crystallisation is interrupted. This is associated with the well-known fact that the crystallisation of amorphous polyethylene terephthalate is made much easier in the case of its molecular orientation [51, 52]. At the same time, at 95°C the crystallisation of the non-oriented polyethylene terephthalate no longer takes place so that the regions of the polyethylene terephthalate, crystallised in the oriented state, are clearly visible on the background of the non-oriented amorphous polyethylene terephthalate (Fig. 7.18 *c*).

It is important to mention that in the areas where the distances between the crazes are small the individual fronts of the oriented crystallisation make contact and in these regions the polymer completely changes to the oriented crystalline state (Fig. 7.18 *c*). At the same time, in the structure of the samples of polyethylene terephthalate, showing self-spontaneous elongation (Fig. 7.6), the number of crazes is considerably greater than in the sample shown in Fig. 7.18. Naturally, in this case, the distances between the individual crazes are very small (~5–50 μm) and they are easily overcome by the front of oriented crystallisation. Consequently, the entire material of the bulk polyethylene terephthalate, situated between the crazes, crystallises in the oriented state. In particular, the oriented crystallisation of the bulk polyethylene terephthalate is also the driving force of self-spontaneous elongation.

7.3. Colloidal swelling

Thus, the previously discussed special features of the mechanical and thermomechanical behaviour of the glassy polymers confirm the important and often controlling role of the surface phenomena in these properties. In fact, the crazed polymers are unique colloidal systems having the set of the properties characteristic of this type systems. It may be expected that the mechanical and thermomechanical behaviour of the crazed polymers does not exhaust the entire range of these special characteristics.

As mentioned previously, the removal of the AAM from the volume of the crazes after deformation of the polymer in the AAM does not restore the initial structure of the bulk glassy polymer. The polymers, subjected to cold drawing in adsorption from active medium, represent a specific high-dispersion colloidal systems (Fig. 7.4). However, as already mentioned, after removing the AAM from the volume of the crazes, the polymer sample does not show any tendency for the change of the dimensions, but nevertheless the resultant systems, like any colloidal systems, are not completely stable. For example, if the polymer whose structure is shown in Fig. 7.4 *b* is again placed in the adsorption-active liquid, then the repeated removal of the polymer from the liquid results in extensive shrinkage [9]. This confirms the unusual and complicated interaction of the liquid AAM and the high-dispersion oriented polymer. The repeated interaction of the high dispersion oriented polymer with the absorption active medium was investigated in detail in [63]. In this study, samples of polyethylene terephthalate were tensile loaded in the AAM to the given degree of elongation and the active liquid was then removed from the porous structure. Some of the specimens were dried in the free state, others with the fixed dimensions. After removing the liquid, the samples were again immersed in the adsorption-active liquid which was evidently not capable of causing the true, volume swelling of the polymer.

Figure 7.19 shows the dependence of the relative variation of the mass of the samples of polyethylene terephthalate on the duration of contact of the samples with the active medium. In particular, it should be mentioned that the samples are capable of absorbing large amounts of n-propanol, and the increase of the mass of the samples in the individual cases reaches 40% of the mass of the initial polymer. This may prove to be expected because the weight method, used in the present study, cannot record any substantial absorption of n-propanol in holding the monolithic non-deformed sample of polyethylene terephthalate in it for even for one month. However, the observed phenomenon can be easily explained if it is regarded as the peptisation process, separation of the coalesced structural elements with subsequent filling of the existing microcavities with liquid media. This explanation is supported by large differences in the amount of the liquid absorbed by the samples dried in the different conditions. Evidently (Figs. 7.19 a and b), the volume of the

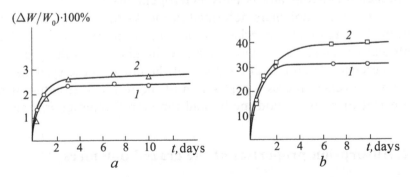

Fig. 7.19. Dependence of the relative variation of the mass of the samples of polyethylene terephthalate subjected to preliminary tensile loading in the AAM (propanol) by 50 (1) and 100% (2) and dried in the free state (a) and with fixed dimensions (b) [63].

micro-cavities in the polyethylene terephthalate, dried after tensile loading in the AAM, in the free state is considerably smaller than in the appropriate specimens dried with fixed dimensions because the opposite walls of the crazes come closer together during shrinkage (see the diagram in Fig. 7.4).

It is important to note that there are available polymer colloidal systems, capable of absorbing a large amount of the liquid component by the mechanism different from the true volume swelling [64]. These systems form during phase separation of the polymer solutions by deposition [65] chemical modification of the dissolved polymer. The removal of the solvent from the resultant porous polymer frame results in a large decrease (2–8 times) of the dimensions of the sample leading to the formation of a transparent and, at the first sight, monolithic material. However, this material 'remembers' its previous history as a result of stored internal stresses. If the resultant 'monolith' is placed in the medium of the solvent, under the effect of the internal stresses the sample increases its dimensions and restores the initial porous structure in which a large amount of the solvent is included. It may be seen that the craze polymers are similar to the previously described Vlodavets–Rehbinder cryptohererogeneous systems [64]. However, there are also large differences between them. The main difference is that the absorption of the liquid by the polymer, tensile loaded in the AAM and subsequently dried, takes place without any significant changes of its geometrical dimensions and is determined by the filling of the internal microcavities already present in the material. In other words, the colloidal system, produced in the drying of the polymer, deformed in the AAM, is not cryptoheterogeneous and is truly heterogeneous.

Thus, the polymers deformed in the AAM are high dispersion colloidal systems characterised by all the properties typical of the colloidal systems. As shown previously, in addition to the processes of coalescence of structural elements of the crazes these systems also show the reverse process – peptisation of the resultant structure under the effect of active liquid media and the stored internal stresses.

7.4. Adsorption properties of the crazed polymers

As already mentioned, crazing is a special type of the inelastic, plastic deformation of the polymer which is accompanied by the rapid development of its interfacial surface. Many aspects of this phenomenon are still unclear but undoubtedly the following is obvious: the polymer as a result of crazing acquires a high level of the interfacial surface. Below, the attempt to characterise the purposes of the crazed polymers from this viewpoint.

As mentioned previously [1, 2] the transition to the oriented state of the polymers during crazing takes place by its transformation to the highly organised structural form, shown schematically in Fig. 7.4 *a*. The main element of such a structure is the asymmetric aggregate of the oriented macromolecules – the fibril. The transverse dimensions of these formations range from several nanometres to several tens of nanometres. Such dimensions are characteristic of the typical highly dispersed colloidal systems. Evidently, the 'disintegration' of the polymer to such small structural elements results in the large interfacial surface of the polymer.

It is natural to assume that any highly dispersed system should have a number of physical–chemical properties, characteristic of the colloidal state of matter. One of these properties is the capacity for extensive adsorption demonstrated by any solid with a highly developed surface. Actually, the solids with a highly developed surface are porous adsorbents and, therefore, they should be studied by the adsorption methods. These methods have been sufficiently developed and can be used to determine the parameters of the porous structure, such as the specific surface area, the size distribution of the pores, the specific volume of the pores, etc [66]. Evidently, if the crazed polymers do indeed have a high level of the interfacial surface, they should be effective porous adsorbents.

To explain this capaciy of the crazed polymers, in [67] the authors studied the adsorption of iodine and a number of organic substance from their aqueous solutions on the surface of polyethylene terephthalate, deformed in the AAM (*n*-propanol). The polyethylene terephthalate samples were tensile loaded to a specific degree of elongation in the AAM and then transferred into the appropriate aqueous solutions and the change of their concentration was studied by the colorimetric method.

The results of the measurement of the concentration of sorptives in the solution were used to construct adsorption isotherms. Figure 7.20 shows the adsorption isotherms of iodine (a) and the organic due rhodamine C (b) from their aqueous solutions by the polyethylene terephthalate samples, deformed in *n*-propanol to various degrees of elongation. The first fact that should be noted is the extensive adsorption of low-molecular substances on the samples of the crazed

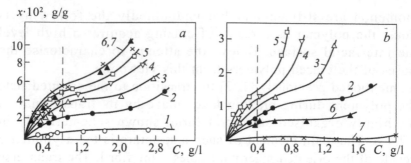

Fig. 7.20. Isotherms of sorption of iodine (*a*) and rhodamine C (*b*) from aqueous solutions by polyethylene terephthalate samples tensile loaded by the crazing mechanism in the AAM to draw ratios: 1) 20%, 2) 50%, 3) 100%, 4) 150%, 5) 200%, 6) 300%, 7) 400%.

polymer. This result directly confirms the often observed presence of the highly developed surface in the crazed polymers. Secondly, sorption from the solutions on the crazed polymer does not depend on the nature of the sorptive or polymer [68–70]. This behaviour indicates the non-specific nature of absorption. In other words, the observed absorption is not determine by the presence in the sorbent and sorptive of specific functional groups, capable of reacting with each other, and is determined exclusively by the presence of the highly developed surface of the crazed polymer.

Finally, it appears that the nature of the adsorption isotherms depends strongly on the draw ratio of the polymer in the AAM. Evidently, the observed differences are determined by the structure of the polymer because all other conditions (temperature, concentration and the nature of the sorptive) were the same. These data are very important because they show directly for the first time that the polymers, deformed in the AAM, actually have a more developed interfacial surface and are effective porous adsorbents. In fact, the resultant samples of polyethylene terephthalate adsorb, for example from iodine solutions with a relatively low concentration (approximately 1.5%) up to 40 wt.% of iodine in relation to the mass of the dry polymer. At the same time, the non-deformed samples of polyethylene terephthalate, and also the samples tensile loaded in air with the formation of a neck do not show any adsorption, recorded by the colorimetric method. In addition to this, these data show that we can use the adsorption methods of investigation for obtaining valuable information on the nanoporous structure of the craze polymers which cannot be or is very difficult to produce by other methods. To obtain the data on the structure of the porous sorbent it is not absolutely necessary to construct the adsorption isotherms in such a wide concentration range as shown in Fig. 7.20. It is sufficient to determine the region of the adsorption isotherms in which the first adsorption monolayer is saturated.

In [67], in addition to iodine and organic dye rhodamine C (molecular mass 478), attention was also given to the adsorption of potassium indigo tetra sulphonate (ITS) (molecular mass 742) and vitamin B12 (molecular mass 157]. It may be seen that the selected sorptives have different molecular dimensions. Evaluation of these dimensions by analysis of the molecular models gives for the rhodamine C molecules the effective diameter of 1.75 nm, for the ITS 2.6 nm, and for the molecule of the vitamin B12 3.8 nm. The effective diameter of the iodine molecule is 0.54 nm. The molecular dimensions of the series of the sorptives investigated in [67] varied in a relatively wide range.

To analyze changes, taking place in the polymer, it is far more efficient to examine the dependence of adsorption at some fixed equilibrium concentration of the sorptive on the degree of deformation of the polymer. Such concentrations in [64] were selected in the region of saturation of the monolayer (indicated in Fig. 7.20 by dashed lines for each family of the adsorption isotherms).

Figure 7.21 shows the dependence of adsorption on the draw ratio of the polymer for the sorptives of different molecular dimensions. It may be seen that the adsorption of iodine at low degrees of elongation increases and then, reaching the maximum, ceases to change, whereas the large volume molecules of rhodamine C in the region of 200% elongation show a large reduction of the degree of adsorption. The potassium indigo tetra sulphonate, having even larger molecular dimensions, also shows the extreme dependence of adsorption on the draw ratio of polyethylene terephthalate in the AAM, and in the region of the high values of the draw ratio this substance does not show any adsorption at all. A similar behaviour was recorded for vitamin B12 for which the maximum adsorption is displaced into the range of 100% elongation.

These experimental results show that the adsorption from the solution is highly sensitive to the changes of the structure of the polymers, deformed in the AAM. Actually, the adsorption data show that in the initial stages of tensile loading the resultant crazes containing large cavities are easily accessible to even the largest (~3 nm) molecules of vitamin B12. The increase of the draw ratio results in the further growth of the crazes and this increases the

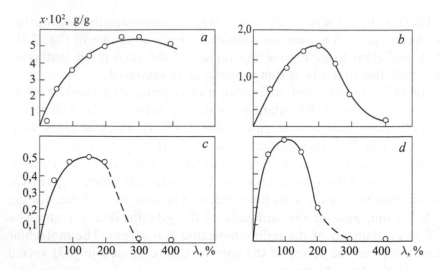

Fig. 7.21. Dependence of adsorption on the draw ratio of polyethylene terephthalate in the AAM: a) the aqueous solutions of iodine at an equilibrium concentration of 0.8 g/l; b) the rhodamine C at an equilibrium concentration of 0.04 g/l; c) ITS at an equilibrium concentration of 0.04 g/l; b) vitamin B12 at an equilibrium concentration of 0.6 g/l.

interfacial surface area of the polymer and, correspondingly, the adsorption. However, tensile loading of the polymer results in the start of the previously mentioned collapse of the high-dispersion material of the crazes leading to the contraction of the structure and a reduction of the effective pore diameter. Starting at 100% elongation, the number of the pores, accessible to the molecules of B12 vitamin, with the molecular size of ~3 nm, decreases from 200% – accessible to ITS (approximately 2.3 nm), from 250% – accessible to rhodamine C (approximately 1.8 nm) and from 300% – accessible to iodine (approximately 0.6 nm). At high draw ratios (300–400%) the polymer is almost completely free from the pores accessible to ITS and vitamin B12, slightly accessible for rhodamine C and easily accessible for iodine. It may be seen that the method of adsorption from the solutions can be used for the highly detailed study of the porous structure of the polymer, deformed in the AAM.

The results are in good agreement with the previously proposed (on the basis of structural and mechanical studies) mechanism of the structural rearrangement of the polymer at high degrees of deformation (Figs. 7.6 and 7.8). Actually, in transition from the porous structure at low degrees of deformation to a more compact structure – at high degrees of deformation, the results show the collapse of structural elements (fibrils) of the crazes which leads unavoidably to a large decrease of the radius of the pores.

The data on adsorption provide accurate information on the structure of the polymer deformed in the AAM. Firstly, they can be used for the quantitative characterisation of the change of the dimensions of the pores in the structure of the crazes depending on the draw ratio of the polymer. Actually, using these data, it may be assumed that the first stages of tensile loading are characterised by the formation of considerably larger pores than the molecular dimensions of the largest of the investigated sorptives, i.e., larger than 3 nm.

Taking into account the fact that the adsorption of iodine and rhodamine C from water was considerably greater than from n-propanol, it may be expected that placing the polyethylene terephthalate sample which reached the adsorption equilibrium in water, in pure n-propanol, will result in the process of desorption in the case in which the adsorbed substance does not react chemically with the surface of the adsorbent. These experiments were carried out in [67]. The experimental results indicate that both iodine and the rhodamine C show extensive desorption when water is replaced by n-propanol, indicating the physical nature of adsorption.

The desorption process is illustrated by the data in Fig. 7.22 which shows the dependence of the relative amount of the desorbed substance in n-propanol to the total amount of the adsorbed substance from water on the draw ratio of polyethylene terephthalate. As indicated by the graph, in the range of small values of deformation, the adsorbent loses 85–95% of the sorptive because of desorption. However, desorption for both rhodamine C and iodine decreases with increasing draw ratio. In the case of rhodamine C this transition is detected in the range of 150–200% elongation of rhodamine, and for iodine at 300–400%. Evidently, the desorption process is also highly sensitive to the structure of the polymer and can be used as a method of investigating this phenomenon.

Actually, the change of the structure of the polymer during its tensile loading in the AAM takes place in such a manner that the size of the resultant pores (the distance between the fibrils) gradually decreases as a result of the coalescence of the high dispersion material of the crazes (Figs. 7.6 and 7.8). When the size of the pores becomes comparable with the size of the molecules of the sorptive, the degree of desorption rapidly decreases because the exit of the sorptive molecules from the relatively wide pores is easier and from

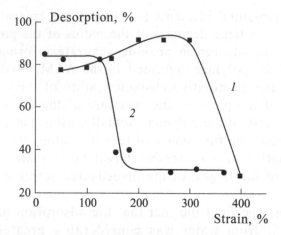

Fig. 7.22. Dependence of the relative desorption of iodine (1) and rhodamine C (2) on the draw ratio of polyethylene terephthalate in the AAM.

the narrow ones it is more difficult, as a result of the overlapping of the dispersion forces of the walls of the pores [66]. As a result of the difference in the size of the iodine and rhodamine C molecules, this transition is detected for them at different draw ratios, in particular earlier for the large rhodamine C molecules and later for the small iodine molecules.

Thus, the adsorption data confirm the existence in the crazed polymers of the highly developed surface and help to describe quantitatively the unique porous structure of the crazes and its evolution in the process of deformation of the amorphous glassy polymer. The results again indicate that the amorphous polymers, subjected to crazing, are unique colloidal systems and many of their properties are determined by the presence in them of the highly developed interfacial surface.

Conclusions

Analysis of the data, presented in chapter 7, shows that, firstly, the polymers, deformed in the AAM, acquire a large interfacial surface as a result of their dispersion to fibrillar aggregates of the oriented macromolecules of the nanometric size. The required excess surface transfers to the glassy polymer an entire set of unusual physical-mechanical, physical–chemical, and other properties. Actually, the crazed oriented polymers demonstrate the low-temperature (below the appropriate glass transition temperature) shrinkage, spontaneous elongation during annealing, and a capacity for colloidal swelling.

As a result of crazing, these polymers acquire a highly developed surface which in fact transforms them by a simple mechanism to porous non-specific adsorbents. All the anomalies in the structural-mechanical behaviour of the crazed glassy polymers are determined by two factors: firstly, by the presence in them of a highly developed interfacial surface and, secondly, the fundamental property of the glassy polymers – a large reduction of their glass transition temperature in the thin-films and interfacial layers of the nanometric thickness.

References

1. Volynskii A.L., Bakeev N.F, Highly dispersed oriented state of polymers, Moscow, Khimiya, 1984, 190 pp.
2. Volynskii A. L., Bakeev N.F., Solvent crazing of polymers, Amsterdam, New York, Tokyo, Elsevier, 1995. P. 410.
3. Volynskii A.L. Bakeev N.F., Vysokomolek. Soed. A, 1975, V. 17, No. 7. P. 1610.
4. Treloar L., Physics of rubber elasticity. Moscow, Mir, 1975.
5. Volynskii A.L. Smirnov V.D., Stoches R.N., Gerasimov V.I., Aleskerov A.G., Bakeev N.F., Vysokomolek. Soed. A, 1976, V. 18, P. 940.
6. Volynskii A.L., Aleskerov A.G., Grokhovskaya T.E., Bakeev N.F., Vysokomolek. Soed. A, 1976, V. 18, No. 9, P. 2114.
7. Volynskii A.L., Loginov V.S., Bakeev N.F., Vysokomolek. Soed. B, 1981, V. 23, No. 4, P. 314.
8. Papkov S.P., Syneresis, Encyclopedia of polymers, Sov. Entsiklopediya, Moscow, 1977, V. 3, P. 409.
9. Volynskii A.L., Gerasimov V.I., Bakeev N.F., Vysokomolek. Soed. A, 1975, V. 17, P. 2461.
10. Kambour R.P., Kopp R.W., J. Polymer Sci. A-2, 1969, V. 7, P. 183.
11. Sinevich E.A., Prazdnichnyi A.M., Bakeev N.F., Vysokomolek. Soed. B. 1988, V. 30, P. 536.
12. Wool R. P., O'Connor K.M., Polym. Engng. Sci., 1981, V. 21, P. 970.
13. Berger L.L., Sauer B.B., Macromolecules, 1991, V. 24, P. 2096.
14. Gent A.N., Thomas A.G., J. Appl. Polymer Sci., 1959, V. 1, P. 10.
15. Volynskii A.L., Aleskerov A.G., Bakeev N.F., Vysokomolek. Soed. A, 1982, V. 24, No. 9, P. 1855.
16. Olf H.G., Peterlln A., J. Polymer Sci. Polymer Phys. Ed., 1974, V. 12, P. 2209.
17. Tan E.P.S., Lim C.T., Composites Science and Technology, 2006, V. 66, P. 1102.
18. Hodges C.S., Advances in Colloid and Interface Science, 2002, V. 99, P. 13.
19. Qingzheng Cheng, Siqun Wang, Composites: Part A, 2008, V. 39, P. 1838-1843.
20. Arnold C.-M. Yang, Materials Chemistry and Physics, 1995, V. 41, P. 295-298.
21. Sprague S., Journal of Macromolecular Science, Part B: Physics, 1973, V. B8, No. 1-2, P. 157-187.
22. Forrest J.A., Dalnoki-Veress K., Adv. Colloid and Interface Sci., 2001, V. 94, P. 167.
23. Forrest J.A., Dalnoki-Veress K., Stevens J.R., Dutcher J.R., Phys. Rev. Lett., 1996, V. 77, P. 2002.

24.   Kajiama T., Tanaka K., Takahara A., Proc. Japan Acad., 1997, V. 73B, No. 7, P. 132.
25.   Tanaka K., Kajiyama T., Takahara A., Acta Polymerica, 1995, V. 46, P. 476.
26.   Kajiyama T., Tanaka K., Takahara A., Macromolecules, 1995, V. 28, No. 9, P. 3482.
27.   Lazurkin Yu.S., Disseration, Dr. Sci. Sciences, S.I. Vavilov Institute of Physical Problems, USSR Academy of Sciences, 1954.
28.   Gul' V.E., Kuleznev V.N., Structure and mechanical properties of polymers, Moscow, Khimiya, 1972.
29.   Askadskii A.A., Deformation of polymers, Moscow, Khimiya, 1973.
30.   Le Grand D.G., J. Appl. Polymer Sci., 1971, V. 16, P. 1367.
31.   Solee G., J. Appl. Polymer Sci., 1971, V. 15, P. 2049.
32.   Grokhovskaya T.E., Volynskii A.L., Bakeev N.F., Vysokomolek. Soed. A, 1977, V. 19, No. 9, P. 2112.
33.   Grokhovskaya T.E., Dissertation, Cand. chemical sciences, M.V. Lomonosov Moscow State University, 1977.
34.   Berger L.L., Buckley D.J., Kramer E.J., J. Polymer Sci. Part. B, Polymer Physics, 1987, V. 25, P. 1679-1697.
35.   Volynskii A.L. Yarysheva L.M., Bakeev N.F., Vysokomolek. Soed., 2001, V. 43, No. 12. P. 2289.
36.   Chernov I.V., Dissertation, Cand. chemical sciences, M.V. Lomonosov Moscow State University, 1977.
37.   Chernov I.V., Yarysheva L.M., Lukovkin G.M., Volynskii A.L. Bakeev N.F., Vysokomolek. Soed.. B, 1989, V. 31, No. 6, P. 404.
38.   DiMarzio E.A., J. Res. Natl. Bur. Stds. A, 1964, V. 68, P. 611.
39.   Volynskii A.L. Yarysheva L.M., Bakeev N.F., Vysokomolek. Soed. A, 2011, V. 53, No. 10, P. 871-898.
40.   Volynskii A.L. Grokhovskaya T.E., Gerasimov V.I., Bakeev N.F., Vysokomolek. Soed. A, 1976, V. 18, No. 1, P. 201.
41.   Volynskii A.L., Aleskerov A.G., Grokhovskaya T.E., Bakeev N.F., Vysokomolek. Soed. A, 1976, V. 18, No. 9, P. 2115.
42.   Volynskii A.L. Dissertation, Doctor chemical. Sciences, M.V. Lomonosov Moscow State University, 1979.
43.   Volynskii A.L. Grokhovskaya T.E., Kulebyakina A.I., Bolshakova A.V., Bakeev N.F., Vysokomolek. Soed. A, 2007, Vol 49, No. 7, P. 1224-1238.
44.   Volynskii A.L. Grokhovskaya T.E., Kulebyakina A.I., Bolshakova A.V., Bakeev N.F., Vysokomolek. Soed. A, 2007, Vol 49, No. 11, P. 1946-1958.
45.   Volynskii A.L. Grokhovskaya T.E., Kulebyakina A.I., Bolshakova A.V., Bakeev N.F., Vysokomolek. Soed. A, 2009, V. 51, No. 4, P. 598-609.
46.   Volynskii A.L., Priroda, 2011, No. 7, P. 14-21.
47.   Berger L.L., Sauer B.B., Macromolecules, 1991, V. 24, No. 8, P. 2096.
48.   Dong J., Xue G., Chang R., Chinese Journ. Polym. Sci., 1994, V. 12, No. 3, P. 266.
49.   Busiko V., Corradini P., Riva F., Makromol. Chem. Rapid Commun., 1980, V. 1, P. 423.
50.   Godovskii Yu.K., Thermophysical methods of study of polymers, Moscow, Khimiya, 1976.
51.   Smith W.H., Sayior C.P. J., Res. Natl. Bur. Stand., 1938, V. 21, P. 257.
52.   Lee S.C., Min B.G., Polymer, 1999, V. 40, P. 5445.
53.   Mandelkern L., Roberts D.E., Diorio A.F., Posner A.S., J. Am. Chem. Soc., 1959, V. 81, P. 4148.
54.   Fomenko B.A., Perepechkin L.B., Vasil'ev B.V., Naimark N., Vysokomolek. Soed. A, 1969, V. 11, P. 1971.

55. Oswald H.J., Turi E.A., Harget P.J., Khanna Y.P., J. Macromol. Sci. Phys.,1977, V. B13. P. 231.
56. Liska E., Kolloid-Z. Z. Polym., 1973, V. 251, P. 1028.
57. Pinnock P., Ward I.M., Trans. Faraday Soc., 1966, V. 62, P. 308.
58. Pereira J.R.C., Porter R.S., Polymer, 1984, V. 25, P. 877.
59. Bosley D.E.J., J. Polym. Sci. C, 1967, V. 20, P. 77.
60. Sinevich E.A., Prazdnichnyi A.M., Tikhomirov V.S., Bakeev N.F., Vysokomolek. Soed. Aa 1989. V. 31, No. 8a P. 1697.
61. Sinevich E.A., Prazdnichnyi A.M., Bakeev N.F., Vysokomolek. Soed. A, 1990, V. 32, No. 2, P. 293.
62. Volynskii A.L., Loginov V.S., Bakeev N.F., Vysokomolek. Soed. B, 1981, V. 23, No. 5, P. 371.
63. Sinitsyna G.M., Vlodavets I.N., Rehbinder P.A., Dokl. AN SSSR. 1967, V. 175, P. 2461.
64. Rehbinder P.A., in: Problems of physics and mechanics of fibrous and porous dispersed structures and materials, Riga, Zinatne, 1967, P. 5.
65. Greg S., Singh K., Adsorption. Specific surface Porosity. Trans. from English, Ed. K.V. Chmutov, Moscow, Mir, 1970, P. 407.
66. Volynskii A.L., Loginov V.S., Plate N.A., Bakeev N.F., Vysokomolek. Soed. A, 1980, V. 22, No. 12, P. 2727.
67. Volynskii A.L., Loginov V.S., Plate N.A., Bakeev N.F., Vysokomolek. Soed. A, 1981, V. 23, No. 4, P. 805.
68. Volynskii A.L., Loginov V.S., Bakeev N.F., Vysokomolek. Soed. A, 1981, V. 23, No. 5, P. 1059.
69. Volynskii A.L., Loginov V.S., Bakeev N.F., Vysokomolek. Soed. A, 1981, V. 23, No. 6, P. 1216.

8

Multiphase nanodispersed systems based on crazed polymers

As already mentioned, the crazing of the polymer in the AAM is accompanied by its transfer to the highly dispersed fibrillised state. However, the structural special features of crazing are not limited to this. Firstly, the unique structure of the crazes can develop only if the active liquid is supplied in a sufficient amount and at the appropriate time into the zone of local transition of the polymer to the oriented state. As a result of this deformation the final product is the polymer whose nanoporous structure is filled with a liquid component in which deformation is carried out. This means that the crazing of the polymer in the AAM is an efficient method of supplying low-molecular liquids (AAM) and any substance dissolved in them to the developing pores of the polymer.

Secondly, since the size of the pores in the crazed polymer does not exceed several nanometres, crazing is an efficient method of not only dispersion of the polymer but also of the substance of a different chemical nature dissolved in the AAM. The latter circumstance not only makes it possible to produce the nanodispersed mixtures with the polymer matrix but also offers simple and effective possibilities of investigating almost any substance in the highly dispersed, colloidal state.

8.1. Interaction of low-molecular substances with the highly developed surface of the crazed polymer

It should be mentioned that the crazing can be used to introduce almost any low-molecular substances into the structure of the polymer. The essential condition for this is the low melting point of the low- molecular component or its solubility in any AAM. In the first case, the low-molecular compound may be used as the AAM at a reduced temperature (but, evidently, not higher than the glass transition temperature (melting point) of the deformed polymer). As a result of drawing the polymer in the melt of this substance crazing takes place and its porous structure is filled with this substance. Subsequent cooling makes it possible to crystallise the low-molecular component *in situ* and produce a polymer nanocomposite in which the low-molecular component crystallises in the highly dispersed state. In the second case, the substance, dissolved in the AAM, is introduced to the volume of the craze. Subsequent evaporation of the volatile component also leads to crystallisation *in situ* of the dissolved substance. In this case, it is evident that low-molecular substances with almost any melting point can be introduced into the volume of the craze [1, 2]. Both these methods are based on the direct introduction of the low-molecular substances into the volume of the crazes so that this method of addition to the structure of the crazes was referred to in [3] as the direct introduction method.

Obviously, there is a large number of substances which cannot be added to the crazed polymer by this method. For example, the substances include metals, many metal salts, oxides and other inorganic substances. In this case, the above-mentioned polymer –low-molecular substance mixtures are produced by the chemical reaction of the corresponding precursors directly in the porous structure of the crazes (*in situ*). In [3] this method of producing the nanocomposite is referred to as the indirect method of adding low-molecular substances to the structure of the crazes.

Special features of structure formation in the crazed polymer– low- molecular crystalline substance system

Initial investigations dealing with the addition of the low-melting substance to the polymer by crazing were carried out on the example of the amorphous polyethylene terephthalate–n-octadecane (OD) ($T_m = 28°C$) system [4]. A polymer sample was tensile drawn in the liquid OD at a temperature of 50°C and cooled down in the clamps of the tensile loading machine. As a result of this procedure, the OD which is an efficient AAM fills the porous structure of the crazes. Cooling the produced material below the melting point of the introduced additive results in its crystallisation in the volume of the

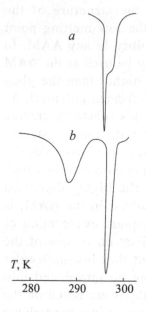

T, K

280　　　290　　　300

Fig. 8.1. DSC data of crystallisation of OD (*a*) and OD, added to the crazes of polyethylene terephthalate (*b*). The draw ratio of polyethylene terephthalate in the OD 50% [4].

crazes. This simple methods can be used to produce nanocomposites with the content of the low-molecular component to 50 wt.% or more. The amount of the added low-molecular component depends on the evolution of the porous structure of the polymer during its tensile loading in the AAM (Figs. 7.6–7.8). Regardless of the draw ratio of the polymer in the AAM, the low-molecular component is dispersed to the finest aggregates with the size not exceeding the diameter of the pores in the structure of the crazes (tens or hundreds of angströms). The production of these nanocomposites not only offers new possibilities for developing a new class of advanced materials but also makes it possible to investigate the properties of the substance in the highly dispersed colloidal state. Some properties of the nanocomposites, produced by crazing in the liquid media, will now be discussed.

Figure 8.1 shows the typical calorimetric curves of crystallisation of the saturated hydrocarbon *n*-octadecane (OD) included in the porous structure of the polyethylene terephthalate and tested in the free state. It may be seen (Fig. 8.1, curve *a*) that in the free state the hydrocarbon crystallises at 24°C, which is reflected in the form of a non-symmetric exothermic peak. Evidently, the latter circumstance is determined by the transition in the melting range from the rhombic to hexagonal packing, typical of the saturated hydrocarbons [5]. At the same time, the OD, situated in the polymer matrix, shows some previously unknown special features in the thermophysical properties

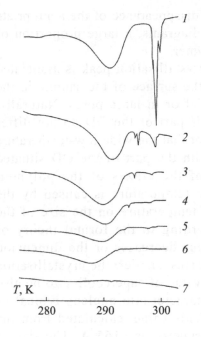

Fig. 8.2. DSC thermograms of crystallisation of polyethylene terephthalate samples containing OD obtained after rinsing the samples in *n*-heptane for 0 (1), 5 (2), 10 (3), 15 (4), 20 (5), 30 (6) and 60 min (7) [4].

As shown in Fig. 8.1, curve *b*, crystallisation of the OD in the polymer matrix appears to take place in two stages. Initially, there is a transition whose temperature coincides with the corresponding transition in the free OD, and subsequently – an exothermic peak forms at a temperature 6–8°C lower than the melting point of the free OD. It may be seen that the main contribution (~80%) to the crystallisation heat is provided by the wide low-temperature peak. The reasons for these differences can be clarified to a large degree by studying the process of washing out the OD from the porous structure of polyethylene terephthalate. The results show that using certain solvents a large part of the added OD is in fact removed by dissolution, but a large part (up to 30%) still remains inside the polymer. Figure 8.2 shows the calorimetric curves of crystallisation of the polyethylene terephthalate samples rinsed in *n*-hexane for different periods of time. It may be seen that the first part is the removal of the part of the added *n*-octadecane with the crystallisation temperature identical with the crystallisation temperature of the free OD (Fig. 8.2, curves 1, 2). Further washing leads gradually to the removal from the polymer of the part of the OD responsible for the low-temperature wide crystallisation peak (Fig. 8.2, curves 2–7). Although long-term rinsing does lead to the complete disappearance of the temperature transitions in the polyethylene terephthalate–

n-octadecane system, and also to the disappearance of the appropriate reflections on the x-ray diffraction diagrams, a large proportion of the hydrocarbon remains in the polymer.

Evidently, the high-temperature crystallisation peak is associated with a small amount of the OD on the surface of the sample in the macroscopic irregularities of its relief or in large pores. Naturally, the thermophysical properties of this part of the OD do not differ greatly from the free OD. It is evident that the wide low-temperature crystallisation peak is associated with the part of the OD situated directly in the porous structure of the crazes of the polymer. The reduction of the crystallisation temperature is caused by the dependence of the phase transition temperature on the size of the resultant crystalline nucleus. According to the formal theory of nucleation, as the size of the nucleus decreases or the dimension of the section of the new phase becomes smaller, the crystallisation temperature also decreases [6]. The value of supercooling to be also used to estimate the size of the characteristic nucleus typical of the crystallisation of the free OD. This value, calculated from the results of differential-scanning calorimetry, is ~165 Å. The size of the nucleus is considerably larger than the size of the large part of the pores in the structure of the polyethylene terephthalate deformed in the AAM. The dimensions of the crystallising phase are limited by the walls of the pores, and the length of the phase is smaller than that of the nucleus in the free OD. This results in a large reduction of the crystallisation temperature of the substance dispersed to such small aggregates. The size distribution of the pores in the polymers is relatively wide because the low-temperature crystallisation peak is greatly stretched over the temperature scale. Consequently, this phenomenon was used for evaluating the size distribution of the pores, typical of the polymer deformed in the given AAM. The procedure used for these calculations is described in [2]. The results of these calculations are presented in Fig. 8.3 in the form of the size distribution of the pores for the polyethylene terephthalate samples tensile loaded in the OD to 50 (1) and 400% (2). It may be seen that the increase of the draw ratio of the polymer results in a large reduction of the effective radius of the pores determined on the basis of the reduction of the crystallisation temperature. This results corresponds to the previously mentioned considerations regarding the evolution of the porous structure of the polymer during its deformation in the AAM (Figs. 7.6–7.8) obtained by other investigation methods.

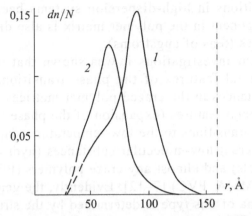

Fig. 8.3. The curves of the size distribution of the pores in the structure of polyethylene terephthalate samples deformed in the OD by 50 (1) and 400% (2). The dashed line shows the size of the critical crystallisation nucleus of OD in the free state [2].

However, the resultant distribution curves can not describe accurately the porous structure of the polymer. As shown previously, after the disappearance of the thermophysical transitions and x-ray reflections as a result of rinsing, a large amount of OD appears in the samples. In [2] it was shown that the fraction of *n*-octadecane, which can be removed from the structure of the craze by rinsing, increases with increasing tensile strain and the reaches high values (~25%). These data indicate that the crazed polymer, deformed in the AAM to high draw ratios, contains a large number of the internal microcavities inaccessible to the solvent so that the added low-molecular substance cannot be extracted from the polymer. This structure forms as a result of the collapse of the structure of the crazes in the manner shown in Figs. 7.6–7.8. At the same time, the shown closed microcavities are small so that the OD present in them can not form an extended crystalline phase so that the DSC thermograms do not show its phase transitions, and the appropriate x-ray diffraction patterns do not show x-ray reflections, characteristic of the crystalline structure of the OD. It is quite interesting to evaluate the size of the aggregate of the molecules below which it can not be described by the thermodynamic concept of the 'phase'. As indicated in Fig. 8.3, in the case of *n*-octadecane the size is ~25 Å.

Thus, the addition to the polymer, deformed in the AAM, of low-molecular organic compounds makes it possible on the one hand to obtain additional data on its porous structure and, on the other hand, experiments can be carried out to investigate the special features of phase transitions in high-dispersion systems because the low-molecular component in the polymer matrix is also dispersed to the finest aggregates (tens of angstroms).

In subsequent investigations it was shown that the previously mentioned special features of the phase transitions of the low-molecular substances in the crazed polymer matrices are of general nature. The special features (expansion of the phase transitions and the shift of the transitions to the low-temperature range) is detected for greatly different low-molecular substances (hydrocarbons, fatty acid and alcohols) and almost any craze polymers (PET, HDPE, PP, PA-6, PTFE, PMMA, PVC) [7–12]. Evidently, the general nature of these phenomena of this type is determined by the structural special features of the fibrillar–porous structure of the crazes.

The above-mentioned special features of the thermophysical behaviour of the low-molecular substances in the structure of the crazes are not unique. For these substances, capable of showing polymorphous transitions, crystallisation in the narrow pores has its own special features. For example, in [9, 11] the methods of differential scanning calorimetry (DSC) were used to investigate the phase composition of low-molecular compounds added to the porous polymer matrices. In all cases, the substance, added to the porous structure of the polymer, is present in the high-temperature modification which is unstable for the same substance in the free state.

The high stability of the modifications of the low-molecular compounds, unstable in the free state, can be explained by taking into account the high dispersion of the microporous structures. In the micropores, all the polymorphous transformations can take place only in the temperature range in which the radius of the critical nucleus of the low-temperature phase is smaller or equal to the radius of the pore. If the temperature at which this condition is satisfied does not fit to the temperature range in which the rate of formation of the new phase is sufficiently high, there are no polymorphous transformations and only the high-temperature polymorphous modification of the low-molecular compound forms.

The low-molecular compounds, solidified in the micropores of the crazed oriented polymer matrices, are in the highly dispersed state. Consequently, analysis of their phase transitions in the investigated systems should be carried out using the surface component of the thermodynamic potentials (free energy, enthalpy, entropy). The thermodynamic potentials of the low-molecular compound will be

Table 1. Parameters of the porous structure of the oriented polymers

System	Fibril diameter, nm	Pore diameter, nm	
		SAXS	DSC
PET/HE	8.0	6.0	6.5
PET/TDA	10.5	7.0	5.0
PC/TDA	29.0	7.0	5.0
Nylon-6/CA	8.3	4.6	7.0

represented as the sum of the volume and surface components. Then, taking into account the fact that the specific surface of the system does not change during the polymorphous transformation of the low-molecular compound in the polymer micropores, it is easy to derive equations describing the variation of the specific surface components of the thermodynamic potentials at the polymer–low-molecular substance interface and the dependence of the thermodynamic parameters of the process of melting of the low-molecular compound on the size of the critical nucleus (the size of the restricting pores).

To verify the applicability of the previously described thermodynamic approach to describing the phase transitions of the low-molecular substances in the porous structure of the polymer matrices, investigations were carried out to determine the parameters of the crazed structure (i.e., the size of the pores and fibrils) by the method of small angle x-ray scattering. Comparison of the results with the DSC data [12, 13] is shown in Table 8.1. There is a good correlation between the values obtained by the various methods and this is a convincing confirmation of the applicability of the thermodynamic analysis for describing the phase transitions of the low-molecular substances in the crazes of the polymers.

Thus, the addition of the low-molecular substances to the porous structure of the craze polymers transfers them to the high-dispersion state. The transition of the low-molecular substances to the highly dispersed state greatly changes the nature of phase transitions in these systems. Analysis of the phase transitions of the low-molecular substances in the polymer matrices and in the free state indicates the strongest effect of the nanoporous structure of the polymer on their thermophysical behaviour. In turn, the detection of this effect provides new information on the structure of the pores in the volume of the crazes.

Orientation effects in crystallisation of low-molecular substances in the volume of the crazes

The previously discussed special features of the thermophysical properties of the low-molecular compounds, added to the structure of the crazes, are determined by the condition of their crystallisation in the narrow (1–30 nm) pores. Such small dimensions of the pores are not the only special feature of the structure of the crazes. Very important is the approximately parallel distribution of the fibrils in the structure of the craze (Figs. 7.6–7.8). This indicates that the narrow asymmetric pores, separating the individual fibrils, are also mutually oriented in relation to the tensile loading axis of the polymer. The distinctive asymmetry of the structure of the craze should also influence the process of crystallisation of the low-molecular substances in their bulk.

This effect was detected and investigated in x-ray studies of a large number of crazed polymer–low-molecular filler systems [13–15]. The results show that regardless of the polymer used and the nature of the added crystallising substance, the low-molecular substance crystallises in all cases with the formation of highly ordered textures. Figure 8.4 shows this phenomenon on the example of polyethylene terephthalate and polycarbonate, containing low-molecular substances of different nature. It may be seen that in all cases the low-molecular substances crystallise with the formation of highly ordered textures and their x-ray diffraction patterns resemble the x-ray diffraction patterns usually obtained from the monocrystals. This phenomenon is of the general nature and is detected also when using, as the matrix, the crazed crystallising polymers (polyethylene terephthalate, polycarbonate) and amorphous polymers (atactic PMMA) [14, 15]. All these special features are retained when adding

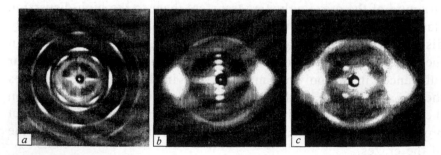

Fig. 8.4. X-ray patterns of the crazed samples of polyethylene terephthalate, containing K1 (*a*) and *n*-octadecane (*b*) and the crazed polycarbonate sample containing the tridecane acid (*c*). The tensile loading axis of the polymers in the AAM is vertical.

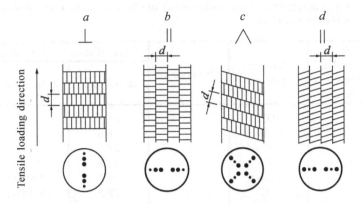

Fig. 8.5. Layer orientation of HE, CS (*a*, *b*) and carbonic acids (*c*, *e*) in the pores of the polymer matrices and the distribution of the appropriate reflections on the x-ray patterns. The abbreviations are explained at the end of Table 8.2.

the matrices of both ionic and molecular crystals to the crazed polymers.

Although the orientation of the low-molecular substances in the structure of the crazes is observed in all cases, the type of orientation depends on the nature of the polymer–low-molecular substance pair. In [8–13, 16, 17] investigations were carried out into a large number of nanocomposites based on a number of crazed polymers, on the one hand, and the long-chain fatty alcohols, hydrocarbons and acids dispersed in them, on the other hand. These compounds crystallise with the formation of ordered layers and this facilitates the interpretation of the resultant x-ray diffraction diagrams. Analysis of these diagrams shows that the set of the point reflexes, distributed on the meridian, equator or diagonal, can be attributed to the scattering of the x-ray beams from the oriented layer planes [12].

The nature of packing of the asymmetric molecules in the oriented pores of the crazes and the appropriate x-ray diffraction reflections are shown schematically in Fig. 8.5. The normal and parallel orientations of the layer planes of the low-molecular substances in relation to the direction of tensile loading of the polymer are indicated by ⊥ and ||, respectively. The oblique order of the layer planes of the *n*-carbonic acids is indicated by Λ.

Table 8.2 shows the interplanar spacings and the nature of the orientation of the low-molecular substances in the pores of different polymers. For comparison, appropriate data are given for the same substances in the free state. Table 8.2 shows that all the investigated molecular substances are oriented during crystallisation in the

Table 8.2. Orientation and phase composition of the low-molecular substances in the crazed polymer matrices

Low-molecular mass compound	Polymer matrix	Orientation	Layer distance, $d \times 10$ nm,	
			in polymer	in free state
HE	HDPE	⊥	28.8	
	PTFE	⊥	28.6	
	PP	⊥	28.7	
	PET	⊥	28.8	
	Nylon-6	⊥	28.8	28.65(A)
	PC	‖	28.5	28.92(R)
	PVC	‖	28.6	
CA	HDPE	⊥	45.5	
	PTFE	⊥	45.4	37.37(γ)
	PET	⊥	45.4	43.83(α)
	Nylon-6	⊥	45.4	44.90(β)
	PC	‖	45.4	
UDA	HDPE	Λ	26.1	
	PTFE	‖	25.0	25.68(C′)
	PC	‖	26.1	30.16(A′)
DDA	HDPE	Λ	27.7	
	PP	Λ, ‖	27.7	
	PET	Λ, ‖	27.7	31.2(A)
	PC	‖	27.7	27.42(C)
TDA	HDPE	Λ	30.1	
	PTFE	Λ	29.8	
	PP	Λ, ‖	30.0	
	PET	Λ, ‖	30.1	
	Nylon-6	Λ	29.7	35.35(A′)
	PC	‖	29.7	30.00(C′)
	PVC	‖	29.8	
	PMMA	‖	29.8	
PDA	HDPE	Λ	35.5	
	PP	Λ, ‖	35.7	40.20(A′)
	PET	Λ, ‖	35.6	31.20(B′)
	PC	‖	35.8	27.42(C′)

Comment. HE – heneicosane; CA – cetyl alcohol; UDA – undecanoic acid; DDA – dodecanoic acid; TDC – tridecanoic acid.

polymer matrices. For the linear hydrocarbon heneicosane (HE) and cetyl alcohol (CA) the orientation of the layers is perpendicular or parallel to the tensile loading axis (the direction of the pores) of the polymer (Fig. 8.5 *a, b, d*). For the long-chain acids, in addition to the parallel orientation, examination also showed the inclination of the orientation of the crystalline layers (Fig. 8.5 *c*).

The orientation of the low-molecular substances accompanying their crystallisation in the narrow (~10 nm) asymmetric pores of the crazed polymer matrices is determined mainly by the thermodynamic stability of the ordered state in comparison with the arbitrary distribution. In ordering of the crystallites of the low-molecular substances in the crazes the intercrystalline surface energy is minimum. The nature of orientation of the low-molecular substances in the polymer matrices is determined by the minimum free energy of the surface component at the polymer–low-molecular substance interface; thus, each type of interaction of the crystalline planes, which depends on orientation, is characterised by its own value of surface energy.

In [13, 18] it was shown that there is some factor which determines the parameters of the crystalline lattice of the low-molecular substance in the polymer matrix. This factor is associated with the existence of internal stresses in the polymer containing the crazes. It is important to note that their size and direction determine the nature of deformation of the crystalline lattice of the low-molecular substances distributed in the volume of the craze.

The orientation effect of the highly developed surface of the structure of the craze influences not only the processes of crystallisation of conventional low-molecular substances. This effect is also observed in the phase transitions, taking place in the liquid crystalline compounds. For example, in [19] investigations were carried out into the phase transitions in a substance capable of existing in the liquid crystalline state (*n*-butoxy benzylidene aminobenzonitrile) in the crazes of a number of polymers. The results show that this compound is also oriented in the narrow pores of the polymer crazes. The orientation of this type is usually recorded by IR dichroism.

To conclude this section, it is necessary to make several assumptions regarding the nature of the phenomenon of orientation of the low-molecular substances in the structure of the crazes reported in [15, 20]. The orientation effect of the crazed polymer matrix on the crystallisation of the added low-molecular compounds cannot be

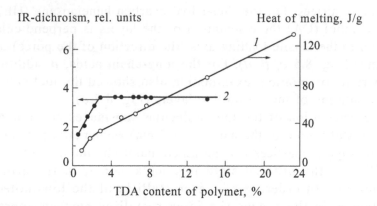

IR-dichroism, rel. units   Heat of melting, J/g

TDA content of polymer, %

Fig. 8.6. Dependence of the heat of melting (1) and IR-dichroism of the absorption band 2925 cm^{-1} of the tridecanoic acid (2) added to the crazed polytetrafluorethylene matrix, on its content in the polymer.

regarded as epitaxy. Epitaxy is the formation of crystals of the same substance uniformly oriented in relation to each other at the phase of a crystal of another substance [21]. As shown previously [14–16], the low-molecular component crystallises with the formation of highly ordered textures irrespective of whether the crazed polymer matrice is crystalline or completely amorphous. In the examined case the main orientation factor is evidently the strictly ordered distribution of the asymmetric pores in the structure of the craze (see the schemes in Figs. 7.6–7.8).

Nevertheless, it is the crystals that are oriented in the structure of the craze and not the individual molecules of low-molecular component. In particular, this is indicated by the data in Fig. 8.6. This figure compares the results of IR-dichroism (band 2925 cm^{-1}) and the heat of melting of the tridecanoic acid (TDA) in the crazed matrix of polytetrafluorethylene in dependence on the content of the low-molecular component. The amount of the TDA in the polymer matrix can be regulated by inducing crazing of the polymer in its solutions with different concentration. As indicated by Fig. 8.6 (curve 1), the heat of melting of the TDA decreases in the entire investigated concentration range. This effect is associated with the fact that with a decrease of the content of the TDA in the crazed polymer matrix the size of the resultant crystallites evidently decreases. The reduction of the size of the crystal is obviously accompanied by the increase of the specific surface area of the substance. In turn, the increase of the size of the specific surface is accompanied by the decrease of the heat of melting because the surface layer of the crystallite does not provide any energy contribution to the melting process because it does not form the tri-dimensional crystal lattice.

At the same time, the IR-dichroism, characterising the degree of orientation of the low-molecular substance in the polymer matrix, changes in a completely different manner (Fig. 8.6, curve 2). The orientation of the TDA remains constant up to its 3% content in the polymer matrix. After reaching this concentration threshold, the IR-dichroism starts to decrease rapidly. Evidently, this result is associated with the fact that up to the content of 3% the TDA is in the polymer matrix mostly in the crystalline state. A further decrease of the TDA concentration results in its molecular dispersion in the volume of the craze. In particular, in this region the amount of the crystalline phase starts to decrease rapidly. The results indicate that the volume of the crazes is characterised by the orientation of the crystallites and not of the individual molecules.

Thus, the crystalline and liquid crystal substances, added to the porous structure of the crazed polymer matrices, are oriented with a high degree of ordering inversely changing in the phase transition. It is important to note that the high level of self-organisation of the crazed polymer also determines the high degree of ordering of the low-molecular component, added to the structure of the crazes. The main factor responsible for these phenomena is evidently the presence of the highly developed surface of the polymer, characteristic of the structure of the crazes.

Special features of the structure formation of the low-molecular component, produced by chemical transformations in the volume of the crazes

Thus, we have examined a number of special features of crystallisation of the low-molecular substances in the nanoporous structure of the crazes. As mentioned previously, the direct method can be used to add only a limited range of the low-molecular substances to the volume of the crazes because of the previously mentioned reasons. The substances insoluble in the AAM and not melting below the glass transition temperature (melting point) of the deformed polymer evidently can not be added to the volume of the crazed by the direct method. Nevertheless, the nanocomposites based on the polymers, containing metals, semiconductors, ferroelectrics and other target additions, are of obvious interest from the applied viewpoint. Therefore, methods have been developed for producing nanocomposites by the chemical reaction of the appropriate precursors directly in the polymer matrix (*in situ*). In fact, in this case, the microscopic pores in the structure of the crazes can be used as 'micro-reactors' for the synthesis and stabilisation of the nanophases at the required level of dispersion and morphology. The formation of nanometric cavities makes it possible to use them as 'micro-reactors' for carrying out different chemical reactions in them (reduction, exchange, etc). This approach to solving the problem of formation of the nanocomposites makes it possible to solve several fundamental problems: stabilisation of the nanophases as a result of the restricting factor of the wall and mixing at the nanolevel of the thermodynamically incompatible components.

The first attempt to realise the chemical reaction *in situ* in the structure of the crazed polymer was evidently made in [15]. In this study, a classic photographic process was performed *in situ* in the structure of the crazed polyethylene terephthalate. For this purpose, a polyethylene terephthalate film was tensile loaded in a water–alcohol solution of potassium iodide. Drawing resulted in the addition of 25 wt.% of KI to the film. The resultant film was then placed in a water–alcohol solution of silver nitrate. As a result of the interaction of KI and AgNO$_3$ the silver iodide crystals precipitated in the structure of the polymer. Finally, the samples were held in the solution of a standard photodeveloper. This type of treatment resulted in the dissociation of silver iodide to metallic silver. All the stages of this process were controlled by x-ray diffraction analysis. A special feature of the processes is the gradual complete disorientation of the low-molecular component in reactions *in situ*. For example, if in the first stage KI crystallises with the formation of a highly ordered texture, then the orientation of silver iodide in the second stage of the process is 'very weak' and the silver, precipitated in the final stage, is absolutely isotropic.

The previously discussed method of adding the metal to the structure of the craze has limited possibilities from the viewpoint of filling the volume of the craze by the low-molecular component. Actually, the increase of the concentration of the solution of the anorganic substance, filling the porous structure of the polymer, also increases the content of the low-molecular component after removing the solvent [22]. Nevertheless, it is evident that this method can not be used to fill the porous structure of the polymer completely as is the case in drawing of the polymer in the melt of the low-molecular substance. Even in the case of substances with high solubility, for example KI in the water–alcohol solution, the amount of the

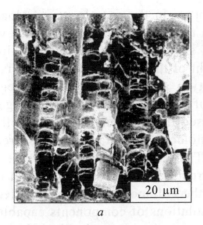
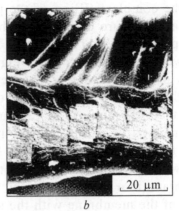

Fig. 8.7. Scanning electron micrographs of the crazed polyethylene terephthalate samples after holding in a saturated solution of potassium iodide and subsequent drying with the fixed dimensions (*a*); the same sample to which AgCl crystals were added by the counter diffusion method (*b*). Explanation in the text.

inorganic component does not exceed 30–50% [22]. Figure 8.7 *a* shows the micrograph of a polyethylene terephthalate sample, tensile loaded by 100%, where the sample acquired a continuous porous structure. The sample was then transferred into a saturated water– alcohol solution of KI and held there for a week. During this time, the AAM in the pores of the volume was evidently replaced by the KI solution. Subsequent drying of the sample produced a polyethylene terephthalate film containing 47 wt.% of KI. Potassium iodide precipitated from the saturated solution in the cavities of the crazes in the form of fine crystals with a size of ~2 – 3 μm which are randomly distributed in the pores. It is characteristic that a relatively large number of such small crystals can form in a single craze. The dimensions of these crystals are non-uniform and this reflects the highly disordered process of their precipitation from the solution. At the same time, Fig. 8.7 *a* shows clearly that although the crazes were filled using the saturated solution of KI, a large part of the structure of the craze is not occupied by the low-molecular component.

It is well-known that the magnitude of porosity, developed in the polymer during drawing in the AAM, can be very high. The glassy polymers are often deformed in the conditions of uniaxial tensile loading in the liquid media in such a manner that the variation of the geometrical dimensions of the specimen is almost completely associated with the development of porosity, and the contribution of other types of deformation is negligible. Simple evaluation indicates that the complete filling of the pores in the sample, tensile loaded by this procedure by 100%, with the low-molecular substance with the specific weight of 3 g/cm^3 should ensure a weight gain of 300 wt.% in relation to the weight of the initial polymer matrix.

To overcome this contradiction and obtain a higher degree of filling the porous structure of the polymer with the inorganic substance, the authors of [23] used the counter diffusion method. This method of filling the low-molecular substance differs greatly from that described previously. The crazed polymer film, i.e., the film with a continuous porous structure, is placed in the form of a membrane in a dialysis cell and the volumes are filled on different sides of the membrane with the solutions of components capable of interacting with each other. In this case, the low-molecular substances diffuse against each other and meet directly in the volume of the pores of the membrane. Consequently, the chemical reaction takes place in the pores so that the porous structure of the polymer can be filled more efficiently.

This method will be described by the classic photoprocess investigated previously in [22]. For this purpose, a crazed polyethylene terephthalate sample was placed in a dialysis cell filled on one side with a water–alcohol solution of AgNO$_3$ and on the other side with a NaCl solution. After 24 hours, the film was extracted from the cell, rinsed in water, dried and examined by scanning electron microscopy.

Figure 8.7 *b* shows the micrograph of the polyethylene terephthalate film containing AgCl. It may be seen that in this case the process of precipitation of the low-molecular component greatly differs from that discussed previously. Figure 8.7 *b* shows that approximately in the middle of each craze there is only one AgCl crystal with a height of ~15 μm and the width equal to the distance between the walls of each craze. Thus, it is in fact possible to fill quite efficiently the volume of the craze with the low-molecular inorganic substance. The resultant nanocomposite contains a large amount of the high dispersion inorganic substance.

In [23] the nanocomposite, whose structure is shown in Fig. 8.7 *b*, was transferred into a standard photodeveloper and held there for two days. Subsequently, the sample was dried and examined in an electron microscope. The results of this examination are presented in Fig. 8.8. It may be seen clearly that the treatment with the AgCl developer leads to the precipitation of fine metallic silver crystals in the volume of the crazes. These crystals have a more porous structure

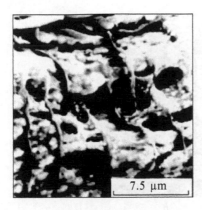

Fig. 8.8. Scanning electron micrograph of the crazed polyethylene terephthalate sample, filled with AgCl, after holding in the standard photodeveloper [23].

than the AgCl crystals from which they formed. The procedure of dissociation of AgCl makes it possible to explain the problem of the interaction of the growing crystal with a fibrillar structure of the craze. As indicated by Fig. 8.8, the silver crystals precipitate an individual fibrils of the craze forming in some cases a granose structure. The results indicate that the fibrils of the craze penetrate the crystalline formation of AgCl (Fig. 8.7 *b*) and are not substituted by them.

This was followed by a large number of investigations of producing by the counter diffusion method different nanocomposites based on a number of polymers (PP, PE, PA-6, etc.) and metals and oxides. In particular, the metallic polymers were produced by counter diffusion of the salts of the appropriate metals and the reduction agent, for example, in a lithium borohydride solution. The results show that the proposed method is universal and can be used to produce a large number of different nanocomposites. The nanocomposites of this type can be conductors or semiconductors, produced in the form of films with high mechanical properties

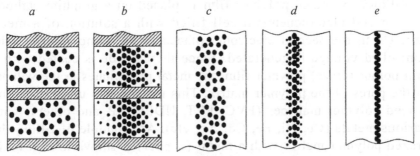

Fig. 8.9. Schematic representation of different variants of adding the low-molecular component of the crazed polymer matrices by the counter diffusion method [32].

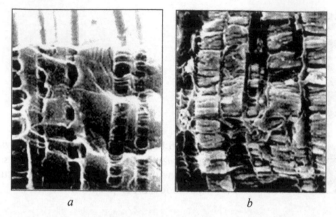

Fig. 8.10. Scanning electron micrograph of the crazed polyethylene terephthalate samples prior to (*a*) and after filling with metallic copper (*b*) [32].

characteristic of the initial polymer matrices. The resultant composite has a large number of morphological forms. In these composites it is possible to regulate not only the total content of the low-molecular dispersed phase, its density and dispersion, but also the distribution inside the polymer phase. Figure 8.9 shows schematically a number of possible variants of the distribution of the low-molecular phase inside the polymer film. Evidently, all these factors influence the macroscopic properties of the produced nanocomposite, such as electrical conductivity, dielectric permittivity, mechanical properties, etc.

In addition to the discussed methods, there is another universal method of precipitating the metals from their compounds – electrolysis. This method was used in [32] for producing mixtures of polymers and metals using crazed polymer matrices. The method of electrolytic production of the metal polymers may be described as follows. A crazed polymer film is placed on a graphite cathode in a special electrochemical cell filled with a solution of a metal salt. Using the second electrode, placed in the same solution, the controlled voltage is generated. Since the cathode is coated with a thin porous (craze) polymer film, the metal precipitates in the volume of the pores of the polymer matrix. This method was used for filling crazed polymer matrices (PVC, PET, HDPE, PP, and others) with various metals (Cu, Ni, Fe, Co, Ag, etc). An example of filling the crazed polymer matrix with a metal (copper) is shown in Fig. 8.10.

Thus, as indicated by the experimental results, it is possible to fill the porous matrix of the crazed polymer with almost any target addition in the highly dispersed state. The amount of the active substance may be larger than 300 wt.% in relation to the unfilled polymer matrix. The production of these compositions opens a path to the formation of new types of combined materials with a polymer base. The crazing of the polymer in the liquid medium is not only a universal method of development of the interfacial surface of the polymer but also a unique method of dispersion of the high- and low-molecular substances in the general volume with the formation of new types of nanocomposites.

This concept is not exhausted only by the formation of the polymer–low-molecular substance high dispersion mixtures. The crazing of the polymers offers also a universal and non-traditional method of producing new types of polymer–polymer mixtures with a high level of mutual dispersion of the components. This aspect of the problem will be described in more detail in the following section.

8.2. Polymer–polymer nanomixtures based on crazed polymers

Recently, a large amount of work has been carried out in the area of production of multicomponent multiphase polymer systems. The main difficulty, preventing progress in this area, is the thermodynamic incompatibility of the majority of the polymers with each other as a result of the low values of the mixing entropy [33]. As a result of this, the mixed polymers break up to various degrees to extended phases with poor adhesion and this has a catastrophic effect on the properties of the resultant compositions.

To overcome the problems, associated with the poor compatibility of the polymers, the best solution is the production of polymer–polymer nanocomposites, i.e., the formation of stable multiphase systems with a high level of mutual dispersion. This type of nanocomposite can be produced in particular by the simultaneous or gradual polymerisation of two monomers in the conditions excluding their co-polymerisation. This approach makes it possible to obtain a high mutual dispersion of the polymer components in the final products. The method has been known for a long time and was developed to produce the so-called interpenetrating polymer networks (IPN) [34]. This class of the polymer–polymer mixtures is synthesized by combining in the same volume two or more monomers, followed by their independent polymerisation. In this case, the phase separation in the system takes place in the process of polymerisation but in these conditions it does not lead to the separation of the system into extended phases. Consequently, the IPNs are two-phase systems with a high level of the mutual dispersion of the components, which are in contact with each other.

There are two main methods of producing IPN: 1) same time IPN – the systems produced by simultaneous polymerisation of two or more monomers; 2) successive IPN – systems produced by swelling the monomer 2 in the ready network of the polymer 1 followed by its polymerisation *in situ*. The main methods of producing the IPN have been generalised in the excellent monograph by Sperling [34].

From the viewpoint of the subject of this section it is important to note that the large majority of the IPNs are typical nanocomposites. In fact, the dimensions of the domains of the polymer 2 in the network of the polymer 1 (D_2) can be suppressed as follows [35]:

$$D_2 = (2\gamma W_2)/RT\nu_1\{[1/(1-W_2)]^{2/3} - 1/2\} \qquad (8.1)$$

where W_2 is the mass fraction of the component 2, γ is the interfacial energy, ν_1 is the effective number of the moles of cross-linked chains in the network of the polymer 1, R is the universal gas constant, T is absolute temperature. Equation (8.1) was verified by a large number of experiments using direct microscopic methods of measuring the phase domains in the IPN. The results show that in the large majority of the cases the dimensions of the phase domains, calculated from equation (8.1) and determined by electron microscopy, are in good agreement with each other. In addition to this, these dimensions equal, for example, for IPN on the basis of butadiene styrol rubber and polystyrene from 48 to 150 nm, depending on the composition and the degree of cross-linking [35], so that these systems can be regarded as typical nanocomposites. Calculations of the phase domain for the IPN on the basis of the caster oil–urethane–polystyrene give the dimensions from 25 to 55 nm, which is fully confirmed by the electron microscopic data [36]. Similar results were obtained in [37–40]. Even the homoIPN produced on the basis of polystyrene (these systems are produced by swelling the cross-linked polystyrene in the styrene, containing an initiating agent, followed by polymerisation *in situ*) also confirmed the presence of domains with the size of 6–10 nm [41].

Recently, the IPNs have been produced in the form of multiphase nanogels, suitable for solving a large number of important applied problems, such as the modulation of light [42], recognition of haemoglobin in water solutions [43], long-term release of low-molecular ligands into the surrounding environment [44], etc [45, 46]. In addition to this, the principle of production of the IPNs is used for the production of classic nanocomposites – the so-called organic – in organic hybrid polymer materials [47–49].

Thus, there is a widely used method of producing polymer–polymer nanocomposites by carrying out to independently occurring processes of synthesis of polymers, accompanied by the phase separation of polymer components. In addition to this, there is also a completely different approach to the production of polymer–polymer nanocomposites. In the most general form this approach is based on the formation of a nanoporous polymer matrix followed by filling the resultant pores with a second polymer component. In this approach, the phase inhomogeneity of the nanocomposite does not form in the process of polymerisation and appears in the stage of formation of the nanoporosity of the polymer matrix.

As shown previously, the universal method of producing the nanosized porosity in the polymer films and fibres is the crazing of the polymers in liquid media [1, 2]. Therefore, the application of nanoporous crazed polymers is highly promising from the viewpoint of production of new types of polymer–polymer nanocomposites.

8.3. Crazing as a method of producing nanosized porosity in polymers

It should be mentioned that the crazing of the polymers in liquid media is one of the fundamental types of inelastic plastic deformation of solid polymers [50, 51]. This type of deformation, which is a unique manifestation of the Rehbinder effect in the polymers [3, 52], results in the dispersion of the solid polymers on the finest aggregates of the oriented macromolecules (fibrils), separated by micro-cavities of approximately the same size.

Deformation of the polymer by the crazing mechanism is easily performed during its tensile loading in the adsorption-active media (AAM) and takes place by the nucleation and development of special zones of the plastically deformed polymer – crazes. During deformation of the polymer, the initial non-oriented material gradually transforms to the oriented fibrillised state. In these conditions, two parts of the polymer coexist constantly and one of them transfers to the highly dispersed fibrillised state, and the other one represents the fragments of the initial non-deformed polymer (Figs. 7.6–7.8). The ratio of these structural components of the polymer changes during deformation. This type of self-regulation of the polymer is referred to as classic crazing. The evolution of the porous structure in the crazing process of the polymer was described in more detail in chapter 7.

There is another type of crazing which also produces the nanoporous structure in the polymer although the mechanism is slightly different – delocalized crazing [53, 54]. In delocalized crazing, the porosity develops place simultaneously throughout the entire volume of the polymer so that in all stages of deformation the polymer is a homogeneous nanoporous material. If classic crazing can take place in both crystalline and amorphous polymers, the delocalized crazing is characteristic only of the polymers with a crystalline structure.

It is important to note that the size of the fibrils and of the micro-cavities that separate them during its crazing in the AAM is equal to ~1–10 nm. In this case, the total porosity reaches 60%, and the specific surface several hundreds of m^2/g. The parameters of the porous structures can be easily regulated by changing the draw ratio of the polymer in the AAM, the nature of the AAM and the temperature–force conditions of deformation of the polymer [55–57]. In deformation of the polymer in the AAM the polymer becomes oriented. However, the orientation of the macromolecules does not take place in the monolithic neck and it takes place in the finest fibrillar aggregates of the macromolecules separated in space.

The formation of the unique fibrillar–porous structure in crazing is possible only if the submicropores, formed during the process, are continuously filled with the surrounding liquid medium in which the polymer is deformed. Consequently, both the polymer and the low-molecular substance which is thermodynamically incompatible with the polymer are not only mutually dispersed to the nanometric level but, most importantly, form a highly dispersed and homogeneous mixture with the polymer. Naturally, these special structural features of the crazing of the polymers in the liquid media create suitable conditions for the development of a new universal method of introducing to the polymer the second component in order to produce the nanocomposite.

8.4. Special features of production of polymer–polymer nanocomposite by polymerisation *in situ* in a crazed polymer matrix

Since the delocalized crazing takes place uniformly throughout the entire volume of the polymer, it is attractive to use it primarily for the development of new types of polymer–polymer nanocomposite. As mentioned previously, delocalized crazing takes place during deformation of only crystalline polymers. At the same time, almost all currently available types of IPN are synthesised on the basis of exclusively amorphous polymers. This is associated with the fact that the production of the polymer nanocomposite by the method of polymerisation *in situ* requires the capacity of the so-called first polymer network to swell considerably in the second monomer [58, 59]. The crystalline elastomers, such as polyethylene or polypropylene, are capable of only limited swelling in the organic liquids which have affinity for them [60]. However, the swelling of this type is usually small and equals several percent, and as the degree of crystallinity of the polymer increases, the magnitude of its equilibrium swelling in the solvents compatible with them decreases. This is caused by the fact that the low-the molecular component is capable of penetrating only into the amorphous regions of the polymer and does not affect the crystallites. It is therefore natural that the highly crystalline polymer, for example, high-density polyethylene, is not capable of sorbing large quantities of the low-molecular liquid. Evidently, this is the reason why the IPNs based on the crystalline polymers can not be synthesized.

The crazing of the polymers in the liquid media greatly increases the amount of the low-molecular component in the deformed polymer. As shown in [61], the tensile loading of the polymer in contact with the plasticising liquid increases its swelling to 100 or more percent in relation to the weight of the dry polymer. In the case of crazing the situation is reversed: as the degree of crystallinity of the polymer increases the porosity developed in the polymer is greater, with other conditions being equal. This circumstance was also used in [62, 63] for producing a number of polymer nanocomposites on the basis of HDPE. In this study the active liquid was in the form of monomers which cause delocalized crazing of the polymer and are actively included in the developing porous structure of the polymer.

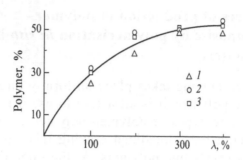

Fig. 8.11. Dependence of the percent content of PMMA (1), polystyrene (2) and polybutylmethacrylate (PBMA) (3) in the composition on the tensile strain (γ) of high-density polyethylene in the appropriate monomer [63].

To produce polymer composites, the film of the extruded HDPE was tensile loaded at room temperature in the monomers, containing 0.3 wt.% of the initiator (benzoyl peroxide) and, in some cases, the cross-linking agent. Experiments were carried out using the monomers compatible with the HDPE – methyl methacrylate, styrene and n-butyl methacrylate. After deformation to the required strain the dimensions of the specimen were fixed in a special frame and the sample was transferred into a thermostatically controlled vessel in which polymerisation was carried out. It can be seen that as regards the procedure, the process of formation of the polymer–polymer nanocomposite in this case is completely identical with the method of production of successive IPNs [34].

Figure 8.11 shows the dependence of the amount of the added polymer by polymerisation of the appropriate monomer on the tensile strain of HDPE in the liquid monomer. It may be seen that using the method described previously it is possible to produce polymer nanocomposites based on HDPE containing up to 50% or more of the second component which, obviously, could not be achieved in simple swelling of the non-deformed polymer in the same monomers. The amount of the added second polymer monotonically increases with increasing tensile strain of the polymer and approximately at a strain of 200–250% it ceases to change, as in the case of the amount of the plasticising liquid in which deformation is carried out [61]. The amount of the second component, added to the HDPE, depends only slightly on the nature of the selected monomers, and the dependence of composition on the preliminary tensile strain of the polyethylene terephthalate in the selected monomers can be approximately described by a single curve. Since both components

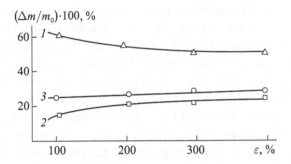

Fig. 8.12. Dependence of the percentage content of HDPE (1), PMMA (2) and the residue insoluble in selective solvents (3) in HDPE–PMMA compositions on the tensile strain (γ) of HDPE in methyl methacrylate.

of the mixture are linear, in cases in which the cross-linking agent was not added it was possible to verify the results by the method of selective washing out of the components. This procedure is often used for analysis of the composition and structure of the polymer compositions, in particular, the interpenetrating polymer networks [64, 65].

For this purpose, the compositions based on HDPE–PMMA were subjected to gradual rinsing in chloroform (for dissolution of PMMA) and n-decane (for dissolution of HDPE). Independent experiments show that the homopolymers completely dissolve in the same conditions. The results of these experiments are presented in Fig. 8.12.

It appears that by selective dissolution it is not possible to separate completely the components of the resultant composite. In all cases after washing out ~25–30 wt.% of the insoluble product remains. Evidently, polymerisation is accompanied either by the chemical grafting of the chains of the components [59] or by their mutual penetration at the molecular level (as observed in the production of the majority of the interpenetrating polymer networks [66]), or by simultaneous occurrence of both processes. Evidently, the process of this mutual penetration of the components is supported by the surface 'loosening' of the nanosized phases of the polymers described in detail previously (chapter 3). Regardless of the high mutual dispersion of the polymer components, the resultant compositions are two-phase systems. Figure 8.13 *a* shows the electron micrograph of the PMMA-frame produced after washing out the HDPE from the nanocomposite. It may be seen that the removal of HDPE results in the formation of a highly porous openwork frame with the dimensions

of the structural elements from several nanometres to several tenths
of a micron. A similar pattern is also obtained if PMMA is washed
out from the composite (Fig. 8.30 *b*). Evidently, the structures shown
in Figs. 8.13 *a* and *b* are complementary and should supplement each
other in the nanocomposite. The two-phase nature of the resultant
systems is also confirmed by the calorimetric data according to which
there is no change of temperature and the heat of melting of HDPE
[67]. Thus, the proposed procedure helps to obtain the required
target: produce a new type of polymer–polymer nanocomposites.

It should be stressed that the polymer nanocomposites based
on HDE, produced by polymerisation *in situ*, greatly differ in the
morphology from identical compositions, produced by the traditional
method (mixing of the polymer melts), regardless of whether the
block copolymer is added to such a mixture for improving the
compatibility or not. For example, in [68–70] in the mixtures based
on PE–PS of any composition there were spherical formations with
the size of 1–10 μm in a continuous matrix. This morphology forms
in the melt under the effect of surface forces due to almost complete
incompatibility of the components. Evidently, on the basis of the
dimensions of the coexisting phases, this mixture can be classified
as a nanocomposite.

In the investigated cases there were absolutely no spherical
formations of one of the components and, as indicated by Fig.
8.13, the level of dispersion of the coexisting phases was relatively

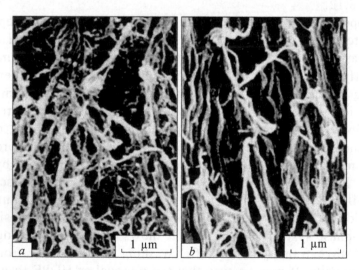

Fig. 8.13. Electron micrographs of the HDPE–PMMA nanocomposite (tensile strain
in the monomer 200%) after selective washing out of HDPE (*a*) and PMMA (*b*).

high and the dimensions of these phases were characteristic of the nanocomposite. This is explained by the fact that the formation of a second polymer phase takes place in the intercrystalline regions of HDPE. In the polymerisation conditions, HDPE has the temperature lower than the melting point and is capable to resist efficiently the effect of surface forces, determining the phase separation, and also stabilise the resultant structure. In other words, the network of the HDPE crystallites restricts the formation of the extended phase of the second component and determines the dispersion of the system. The situation is identical with the case observed in the synthesis of interpenetrating polymer networks where the increase of the density of cross-linking of the first network increases the phase dispersion of the resultant second network [71]. In the examined case, the role of cross-linking in the first network is played by the crystalline structure of the HDPE.

On the basis of these results it may be concluded that in the synthesis of the described nanocomposites the resultant structure is characterised by double phase continuity because the selective washing out of each component results in the formation of a continuous porous frame and not in the breakdown of the components into individual parts.

It should be mentioned that the increase of the draw ratio of HDPE in the liquid monomer causes not only the increase of the amount of the second component in the nanocomposite. According to the data presented in [67], tensile loading is accompanied by the extensive orientation of HDPE. At the same time, the addition of the second polymer component to HDPE is not accompanied by its molecular orientation. Evidently, the molecular orientation of one of the components should be reflected in the properties of the final products. In other words, in contrast to the currently available approaches to produce in the polymer–polymer composites, the crazing of the polymers is another factor which can be used to influence the properties of the produced nanocomposites in the required direction.

Thus, the radical polymerisation of the monomers in the HDPE matrix, deformed in the medium of the monomers, produces a number of interesting polymer–polymer nanocomposites characterised by the high mutual dispersion of the components. By drawing the polymer in the liquid monomer, it is possible to include in the number of objects suitable for producing the nanocomposites by *in situ* polymerisation, a large range of crystalline polymers, such as HDPE [67, 72], polypropylene [63, 73, 74], PA-12 [75], and mixtures of these polymers were produced exclusively by mixing their melts. Naturally, it was not possible to obtain high degrees of mutual dispersion of the components and this had an undesirable effect on the properties of the compositions. In addition to this, using crazing it is possible to produce nanocomposites based on polymers greatly differing in polarity, for example polytetra fluoroethylene and polyacryl amide [76] or polypropylene and polyacrylamide [77] which evidently creates additional difficulties of combining them in the same material by the conventional methods.

Mechanical properties of polymer–polymer nanocomposites, produced by polymerisation *in situ* in a crazed polymer matrix
Prior to studying the mechanical properties of the nanocomposites produced by polymerisation *in situ*, we discuss the change of the properties of the pure crazed polymer matrix, deformed to various strains the active liquid. Figure 8.14 shows the stress–strain properties of HDPE subjected to drawing in *n*-heptane to various strains by delocalized crazing. It may be seen that the polymer matrix, used for producing new types of nanocomposites, shows the behaviour characteristic of the oriented polymers. Drawing increases the modulus, the yield stress and strength of the polymer and reduces its strain at break [74].

We now examine the variation of the mechanical properties of the nanocomposites if polyethylene undergoes polymerisation by the procedure described in the previous section. Figure 8.15 shows the tensile loading curves of the nanocomposites based on HDPE and PS or PMMA. It may be seen that the resultant nanocomposites

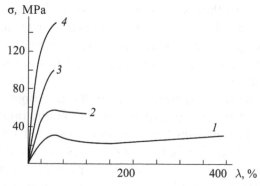

Fig. 8.14. Curves of tensile loading of HDPE (*a*) after preliminary orientation in *n*-heptane, with strains 100 (1), 200 (2), 300 (3) and 400% (4) [74].

'remember' that the polyethylene phase is oriented [72]. Actually, the increase of the preliminary drawing strain of HDPE in a certain monomer increases regularly the modulus and the yield stress of the produced nanocomposite. As indicated by Fig. 8.15 *a, b*, the second component, which was in the glassy state, also has a strong effect on the properties of the composite which is expressed primarily in the increase of the initial modulus of the produced material. At the same time, in all cases the synthesized nanocomposite has the properties not characteristic of any of its components. It also may be seen that the nanocomposites based on HDPE and PS, and also HDPE–PMMA, are capable of high plastic strains. Actually, the addition of PS and PMMA to HDPE may result in very high strains at break even for the composites based on HDPE, deformed by 300 and 400% which is not capable in the real form of large strains (Fig. 8.14). It is well-known that neither PS nor PMMA at room temperature is capable of large inelastic strains and fail at a strain of 3–5%. In addition, even cross-linking of the glassy component in the structure of the nanocomposite (Fig. 8.15 *c*) does not suppress its capacity for high plastic strains. Thus, there is some synergism in

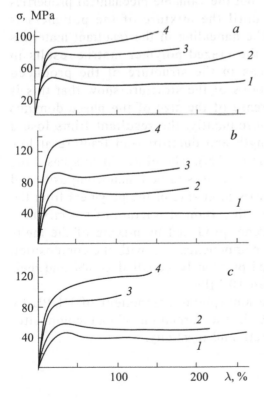

Fig. 8.15. Stress–strain curves of composites based of HDPE–PS (*a*) and HDPE–PMMA (*b, c*) produced by polymerisation *in situ* by stretching HDPE in styrene (*a*) or methyl methacrylate (*b, c*) by 100 (1), 200 (2), 300 (3) and 400 (4) % without the cross-linking agent (b) or in the presence of 20 wt% of dimethacrylate ethylene glycol (*c*) [172].

the mechanical behaviour when two polymers, not capable separately of high strains, are combined in the structure of the nanocomposite and show this capacity.

The reasons for the observed synergism are evidently based on the high dispersion of the glassy polymer, added to the PE matrix. As shown in chapter 3, in the investigations carried out in the last decade it was shown that the thin layers of the glassy polymers (nanosized range) are characterised by a large (by tens or hundreds of degrees) reduction of their glass transition temperature (see, for example [78, 79]).

In other words, the transition of the glassy polymer to the nanostate is accompanied by a large change of its properties and, in particular, a large decrease of T_g. The decrease of T_g below the test temperature (the room temperature) indicates that the amorphous polymer in the HDPE matrix is in the rubbery and, consequently, loses its brittleness. The material of the matrix is also disintegrated to the nanostate and, evidently, changes its properties (see the data in chapter 5). Consequently, the resultant nanocomposite acquires higher plasticity than the bulk polymers forming it.

It is important to mention that the valuable mechanical properties of the system are retained until the mixture of the polymers is dispersed to the nanostate. The annealing of the resultant materials above the melting point of the crazed polymer matrix results in the irreversible consequences in the structure of the produced polymer mixtures. Investigations of the structure show that this is accompanied by a large increase of the size of the phase domains of the components [80]. Consequently, the resultant films lose a large proportion of the strength and ductility and fracture at low stresses (~5 MPa) and strains (2–5%). Therefore, in this case, the resultant mixture of the polymers ceases to be a nanocomposite and a structure which usually appears in mixing of the polymers from the melt forms [81]. For example, the strain at fracture of the mixtures of polyethylene and polystyrene, produced by mixing of the melts and containing up to 50% of components, i.e., with the composition identical with that investigated previously, equalled 2– 8% and their fracture strength was less than 10 MPa.

Thus, the addition of the amorphous component of HDPE by polymerisation *in situ* results in the formation of nanocomposites characterised by higher strength and plasticity.

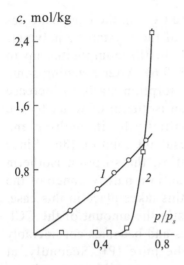

Fig. 8.16. Sorption isotherms of CCl_4 in the IPP sample, deformed in n-heptane by 200% (1) and in the IPP–PMMA composites, produced by deformation of IPP in MMA by 200% (2).

Sorption properties of the polymer–polymer nanocomposites, produced by polymerisation *in situ* in a crazed polymer matrix

The investigations of the diffusion and sorption of the low-molecular components in the polymers provides important information not only on the mechanisms of their mass transfer but also the structure of the material. Whilst the transfer processes in the single-component polymers have been studied quite sufficiently [82], there has been only a small number of studies of this type in the multicomponent heterophase mixtures of the polymers and the results have not been widely published. The sorption properties of the nanocomposites, produced by crazing, have been studied only in [83, 84] which, naturally, could not completely clarify their properties.

Attention will now begin to several results of the investigation of the diffusion and sorption of a selective low-molecular solvent in the structure of the nanocomposite, produced by the previously mentioned methods and based on isotactic polypropylene (IPP) and PMMA [83].

Figure 8.16 shows the typical sorption isotherms of CCl_4 in the IPP, deformed in n-heptane by 200%, and of the IPP–PMMA composites produced by tensile loading of the IPP in MMA by 200% followed by polymerisation *in situ*. It may be seen that the pure IPP has the conventional sorption isotherm, characteristic of the crystalline polymer, with a temperature higher than the glass transition temperature [84]. This isotherm is almost completely identical with the sorption isotherm of the initial non-oriented IPP. At the same time, the sorption isotherm of the CCl_4 in the composite has a number of special features not typical of the crystalline polymer.

Firstly, there is no sorption of CCl_4 to the composites up to high vapour pressures of CCl_4 (p/p_s ~0.5–0.6). After reaching some threshold value of the vapour pressure, sorption starts to increase rapidly. This curve resembles the sorption isotherm of water by the low-molecular sugar which is caused initially by its melting and subsequent absorption of water by the resultant solution [86]. Since the water and sugar are mixed at all ratios, the sorption isotherm of this system shows a large rising section in a narrow range of the vapour pressure. Evidently, nothing like this takes place in this case. Nevertheless, as shown by Fig. 8.16, firstly the amount of the CCl_4 sorbed by the composites rapidly increases and becomes considerably greater than its amount absorbed by the pure IPP. Secondly, at the vapour pressures exceeding 0.6, it is not possible to reach the equilibrium value of sorption because sorption continues for many days. To analyse such unusual sorption behaviour of the IPP–PMMA composites, we examine the data for the sorption kinetics. In the case of the initial non-oriented PP, also stretched in the active liquid, the sorption curve has the usual form which reaches the equilibrium value at approximately 1 h.

The addition of PMMA to the IPP matrix greatly changes the kinetics of the sorption process (Fig. 8.17 *a*, *b*). As indicated by Fig. 8.17 *a*, after a short incubation period the sorption starts to increase in accordance with the linear law. This nature of sorption is maintained over many days (Fig. 8.17 *b*) or even weeks. The equilibrium value of sorption is not reached even after 1 month.

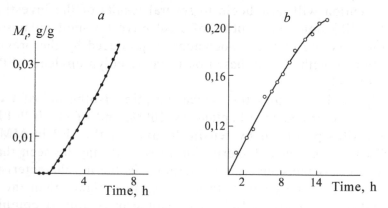

Fig. 8.17. Kinetic curves of sorption of CCl_4 in the IPP–PMMA composite, produced in tensile loading of IPP in MMA by 200%: *a*) initial section, *b*) stationary section (at $p/p_s = 0.69$).

The amount of sorbed CCl_4 reaches 50 wt.% and more, although the pure initial IPP, which is the only component with affinity for CCl_4 is sorbed by less than 15%.

The behaviour of composites in the desorption of CCl_4 is also unusual. With the reduction of the vapour pressure, the samples of pure IPP show the reversed variation of sorption, and in vacuum treatment they lose almost completely the entire sorbed amount of CCl_4. In the case of the IPP–PMMA composites, holding of the specimens which sorbed CCl_4, at a pressure of 0.6 for two days did not lead to any loss of the sorbate, which could be detected by the quartz balance. Holding the same specimen in air for 20 days at room temperature also does not lead to any significant desorption of CCl_4. Partial desorption of CCl_4 can be caused only by the combined effect of vacuum of $\sim 10^2$ mm Hg and increase of the temperature from room (23°C) to 40°C.

Thus, the addition of the glassy PMMA to the IPP by polymerisation in the polymer matrix greatly changes its sorption characteristics. There is a number of questions to be answered. What explains the unusual form of the sorption isotherm and why there is the threshold pressure of the CCl_4 vapours below which there is no sorption? What explains the unusual sorption kinetics and the capacity of the composites to absorb such large quantities of CCl_4, although the pure IPP is capable of sorbing much smaller amounts of CCl_4 and the pure PMMA does not sorb CCl_4 at all? Why is the desorption of CCl_4 from the composite irreversible to a large degree or very difficult?

It was established in [83] that such unusual behaviour of the produced nanocomposite is determined by structural rearrangement taking place in the material during the penetration of the low-molecular component into the material. It appears that (Fig. 8.18) on both surfaces of the specimens, i.e., where the nanocomposite interacted with the CCl_4, its structure appears to be more porous than

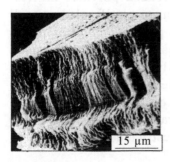

Fig. 8.18. Scanning electron micrograph of brittle cleavage of a IPP–PMMA nanocomposite sample after sorption of CCl_4 from the liquid phase for 30 days.

the dense core. There is a very sharp boundary between the porous shell and the dense core. In particular, the formation of the porous structure also determines the large quantities of the CCl_4 absorbed by its condensation in the resultant micro-cavities.

The calculations together with the electron microscopic data can be used to propose the following mechanism of this phenomenon. The effect of low vapour pressure does not cause any extensive swelling of the composite owing to the fact that the IPP, capable of swelling in the CCl_4, is located in the rigid frame of the PMMA which prevents any change of the dimensions of the IPP phase. This continues until condensation of CCl_4 starts to take place in narrow surface pores, characteristic of the structure of the composite. The formation of such thin films of CCl_4 results in the formation of a wedge force, playing a significant role in the stabilisation of the dispersed systems [87]. The formation of the wedge force results in the start of the unique self-spontaneous dispersion of the structure of the nanocomposite consisting of the thermodynamically incompatible polymer components. The spontaneous dispersion of this type is a result of the penetration of the liquid along the boundaries between the grains is well-known and has been described in detail for the low-molecular disperse systems [88].

Thus, the addition of the glassy component to the IPP by the polymerisation method in the polymer matrix greatly changes the nature of the sorption of the selective low-molecular solvent. A unique synergism is observed in sorption where its magnitude rapidly increases with the addition of the second component which does not interact with the sorbate. The unusual sorption behaviour of the nanocomposite is associated with its unique self-spontaneous dispersion leading to the breakdown to individual phases and a number of phenomenon typical of any of the pure components.

Electrically conducting polymer–polymer nanocomposite, produced by polymerisation *in situ* in a crazed polymer matrix
In the context of this review, it is important to clarify also a small number of attempts to develop polymer–polymer composites in which one of the components is an electrically conducting polymer. The formation of new types of electrically conducting polymers is the most important scientific and applied task of the advanced physical chemistry of polymers. Regardless of considerable advances achieved in recent years in this area, the practical application of the electrically conducting polymers is quite restricted because usually

they are in the form of non-melting, insoluble powders, are of little avail for processing [89].

In [90–92] the authors carried out investigations to develop approaches to the preparation of nanocomposites in which the second polymer, added to the crazed polymer matrix, has high electrical conductivity. In these studies, the procedures described below were developed. The polymer films (high density polyethylene (HDPE) and polyethylene terephthalate (PET)) were subjected to crazing during which the appropriate monomers (acetylene and aniline) were added to the developing nanoporous structure. Subsequent polymerisation, addition of the monomers in situ and doping of the resultant products made it possible to produce a number of advanced nanocomposites which combine efficiently the excellent electrical conductivity, characteristic of polyacetylene and polyaniline, with high mechanical parameters, characteristic of HDPE and PET.

It appears that the addition to HDPE of even a relatively small amount of polyacetylene (8%) and subsequent doping of the polymer mixture with iodine increase the bulk electrical conductivity by 14–16 orders of magnitude, i.e., to $(0.2–1.4)\cdot10^2$ ohm^{-1}·cm^{-1}. This fact, taking into account that the bulk specific electrical conductivity of the pure polyacetylene equals ~10^2 ohm^{-1} cm^{-1} [89], indicates that the structure of the produced polymer mixture results in a low percolation threshold of electrical conductivity. It is important to note that the electrically conducting polymer, added to the crazed matrix, acquires the distinctive molecular orientation which, as is well-known [89], greatly increases its electrical conductivity. Similarly successful were the attempts to add a second electrically conducting polymer – polyaniline – to the crazed matrices based on PE and PET [91, 92].

8.5. Direct addition of the second polymer component to the crazed polymer matrix

To produce a polymer–polymer nanocomposite, in the cases examined above a monomer was added to the porous structure of the craze polymers followed by its polymerisation in situ. This approach is determined by the assumption that the direct addition of the macromolecule to the crazed polymer is not possible because of steric reasons. In fact, the non-perturbed dimensions of the macromolecule with a molecular mass of 1 mln of even one of the most flexible-chain polymers (PE) equals approximately 200 nm, whereas the size of the nanopores in the structure of the crazes is in the range from

1 to 50 nm. Evidently, if it would be possible to include by some procedure a macromolecule in the structure of the crazed polymer, this would greatly simplify the procedure for producing the polymer–polymer nanocomposite of this type because it would be possible to exclude the relatively labour-consuming and ecologically dangerous stage of polymerisation from the process of production of these nanocomposites.

For this reason, investigations were started recently to determine the possibilities of direct introduction of the macromolecules to the nanoporous structure of the crazed polymers. The most attractive proposal for the explanation of this possibility is the development of nanocomposites based on polymers with greatly differing properties, for example hydrophilic and hydrophobic. In this case, one can expect the most interesting and unexpected properties in the resultant polymer products. The combination of components with such different polarities is itself a serious task. For example, the non-polar polymer polyethylene terephthalate and the polar polymers polyethylene glycol (PEG) or polypropylene glycol (PPG) can not be combined with the formation of a highly dispersed mixture by any of the currently available methods. Actually, the melting point of PET is 245°C, and the temperature of chemical degradation of PEG and PPG is 180 and 220°C, respectively. Evidently, it is not possible to produce a mixture of these components by mixing the melts. Mixing such greatly differing polymers through the solution is also not possible because these polymers do not have common solvents. Thus, the application of crazing is evidently the only method of producing nanocomposites based on these components.

It has been established [93] that the oligomers PEG and PPG are effective crazing agents for PET and HDPE. The amount of the added second component is relatively large (up to 60 wt.% or more) so that it may be expected that it will have a strong effect on the properties of the final product. However, in this case, the liquid component is capable of migrating from the volume of the produced nanocomposites.

For this reason, attempts have been made to add the second polymer component with a higher molecular mass (400 000 and 1 million). The PEGs with these molecular weights are solid products at room temperature, so that they were added to HDPE and PET [94–96] using their solutions in crazing solvents. For this purpose, the PET films were deformed in solutions of PEG in ethanol–water mixtures. The composition of the solvent was selected to ensure

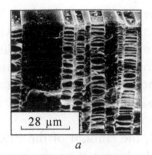 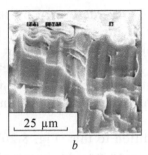

28 μm

25 μm

a

b

Fig. 8.19. Electron scanning micrographs of the polyethylene terephthalate samples deformed by 100 % in pure AAM (*a*) and in a 5% solution of PEG with a molecular mass of 1 000 000 (*b*) [94].

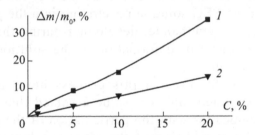

Fig. 8.20. Dependence of the amount of the PEG with a molecular mass of 1 000 000, included in the polyethylene terephthalate in the process of its tensile loading in the water–alcohol solution on the concentration of this solution (1); theoretical calculations were carried out to determine the dependence of the amount of the PEG which can be added to polyethylene terephthalate on the basis of the assumption of the maximum filling of the porous structure of the crazes by the PEG solution with a concentration of 5% (2) [95].

that, on the one hand, it dissolves a large amount of the PEG (no less than 5 wt.%) and, on the other hand, the same solvent results in efficient crazing of PET.

It was observed unexpectedly that the polymers with such high molecular masses (up to 1 000 000) efficiently penetrate into the nanoporous structure of the crazed PET, forming an appropriate nanocomposite. As indicated by Fig. 8.19, the porous structure of the crazed PET is filled with the second polymer component (PEG 1 million). As shown by the data in Fig. 8.20 (curve 1), deformation of the PET in the solution of PEG with the molecular mass (MM) 1 million produces mixtures with a high content of the second component. It should be remembered that the penetration of the PEG into the porous structure of polyethylene terephthalate takes place in the conditions in which the dimensions of the coil of the macromolecules greatly exceed the dimensions of the pores in the polymer, deformed by the crazing mechanism. The effective diameter of the pores in PET in drawing in the AAM, determined by the method of penetration of the liquid under the effect of a pressure gradient or by the method of small angle x-ray scattering, is 5–10 nm, whereas the root-mean-square radius of the coil of the macromolecules PEG MM from 40 000 to 1 000 000 is 9.2–63 nm [96].

In addition to this (Fig. 8.20), the amount of the PEG penetrating in the crazing process is greatly larger than the possible amount which can be calculated assuming the complete filling of the porous structure with the solution of the polymer with the concentration used as the crazing medium. In other words, there is some mechanism of enrichment of the PEG solution penetrating into the porous structure of polyethylene terephthalate, developing during the process of its crazing in the course of deformation in the solution of the second polymer component.

Many special features of this process have not as yet been completely explained, but it may be assumed that the production of the nanocomposites with the participation of the crazed polymer matrices by direct addition of the second polymer component is fully feasible. Stressing the results, it may be concluded that the application of crazing of the polymers in the liquid media makes it possible to produce a large number of new types of polymer–polymer nanocomposites.

Conclusions

The experimental results presented here can be used to regard the crazing of the polymer in the AAM not only some type of self-dispersion of the polymer under the combined effect of the mechanical stress and the active liquid media. Crazing is also a method of colloidal dispersion of low-molecular substances. During crazing, the active liquid fills the porous structure of the crazes and, at the same time, supplies greatly differing low-molecular substances to the bulk of the polymer. As a result of the colloidal dispersion of this type, the low-molecular substances acquire unique properties. The temperature and duration of the phase transitions in the substances change. There are also temperature ranges in which specific polymer modifications of the given substance are stable. Of special interest is the capacity of the low-molecular substances, added to the bulk of the crazes, to orient themselves in relation to the axis of tensile loading of the polymer. This orientation is universal and is determined by the nature and intensity of the interaction of the low-molecular substances with the material of the polymer matrix. The crazing of the polymers in the liquid media offers new possibilities for producing highly dispersed mixtures of the polymers with greatly different low-molecular substances, such as metals, oxides, salts, etc. The studies concerned with the development of a new universal method of producing polymer–polymer mixtures have been reviewed. The method is based on the crazing phenomenon which occurred during the deformation of the polymers in the adsorption-active liquid media. It is very important to note that during crazing a unique fibrillar–porous structure develops in the polymer and the dimensions of the pores and fibrils of the structure are not greater than several tens of nanometres. Two approaches to developing polymer–polymer mixtures have been described. According to the first approach, the nanoporous structure of the crazed polymer is filled with the monomer and its subsequent polymerisation *in situ* results in the formation of polymer–polymer mixtures. The second approach, discovered recently, is based on the direct penetration of the macromolecules into the fibrillar–porous structure of the deformed polymer developing during drawing in the adsorption-active medium. In both cases it is possible to produce a wide range of the polymer mixtures with a high degree of dispersion of the components. It has been shown that the materials, produced by this procedure, have a unique set of the mechanical, electrical and sorption properties, determined by the specific fibrillar–porous structure of the crazed polymer. The crazing mechanism and the deformation conditions of the initial polymer make it possible to regulate the composition, structure and properties of the mixtures.

References

1. Volynskii A.L.,Yarysheva L.M., Bakeev N.F., Rossiiskie Nanotekhnologii, 2007, Vol 2, No. 3-4. P. 5812.
2. Volynskii A.L.,Grokhovskaya T.E., Lukovkin G.M., Godovskii Yu.K, Bakeev N.F., Vysokomolek. Soed. A, 1984, V. 26, No. 7, P. 1456.
3. Yarysheva L.M. Lukovkin G.M., Volynskii A.L., Bakeev N.F., in: Successes of colloid chemistry and physicochemical mechanics, Ed. Shchukin E.D., Moscow, Nauka, 1992.
4. Mandelkern L., Polymer crystallization, Moscow and Leningrad, Khimiya, 1966. 336 pp.
5. Moskvina M.A., Volkov A.V., Grokhovskaya T.E., Volynskii A.L., Bakeev N.F., Vysokomolek. Soed. A, 1984, V. 26, No. 11, P. 2369.

6.   Moskvina M.A., Volkov A.V., Volynskii A.L., Bakeev N.F., Vysokomolek. Soed. A, 1985, V. 27, No. 8, P. 1731.
7.   Moskvina M.A., Volkov A.V., Volynskii A.L. Bakeev N.F., Vysokomolek. Soed. A, 1985, V. 27, No. 12, P. 2562.
8.   Taylor G.I., Proc. Roy. Soc. London, A, 1950, V. 201, P. 192.
9.   Volkov A., Moskvina M.A., Volynskii A.L., Bakeev N.F., Vysokomolek. Soed. A, 1987, V. 29, No. 7, P. 1447.
10.   Moskvina M.A., Volkov A.V., Volynskii A.L., Bakeev N.F., Vysokomolek. Soed. A, 1987, V. 29, No. 10, P. 2115.
11.   Moskvina M.A., Volkov A.V., Efimov A.V., Volynskii A.L., Bakeev N.F., Vysokomolek. Soed. B, 1988, V. 30, No. 10, P. 737.
12.   Kitaigorodskii A.I., Molecular crystals, Moscow, Nauka, 1971, 392 pp.
13.   Volkov A.V., Moskvina M.A., Arzhakova O.V., Volynskii A.L., Bakeev N.F., J. Therm. Anal., 1992, V. 38, P. 1311.
14.   Volynskii A.L., Grokhovskaya T.E., Shitov N.A., Bakeev N.F., Vysokomolek. Soed. B. 1980. V. 22, No. 7. P. 483.
15.   Volynskii A.L., Shitov N., Chegolya A.S., Bakeev N.F., Vysokomolek. Soed. B, 1983, V. 25, No. 6, P. 393.
16.   Moskvina M.A., Volkov A.V., Grokhovskaya T.E., Volynskii A.L., Bakeev N.F., Vysokomolek. Soed. A, 1984, V. 26, No. 11, P. 2369.
17.   Moskvina M.A., Volkov A.V., Volynskii A.L., Bakeev N.F., Vysokomolek. Soed. A, 1984, V. 26, No. 7, P. 1531.
18.   Volynskii A.L., Moskvina M.A., Volkov A.V., Bakeev N.F., Vysokomolek. Soed. A, 1997, V. 39, No. 11, P. 1833.
19.   Moskvina M.A., Volkov A.V., Volynskii A.L., Bakeev N.F., Vysokomolek. Soed. A, 1989, V. 31, No. 1, P. 160.
20.   Volynskii A.L., Grokhovskaya T.E., Lukovkin G.M., Godovskii Yu.K, Bakeev N.F., Vysokomolek. Soed. A, 1984, V. 26, No. 7, P. 1456.
21.   Shubnikov A.V., Epitaxy. Physical encyclopedic dictionary, Moscow, Publishing house Sov. Entsiklopediya, 1966, V. 5, P. 534.
22.   Volynskii A.L., Grokhovskaya T.E., Shitov N., Bakeev N.F., Vysokomolek. Soed. A, 1982, V. 24, No. 6, P. 1266.
23.   Volynskii A.L., Yarysheva L.M., Arzhakova O.V., Bakeev N.F., Vysokomolek. Soed. A, 1991, V. 33, No. 2, P. 418.
24.   Volynskii A.L., Moskvina M.A., Volkov A.V., Zanegin V.D., Bakeev N.F., Vysokomolek. Soed., 1990, V. 32, No. 12, P. 933.
25.   Stakhanova S.V., Nikonorova N.I., Zanegin V.D., Lukovkin G.M., Volynskii A.L., Bakeev N.F., Vysokomolek. Soed. A, 1992, V. 34, No. 2, P. 133.
26.   Stakhanova S.V., Nikonorova N.I., Lukovkin G.M., Volynskii A.L., Bakeev N.F., Vysokomolek. Soed. B, 1992, V. 33, No. 7, P. 28.
27.   Stakhanova S.V., Trofimchuk E.S., Nikonorova N.I., Rebrov A.V., Ozerin A.N., Volynskii A.L., Bakeev N.F., Vysokomolek. Soed. A, 1997, V. 39, No. 2, P. 318.
28.   Kuleznev V.N., Mixtures of the polymers. Moscow, Khimiya, 1980.
29.   Stakhanova S.V., Nikonorova N.I., Volynskii A.L., Bakeev N.F., Vysokomolek. Soed. A, 1997, V. 39, No. 2, P. 312.
30.   Nikonorova N.I., Stakhanova S.V., Volynskii A.L., Bakeev N.F., Vysokomolek. Soed. A, 1997, V. 39, No. 8, P. 1311.
31.   Nikonorova N.I., Stakhanova S.V., Chmutin I.A., Trofimchuk E.S., Chernavskii P.A., Volynskii A.L., Ponomarenko A.T., Bakeev N.F., Vysokomolek. Soed. B, 1998, V. 40, No. 3, P. 487.

32. Nikonorova N.I., Trofimchuk E.S., Semenova E.V., Volynskii A.L., Bakeev N.F., Vysokomolek. Soed. A, 2000, V. 42, No. 8, P. 1298.
33. Volynskii A.L.,Yarysheva L.M., Lukovkin G.M., Bakeev N.F., Vysokomolek. Soed. A, 1992, V. 34, No. 6, P. 24.
34. Sperling L. Interpenetrating polymer networks and related materials, Moscow, Mir, 1984.
35. Donatelly A.A., Sperling L.H., Thomas D.A., J. Appl. Polym. Sci., 1977, V. 21, No. 5, P. 1189.
36. Yenvo G.M., Sperling R.H., Pulido J., Manson J.A., Conde A., Polym. Eng. Sci., 1977, V. 17, No. 4, P. 251.
37. Huelk V., Thomas D.A., Sperling R.H.., Macromolecules, 1972, V. 5, P. 340.
38. Allen G., Bowden L.J., Bundell O.J., Jeffs G.M., Vyvoda J., White T., Polymer, 1973, V. 14, P. 604.
39. Sionakidis J., Sperling L.H., Thomas D.A., J. Appl. Polym. Sci., 1979, V. 24, No. 5, P. 1179.
40. Yenwo G.V., Manson J.A., Pulido J., Sperling L.H., Conde A., Devia-Manjarres N., J. Appl. Polym. Sci., 1977, V. 21, No. 7, P. 1531.
41. Siegfried D.L., Sperling R.H., Manson J.A., J. Polym. Sci. Phys. Ed., 1978, V. 16, P. 583.
42. Tsutsui H., Moriyama M., Nakayama D., Ishii R., Akashi R., Macromolecules, 2006, V. 39, No. 6, P. 2291.
43. Yong-qing Xia, Tian-ying Guo, Mou-dao Song, Bang-hua Zhang, Bao-long Zhang, Biomacromolecules, 2005, V. 6, No. 5, P. 2601.
44. Gitsov I., Chao Zhu., J. Am. Chem. Soc., 2003, V. 125, No. 37, P. 11228.
45. Kwok A.Y., Qiao G.G., Solomon D.H., Chem. Mater., 2004, V. 16, No. 26, P. 5650.
46. Dhara D., Rathna G.V.N., Chatterji P.R., Langmuir, 2000, V. 16, No. 6, P. 2424.
47. Ogoshi T., Itoh H., Kim K.-M., Chujo Y., Macromolecules, 2002, V. 35, No. 2, P. 334.
48. Zecca M., Biffis A., Palma G., Corvaja C., Lora S., Jrabek K., Corain B., Macromolecules, 1996, V. 29, No. 13, P. 4655.
49. Jackson C.L., Bauer B.J., Nakatani A.I., Barnes J.D., Chem. Mater., 1996, V. 8, No. 3, P. 727.
50. Kambour R.P., J. Polym. Sci., Macromol. Rev., 1973, V. 7, P. 1.
51. Narisawa J., Jee A.F., Material Science and Technology. Structure and Properties of Polymers, 1993, V. 12, P. 701.
52. Volynskii A.L., Priroda, 2006, No. 11, P. 11.
53. Yarysheva L.M., Shmatok E.A., Ukolova E.M., Arzhakova O.V., Lukovkin G.M., Volynskii A.L., Bakeev N.F., Dokl. AN SSSR, 1990, V. 310, No. 2, P. 380.
54. Volynskii A.L.,Yarysheva L.M., Shmatok E.A., Ukolova E.M., Lukovkin G.M., Bakeev N.F., Vysokomolek. Soed. A, 1991, V. 33, No. 5, P. 1004.
55. Volynskii A.L., Mikushev A.E., Yarysheva L.M., Bakeev N.F., Ros. Khim. Zh., (Zh-VKhO im. D.I. Mendeleeva), 2005, V. 49, No. 6, P. 118.
56. Volynskii A.L., Rukhlya E.G., Yarysheva L.M., Bakeev N.F., Rossiiskie Nanotekhonologii, 2007, Vol 2, No. 5-6, P. 44.
57. Yarysheva L.M., Rukhlya E.G., Yarysheva L.M., Volynski A.L., Bakeev N.F., Obzornyi zhurnal po khimii, 2012, V. 2, No. 1, P. 3-21.
58. Lipatov Yu.S., Sergeeva L.M., Interpenetrating polymer networks, Kiev, Naukova Dumka, 1979.
59. Thomas D., Sperling L., In: The polymer blends, Ed. D. Paul, C. Newman, Moscow, Mir, 1981, V. 2, P. 5.

60. Papkov S.P., Encyclopedia of polymers, Moscow, Sov, entsiklopediya, 1974. V. 2, P. 320.
61. Efimov A.V., Bondarev V.V., Kozlov P.V., Bakeev N.F., Vysokomolek. Soed. A, 1982, V. 24, No. 8, P. 1690.
62. Volynskii A.L., Shtanchaev A.Sh., Bakeev N.F., Vysokomolek. Soed. A, 1984, V. 26, No. 11, P. 2374.
63. Volynskii A.L., Bakeev N.F., Advances in Interpenetrating Polymer Networks, Ed. by D.Klempner and KC Frisch, Technomic Publishing, Lancaster, Pennsylvania, 1991, V. 3, P. 53–74.
64. Widmaier J.M., Sperling L.N., Macromolecules, 1982, V. 15, No. 2, P. 625.
65. Widmaier J.M., Sperling L.H., J. Appl. Polymer Sci., 1982, V. 27, No. 12, P. 3513.
66. Frisch H.L., Frisch K.S., Klempner D., Pure and Appl. Chem., 1981, V. 53, No. 8, P. 1557.
67. Volynskii A.L., Shtanchaev A.Sh., Bakeev N.F., Vysokomolek. Soed. A, 1984, V. 26, No. 8, P. 1842.
68. Heikens D., Barentsen W.M., Polymer, 1977, V. 18, No. 1, P. 69.
69. Sjoerdsma S.D., Dalmolen J., Bleijenberg A.S., Heikens D., Polymer, 1980, V. 21, No. 12, P. 1469.
70. Heikens D., Kern. Ind., 1982, V. 31, No. 4, P. 165.
71. Donatelly A.A., Sperling L.H., Thomas D.A., Macromolecules, 1976, V. 9, No. 4, P. 671.
72. Volynskii A.L., Shtanchaev A.Sh., Zanegin V.D., Gerasimov V.I., Bakeev N.F., Vysokomolek. Soed. A, 1985, V. 27, P. 831.
73. Volynskii A.L., Lopatina L.I., Bakeev N.F., Vysokomolek. Soed. A, 1986, V. 28, No. 2, P. 398.
74. Volynskii A.L., Lopatina L.I., Yarysheva L.M., Bakeev N.F., Vysokomolek. Soed. A, 1997, V. 39, No. 7, P. 1166.
75. Volynskii A.L., Lopatina L.I., Yarysheva L.M., Bakeev N.F., Vysokomolek. Soed. A, 1998, V. 40, No. 2, P. 331.
76. Shtanchaev A.S., Sergeev V.G., Baranovsky V.Yu., Lukovkin G.M., Volynskii A.L., Bakeev N.F., Kabanov V.A., Vysokomolek. Soed. B, 1983, V. 25, No. 9, C. 642.
77. Bykova I.V., Prazdnikova I.Yu., Brooke M.A., Isaev G.G., Bakeev N.F., Teleshov E.N., Vysokomolek. Soed. B, 1989, V. 31, No. 1, P. 67.
78. Forrest J. A., Eur. Phys. J. E., 2002, V. 8, P. 261.
79. Volynskii A.L., Bakeev N.F., Vysokomolek. Soed. B, 2003, V. 45, No. 7, P. 1210.
80. Volynskii A.L., Shtanchaev A.Sh., Bakeev N.F., Vysokomolek. Soed. A, 1984, V. 26, No. 11, P. 2445.
81. Heikens D., Kern. Ind., 1982, V. 31, No. 4, P. 165.
82. Reitlinger S.A., Permeability of polymer materials, Moscow, Khimiya, 1974, P. 102.
83. Volynskii A.L., Barvinskii I.A., Lopatina L.I. Volkov A.V., Bakeev N.F., Vysokomolek. Soed. A, 1987, V. 29, No. 7, P. 1382.
84. Volynskii A.L., Lopatina L.I., Arzhakov M.S., Bakeev N.F., Vysokomolek. Soed. A, 1987, V. 29, No. 4, P. 823.
85. Kargin V.A., Gatovskaya T.V., Pavlyuchenko G.M., Berestnev V.A., Dokl. AN SSSR, 1962, V. 143, No. 3, P. 590.
86. Kargin V.A., Usmanov Kh.U., Zh. Fiz. Khimii, 1954, V. 28, No. 2, P. 224.
87. Shchukin E,D., Pertsov A.V., Amelina E.A., Colloid chemistry, Moscow, MGU, 1982. 35 pp.
88. Pertsov A.V., Physico-chemical mechanics and lyophilic properties of dispersed systems, No. 13, Kiev, Naukova Dumka, 1981.

89. Conducting polymers: Special applications, Ed. by Alcacer L., Dordrecht, Reidel, 1987, P. 65.

90. Electronic properties of conjugated polymers, Ed. by Kuzmany H., Berlin, Springer, 1987, P. 128.

91. Yarysheva L.M., Saifullina S.A., Rozova E.A., Sizov A.I., Bulychev B.M., Volynskii A.L., Bakeev N.F., Vysokomolek. Soed. A, 1994. V. 36, No. 2. P. 363.

92. Saifullina S.A., Yarysheva L.M., Volkov A.V., Volynskii A.L., Bakeev N.F., Vysokomolek. Soed. A, 1996, V. 38, No. 7, P. 1172.

93. Saifullina S.A., Yarysheva L.M., Volynskii A.L., Bakeev N.F., Vysokomolek. Soed. A, 1997, V. 39, No. 3, P. 456.

94. Yarysheva A.Yu., Polyanskaya V.V., Rukhlya E.G., Dement'ev A,I,, Volynskii A.L., Bakeev N.F., Kolloid. Zh., 2011, V. 73, No. 4, P. 565.

95. Rukhlya E.G., Litmanovich E.A., Dolinnyi A.I., Yarysheva L.M., Volynskii A.L., Bakeev N.F., Macromolecules, 2011, V. 44, P. 5262–5267.

96. Volynskii A.L., Rukhlya E.G., Yarysheva L.M., Bakeev N., Dokl. AN SSSR, 2012, V. 447, No. 2, P. 176–178.

97. Yarysheva A.Yu., Bagrov D.V., Rukhlya E.G., Yarysheva LM, Volynskii A.L., Bakeev N.F., Vysokomolek. Soed. A, 2012, V. 54, No. 10, P. 1507-1515.

98. Kawaguchi M., Mikura M., Takahashi A., Macromolecules, 1984, V. 17, P. 2063.

Instability and self-organisation of polymer surfaces

As mentioned previously, almost any effect on the polymer results in a change of the interfacial surface area of the polymer and, consequently, these effects are accompanied by the mass transfer of material from the surface into the bulk and vice versa. Generally speaking, the rearrangement of this type is typical of all solids, not only polymers. However, in the polymers the phenomena of this type are far more distinctive and their intensity is higher because of the unique ability of the polymers for high inelastic strains. Evidently, high strains are accompanied by the mass transfer of large quantities of the material and this is also reflected in their structural–mechanical behaviour.

Attention will be given to the most characteristic examples of the reaction of the polymer systems by the development (healing) of the interfacial surfaces as a response to perturbing effects of different type.

9.1. Special features of the development of interfacial surfaces during the flow of polymer melts and solutions

This monograph is concerned with the surface phenomena taking place under effects of different type on solid polymers. Nevertheless, the development of the interfacial surfaces is also characteristic of the polymer melts and solutions. The amorphous polymer systems characterised by the highest degree of disorder at the molecular level are their melts and solutions. It is well known [1] that the deformation (flow) of the polymer solutions and melts is uniform as in the case of the low-molecular liquids. Nevertheless, at sufficiently high, critical values of shear rates and stresses there is a transition to the unstable flow mode [2, 3]. The unstable flow of the polymers is manifested in the appearance of large periodic oscillations of the flow accompanied by sound effects. The start of the unstable flow is associated with the formation of new interfaces so that they are easily observed visually. The stream, leaving smooth capillaries, is uneven and rough. The unstable flow of the polymer solutions and melts is often accompanied by their regular structurisation. There are both regular (helical, bamboo like, flaky) and irregular forms of the extrudates. In fact, the homogeneous laminar jet is dispersed into aggregates and this is evidently accompanied by the development of new interfacial surfaces.

It is assumed that the transition to the unstable flow is associated with the critical nature of the buildup of elastic (rubbery) strains of the polymer streams. As a result of the relaxation of the stored rubber-like elastic strains, the polymer loses adhesion to the walls which are in contact with the flow and starts to slide on them. If the cohesion strength of the melt is lower than the adhesion strengths, fractures formed in the materials, if not – periodic and other structures appear. The critical conditions of transition to the unstable flow are determined by many factors, including the molecular mass and molecular–weight distribution of the polymer [4]. A detailed analysis of the instability phenomena, taking place in the solutions and melts of the polymers, has been presented in particular in the review [5].

9.2. Loss of stability and dispersion in flow and during phase separation in polymer systems

The phenomena of the loss of stability and the associated mass transfer are highly characteristic also of liquid polymer systems which are not subjected to the effect of the external mechanical force. The destabilisation of the thin layers of the polymers and their solutions may take place under the effect of molecular forces, responsible for the phenomena of wetting and spreading of the liquids on the solid surface. The authors of studies [6–8] investigated the loss of stability of a thin-film of the solution of polystyrene in benzene during its spontaneous spreading on a hydrophilic surface (glass, quartz, mica). When a thin-film of the hydrophobic solution is deposited on the hydrophilic surface, the liquid film is destabilised

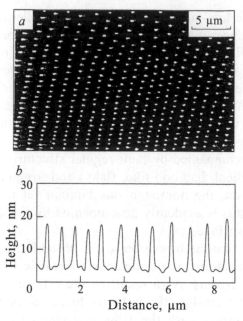

Fig. 9.1. Atomic force microscopy image (*a*) and the corresponding cross section (*b*) of the solution of polystyrene in benzene destabilised by the dewetting mechanism on the surface of mica after evaporation of the solvent [6].

under the effect of surface forces, fractures and forms droplets.

This phenomenon, inverse to the spreading of the working liquid on the surface, is referred to as dewetting. The dewetting process is in fact the loss of stability of the thin liquid film in the field of the effect of surface forces.

It is important to note that the destabilisation is usually accompanied by the formation of ordered structures. In the investigated dewetting case, there is the spontaneous formation of a regular order structure. The liquid film breaks up into individual microscopic closely sized droplets distributed surprisingly regularly on the surface of the hydrophilic substrate. After evaporation of the liquid, the substrate surface retains digitate aggregates of the polymer organised into a regular network (Fig. 9.1). The aggregates have the form of cylinders with a height of 300 nm and a diameter of 10 nm; the spacing of the aggregates is approximately 1 μm. The proposed mechanism of this type of structure formation is shown schematically in Fig. 9.2.

Identical regular surface structures can also be produced by another method [9]. A hydrophilic substrate (oxidised silicon [001]) is immersed in a solution of some diblock copolymers, containing

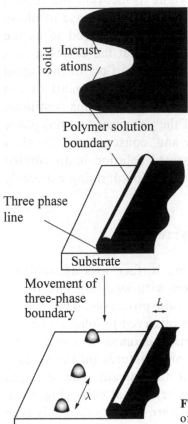

Fig. 9.2. Schematic representation of destabilisation of a thin liquid layer by the dewetting mechanism [6]: *a*) top view; *b*) side view.

rigid and flexible blocks [10]. After extraction of the substrate from the solution and evaporation of the solvent, the regularly distributed ensembles of digitate polymer aggregates with a height of ~7 nm form on the surface. Depending on the concentration of the solution, the nature of the polymer and the holding time of the substrate in the solution, a large number of types of regular surface nanostructures can form (cellular, honeycomb-like, etc). It should be mentioned that the loss of stability of the polymer systems during their phase separation and the associated structure formation are evidently a typical phenomenon. This phenomenon usually accompanies the processes of phase separation in the solutions and mixtures of the polymers, especially in cases in which they take place in the thin films [11–13].

To conclude this section, it is important to mention the results of the study [14] which describes a relatively unusual case of phase separation in the solution of the polymer, also referred to as the loss of its stability. In the study, it is shown that the addition of a horizontal temperature gradient generates a spiral flow of the liquid along its surface from the warm to cold area as a result of the thermal capillarity phenomenon. Under certain conditions (optimum mixture and the thickness of the layer of the polymer solution) phase separation takes place in the spiral flow and, consequently, the flow separates in a regular manner. The separated solution is distributed along continues lines of the resultant periodic flow forming extremely effective regular patterns.

9.3. Inhomogeneous swelling of polymers

The self-organisation, associated with the surface loss of stability, is also observed in the phase separation with swelling and phase separation accompanying the gel formation processes. The critical case of the loss of stability of this type is observed in inhomogeneous swelling of some polymers [15–18]. This phenomenon is expressed in the formation of very beautiful regular structures on the surface of the polymer brought into contact with the solvent. The resultant the relief has the form of folds forming hexagonal cells (Fig. 9.3). The authors assume that the surface structure formation of this type is associated with the loss of the mechanical stability of the thin

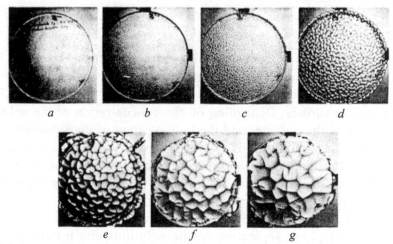

Fig. 9.3. External appearance of the surface of an ionised acrylamide gel, brought into contact with the swelling agent (water) [19]: *a–g*) different contact times between the polymer and the solvent.

surface layer of the polymer, brought into contact with the solid, capable of causing its swelling. The loss of mechanical stability arises when the swollen surface layer of the polymer and the non-swollen core coexist. Since the swollen surface layer of the polymer cannot increase its area as a result of the strong bond with the non-swollen core, it is in the planar compression state. As a result of planar compression, the swollen surface layer of the polymer loses stability and acquires a unique regular relief. Evidently, in this case, the conditions of the planar compression of the swollen polymer layer, strongly bonded with the non-swollen core, are the reason for the observed loss of stability.

The surface loss of stability of the swelling gels is of the exclusively kinetic nature. The above-mention relief is observed only in the process of swelling of gels, i.e., while the surface swollen layer and the non-swollen core (two-layer system) coexist. The size of the observed cells is not constant and depends in particular on the thickness of the swollen layer. Since the thickness of the swollen layer in the conditions of contact of the polymer and the solvent increases continuously, the cell size also increases continuously (Fig. 9.3 a–e). When the swelling process is completed, i.e., when the solvent concentration throughout the entire volume of the gel is equalised, the system ceases to consist of two layers; consequently, the surface of the gel becomes smooth and the resultant regular microrelief disappears.

It should be mentioned that the polymer gels also show the relatively distinctive structural transitions with the reverse sign (shrinkage, deswelling). These processes develop in the swollen gels in response to external effects, such as the change of the pH value, the temperature or quality of the surrounding solvent. It appears that these processes also take place at a relatively high rate (in relatively small ranges of the variation of the external parameters) and also with the formation of a regular structures [20–23]. Figure 9.4 shows schematically the structural rearrangement taking place in a cylindrical sample of the acrylamide gel, swollen in water and placed in the precipitate (water–acetone mixture). The dark areas indicate the regions of the polymer gel which lost a large part of the solvent and their density is relatively high. It may be seen that the phase separation in the samples is highly non-uniform and regular. The separation of the polymer from the gel takes place with the formation of 'bead-like' (Fig. 9.4 a) or 'bamboo-like' (Fig. 9.4 b) periodic structures. The previously examined processes of

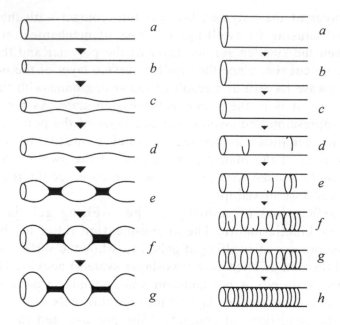

Fig. 9.4. 'Bead-like' (*a*) and 'bamboo-like' (*b*) morphology formed in the swollen acrylamide gels, placed in a precipitant (water–acetone mixture) [20]: *a–h* corresponds to the increasing holding time of the gel in the precipitant.

inhomogeneous swelling and shrinkage of the polymer gels determine the special features of their mechanical instability, formed under the effect of osmotic forces. This is indicated in particular by the fact that the investigated transitions depend more strongly on the external mechanical effects [24]. It should be mentioned that the phenomenon of the surface loss of stability has recently been applied in practice [25].

9.4. Electrodynamic and thermomechanical instability of polymer surfaces

The structural self-organisation of the polymer surfaces is an universal phenomenon which may occur under the effect of the external forces of greatly different nature. A direct confirmation of this possibility are the studies of the occurrence and investigation of the loss of stability and self-organisation of a thin layer of a rubbery polymer under the effect of an electrical field [26–28]. These studies investigated especially surface layers because the thin films (100–200 nm) of an amorphous polymer (polystyrene) were studied.

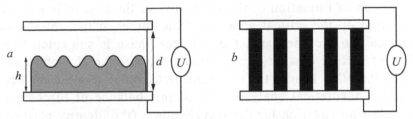

Fig. 9.5. Schematic representation of the device for electrodynamic destabilisation of surface polymer layers [26]: *a* and *b* – different stages of destabilisation.

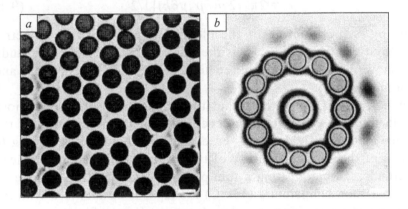

Fig. 9.6. Light micrographs of thin (93–193 nm) polystyrene layers electrodynamically destabilised under different conditions [26].

These films were deposited on a silicone substrate which was used as one of the plates of the condenser. The second plate was positioned at a small distance (100–1000 nm) from the first plate (Fig. 9.5). Subsequently, the entire device was heated to a temperature higher than the glass transition temperature of the polymer (in the case of PC 170°C) and a low relatively voltage (20–50 V) was applied to the condenser plates. The small distance between the plates of the condenser and the small thickness of the polymer determine the high values of the perturbing electrical field (10^7–10^8 V/m²) which in turn resulted in a loss of stability of the polymer surface. As a result, a surprisingly regular structure (Fig. 9.6) formed in the polymer layer. The cooling of the system to temperatures below the glass transition temperature of the polymer fixed the resultant structure and made it possible to study the structure in direct microscopic experiments. As indicated by Fig. 9.6, the loss of stability of the polymer surface results in the spontaneous formation of the system of strictly regularly (periodically) distributed thinnest digitate aggregates. The

mechanism of formation of the instability of the system in this case
was based on the critical balance of the forces at the polymer–air
interface. On the one hand, the polymer phase is subjected to the
effect of surface tension γ, minimising the interfacial surface area,
and on the other hand – the electrical field, polarising the dielectric
[29]. The theoretical analysis [30] of this balance of forces gives
the following equation for the wavelength λ (the identity period) in
the distribution of the resultant structural elements:

$$\lambda = 2\pi \ \{2\gamma/(\partial p_{el})/\partial h)\} 1/2 \qquad (9.1)$$

where p_{el} is the function of the electrical field and the dielectric
permittivity of the polymer, h is the thickness of the film. The study
[26] showed the excellent agreement between the theoretical and
experimental data.

   In [31, 32] it was shown that the phenomena similar to those
described above can also be induced without the application of the
external electrical field. In this case, the perturbing factor leading to
the loss of the surface stability of the system, is thermomechanical.
Nevertheless, when heating a thin polymer layer (PMMA) above its
glass transition temperature (130°C) the layer is destabilised with the
formation of surprisingly regular structures (Fig. 9.7).

   Light (Fig. 9.7 a) and atomic force microscopy (Fig. 9.7
b) provide information on the nature of the resultant structures.
According to the authors of [27, 28], this phenomenon is similar to
that discussed previously but in this case electric charges form in
the surface polymer layer as a result of the upper electrode situated
in the vicinity. Evidently, the formation of surface regular structures
under the effect of these or other destabilising factors is possible

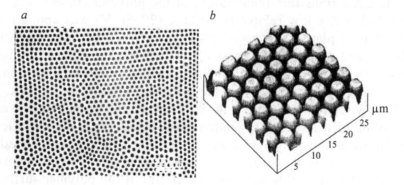

Fig. 9.7. Light micrograph (a) and atomic force microscopy image (b) of the thin
layer (94 nm) of PMMA destabilised at a temperature of 130°C [28].

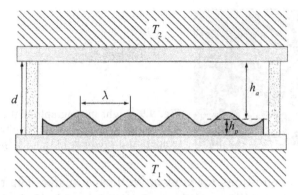

Fig. 9.8. Schematic representation of the cell for destabilisation of the thin polymer film under the effect of a temperature gradient [33].

only in the presence of extensive mass transfer of the material from the bulk to the surface of the polymer.

To conclude this section, it is important to mention the study [33] dealing with another case of destabilisation of the polymer surface – the instability caused by the temperature gradient along the film with restricted geometry. The theoretical model of this type of the surface destabilisation of the thin polymer film was proposed previously in [34].

The morphological instability of the thin-layer (80–110 nm) of polystyrene, placed on a silicone substrate, was investigated using the device is shown in Fig. 9.8. A thin polystyrene film was placed on a silicon substrate which was heated to 170°C (T_1). At a small distance from the surface of the polystyrene film (100–600 nm) there was another similar plate but without the polymer coating and with temperature T_2. Temperature T_2 was selected to fulfil the following condition: $\Delta T = T_1 - T_2$ was in the range from 10 to 55°C. Consequently, the temperature gradient in the polystyrene film reached ~180°C/m. It may be seen that the scale of the investigated object is such that here we are concerned with the study of the interfacial surface of the polymer.

The resultant temperature gradient causes the stabilisation of the thin polystyrene film which consequently acquires either a columnar or banded (Fig. 9.9) periodic morphology. It is important to mention that in the absence of the upper thermostatically controlled plate destabilisation of the polystyrene film does not take place and regular structures do not form. It is assumed that the morphological transition, detected in this study, cannot be associated with the convective instability because the film thickness is too small to

Fig. 9.9. Light micrograph of the destabilised thin polystyrene layer under the effect of a temperature gradient [33].

offer suitable conditions for convective heat transfer. The reasons for the observed instability and the associated morphological transition are accoaited with the diffusion of heat transfer, as indicated by the detected relationship of the wavelength of the resultant structure and the power of the heat flow. According to the authors, the configuration of the resultant morphology of the polymer film indicates the equalisation of the balance of forces between the plates with one of the plates being the heat source and the other one heat absorber.

9.5. Polymers with thin rigid coatings

In the previous section, it was shown that the very thin polymer films (in fact, surface layers) are 'susceptible' to the loss of stability and, consequently, structural self-organisation resulting in the formation of surprisingly regular periodic structures (Figs. 9.6, 9.7). At the same time, the deformation of the thick polymer specimens which are usually used in studies of the structural–mechanical behaviour [35–37] is not accompanied by any structural arrangement which could be recorded in experiments in which could be causes by the processes of mass transfer of the material from the surface of the volume and/or vice versa.

Previously (chapter 1) it was shown that to visualise the structural rearrangements taking place in the default polymers, a large amount of information is obtained using a relatively simple procedure – deposition of a thin (nanometric) coating on the deformed polymer [38–41]. This procedure produces important information on the mass transfer in the deformed polymer under different conditions. At the same time, the deformation of the polymer with thin coatings, deposited on the polymer surface, is also accompanied by the formation and development on the surface of the polymer of regular microstructures which are surprisingly similar to those detected in the thin polymer films, subjected to perturbing influences of different type. We believe that the coincidences of this type cannot be random and, evidently, are based on identical physical processes. Attention will be paid to several cases of the formation of surface regular structures under the mechanical effect on the polymers with thin rigid coatings.

Planar compression conditions

One of the approaches to the formation of structures of this type is the formation of a hard coating on the surface of a amorphous rubbery polymer heated to a high temperature. After depositing the hard coating on the surface, the resultant system should be cooled down and, as a result of thermal shrinkage, the deposited coating will be in the planar compression state accompanied by the loss of its stability and the appropriate structure formation.

This approach was used in studies [42–44] in which thin rigid coatings were deposited on the surface of films of polydimethyl siloxane rubber (PDMS) heated to 50–250°C. The coatings were deposited either by vacuum sputtering (gold) or by oxidation of the surface layer of PDMS in oxygen plasma (deoxidised rigid layer of PDMS). In both cases, the cooling of the PDMS film at room temperature, accompanied by its thermal shrinkage, resulted in the formation of a surprisingly attractive regular microrelief on the surface of the polymer (Fig. 9.10). It is important to note that in contrast to the phenomenon of inhomogeneous swelling of the polymers, discussed previously [15 –21], in this case, the resultant system is fully stable and does not change its structure and properties with time. Consequently, the polymer film with the regular relief can be extensively studied. The mechanism of formation of the regular folds is associated with the loss of stability of the thin coating in

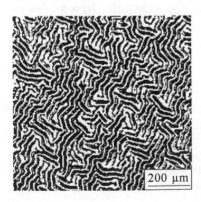

Fig. 9.10. Microscopic image of a PDMS sample with a thin gold coating (50 nm) deposited on its surface at 110°C. The photograph was taken after cooling the sample to room temperature [42].

compression on a soft substratum. The wavelength (the length of the fold) λ of the regular microrelief is associated with the properties of the material of the substrate and the coating [45]:

$$\lambda \approx 4.4h \, (E_1/E_2)^{1/3} \tag{9.2}$$

where E_1 and E_2 are the moduli of the coating and the substrate, respectively, h is the thickness of the coating.

Similar results were also obtained in [46]. Very thin (30–121 nm) polystyrene films were produced in this study which were subsequently coated on both sides with even thinner (6 nm) silicon oxide layers by thermal vacuum sputtering. High-temperature (210°C) annealing of these films resulted in the formation of a microrelief completely identical with that observed in [42–44] (compare Figs. 9.10 and 9.11).

It is assumed that the mechanism of formation of the relief is the same as in the case discussed previously. Detailed analysis of this phenomenon enabled the authors to obtain the following expression for the wavelength γ of the microrelief

$$\lambda = 2\pi \left\{ \pi \frac{E}{4A}(1-\mu 2) \right\} \frac{1}{4} h \frac{3}{4}(L+2h) \tag{9.3}$$

Here E and μ is the Young's modulus and Poisson's coefficient of the coating, respectively, A is the Hamaker constant associated with the coating–air interface, h is the thickness of the coating layer, L the thickness of the polystyrene film. Equation (9.3) describes satisfactorily the experimental data.

The loss of stability and the associated surface relief formation of the rigid coating on the polymer substrate in the conditions of planar compression of the coating may be realised when using the polymer films with planar orientation as the substrate. In a series of articles [47–49], the rigid coatings with a thickness of 10–100 nm

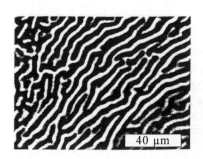

Fig. 9.11. Optical micrograph of a thin (121 nm) polystyrene sample with both PS surfaces coated with thin (6 nm) layers of SiO_2, after annealing at 210°C [46].

40 μm

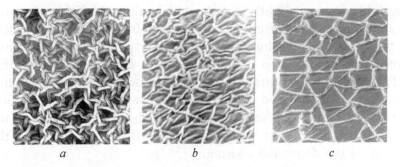

a   b   c

Fig. 9.12. Scanning electron micrographs of the polymer films with the thin coating after their thermal shrinkage: a) PP–Al; b) PET–Al; c) PET–Pt [47].

(gold, platinum, copper, stainless steel, carbon, etc.) were deposited on the surface of polymer films with biaxial orientation (polyethylene terephthalate, polypropylene).

Subsequent annealing of these films resulted in planar shrinkage and appearance of a microrelief similar to that discussed previously (Fig. 9.12). The advantage of this method is the fact that the magnitude of the compressive strains can be regulated and changed in a relatively wide range because the shrinkage of the films oriented in the plane reaches 30% and greater (thermal shrinkage of PDMS did not exceed 2–3%).

Figure 9.12 shows that as a result of annealing and the associated planar shrinkage of the polyethylene terephthalate films, the coating actually loses stability and acquires a unique extremely attractive microrelief. It should be mentioned that the nature of this microrelief depends only slightly on the nature of the material of the coating–substrate pair.

In [47–49] it was established that the resultant relief has at least two levels. The first, more distinctive level is represented by the relatively irregular closed cells with almost the hexagonal shape, formed by folds of the coating with a relatively large thickness. When using the polypropylene as the substrate and the aluminium as the coating, only this type of relief forms (Fig. 9.12 *a*). Evidently, the structures of this type are similar to those which are characteristic of the loss of stability of the inhomogeneously swelling gels, examined previously (Fig. 9.3) [15–19].

However, in the planar shrinkage of the PET–aluminium system, in addition to the structural elements of the type of multifaceted cells there is also another level of the microrelief. It may be clearly seen that inside each cell there is a unique folded structure (Fig. 9.12 *b*).

These folds have a slightly smaller height than the folds, forming the cellular structure, discussed previously. Inside each cell of this type the folds are distributed approximately parallel to each other. In some cases (Fig. 1.16 c) the folds form surprisingly attractive reliefs resembling an entangled cluster filaments. In [47–49] this phenomenon was studied in detail and it was shown in particular that the size of the cell and the period of the folds depend on the magnitude of the planar shrinkage of the film–substrate and the thickness of the deposited coating.

Uniaxial tensile drawing

Even more spectacular regular microreliefs form when the stress, applied to the substrate, has a specific orientation. In [50–53], investigations were carried out on different polymer films coated with thin (tens or hundreds of angstroms) rigid coatings (different metals, oxides, carbon, etc.). The systems, produced by this procedure, were subjected to uniaxial deformation and this was followed by examination of the structure of the produced materials (Fig. 9.13).

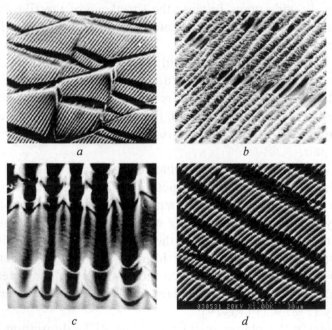

Fig. 9.13. Scanning electron micrographs of PET samples with a thin (15 nm) carbon coating (a), polystyrene with an aminated surface layer (b), PET with a thin (15 nm) quartz coating (c) and PVC with a thin (21 nm) platinum coating (d) after tensile drawing by 100% at 100°C.

Actually, regardless of whether the coating is brittle (carbon) or ductile (platinum), the nature of the resultant relief has many common features in all cases. In addition, if polystyrene is subjected to chemical treatment producing a thin layer of aminopolystyrene on the surface [54, 55], subsequent deformation results in the formation of a microrelief completely identical to that observed in all other cases (Fig. 9.13 b).

Figure 9.13 shows that the uniaxial deformation of the polymer films with the rigid coating is accompanied by two general phenomena: cracking of the coating to a large number of 'islands' (fracture cracks grow normal in relation to the direction of the tensile drawing axis of the polymer–substrate), and the formation of a microrelief. Surprising is the regular nature of the resultant microrelief and its strict orientation in relation to the tensile drawing axis. Depressions and tips of the resultant microrelief are always oriented strictly along (parallel) the tensile drawing axis.

We examine in greater detail the main conditions which determine the formation of a regular microrelief during deformation of the polymer films with a thin rigid coating. Evidently, the only driving force of the previously described surface structure formation is the mechanical stress because no other forces take part in this process. Therefore, it should be remembered that the uniaxial tensile drawing of the rubbery polymers is accompanied by considerable lateral contraction (compression) of the polymers. This means that the hard coating on the surface of the polymer is subjected simultaneously to two types of deformation – tensile drawing and compression.

It may easily be shown that in particular compression is responsible for the formation of the regular microrelief. In [56] it was shown that the structure similar to the structures shown in Fig. 9.13 can be produced not only in tensile drawing of the polymer film with a thin rigid coating. The polymer film can be initially stretched and this can be followed by the deposition of the rigid coating on its surface, with the removal of the stress from the stretched film so that the film can restore its initial dimensions. It appears that in both cases a relief, having all the previously mentioned special features, forms on the surface. The identical features of the structure also suggest identical mechanisms of the formation of these structures. However, there are also large differences between the microreliefs, produced by these methods. The point is that the discussed microreliefs is characterised by different orientations in relation to the tensile drawing axis of the polymer. The microrelief, produced in the shrinkage of the tensile loaded polymer is rotated by 90° in relation to the microrelief produced in direct tensile drawing. This important experimental result is explained by the fact that the essential condition for this phenomenon to take place is the stress leading to the compression of the rigid coating which, evidently, is determined by the compression of the deformed polymer base. It should be remembered that in tensile drawing of the rubbery polymers which do not change their volume there is a considerable lateral contraction which also leads to the compression of the coating in the direction normal to the tensile drawing axis in the formation of the appropriate relief. In shrinkage of the sample subjected to preliminary tensile drawing the directions of tensile drawing and compression are identical. In other words, the directions of compression of the surface of the polymer in the two investigated cases are mutually normal in relation to the axis of tensile drawing of the polymer and, therefore, have the mutually perpendicular resultant microreliefs.

Thus, the controlling role in the mechanism of formation of the regular microrelief is played by the compression stress which forms in the rigid coating in deformation of the compliant polymer matrix. Below, we examine in greater detail the mechanism of formation of the regular microrelief and will pay attention to the role of the main factors which determine the parameters of the resultant regular microrelief.

Effect of the magnitude of compression deformation on the period of the resultant microrelief

Evidently, the period of the resultant microrelief depends in particular on the magnitude of the compression strain of the rigid coating on the soft substratum. It is clear that as the degree of compression of the coating on the soft substratum increases, the period of the microrelief decreases. However, the mechanism of this reduction is not evident and requires experimental justification. Therefore, in [57] the resultant microrelief was compared with the contraction of the deformed substrate in tensile drawing of polyethylene terephthalate films with an aluminium coating. The results of these investigations are presented in Fig. 9.14. It may clearly be seen that the period of the resultant microrelief decreases linearly with an increase of the degree of lateral compression.

It is important to determine the mechanism of the linear decrease of the period of the relief (or, in other words, how the contraction

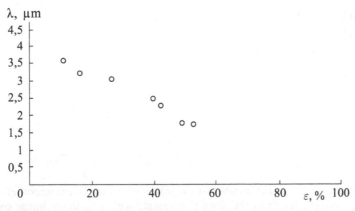

Fig. 9.14. Dependence of the period of the regular microrelief, formed on the surface of the polyethylene terephthalate film with a thin aluminium coating, on the compression strain of the polymer substrate [57].

of the coating takes place) in the process of tensile drawing of the sample because this may take place by many mechanisms. According to one of the theories, in lateral compression the tips and depressions of the relief always remain tips and depressions, whereas the number of the relief waves in the cross-section of the sample remains constant. In this case, the compression of the coating is similar to the compression of the bellow of an accordion. According to other mechanisms, the number of the relief waves in lateral compression changes; the periodic microrelief is rearranged in such a manner that its tips and/or depressions form in new areas.

Studies of the processes of relief formation in the polyethylene terephthalate–aluminium system in [57] showed the dependence of the number of the waves of the relief on the degree of lateral compression. This evaluation was made by the direct measurement of the period of the relief on the resultant micrographs, and the number of the waves was calculated by dividing the width of the specimen by the period of the microrelief.

Figure 9.15 shows the dependence of the number of waves (N) on the compression strain (ε) for the sample with a coating 9 nm thick, tensile drawn at 90°C at a rate of 0.5 mm/min. The data, presented in Fig. 9.15, can be used to make an important conclusion that the number of the relief waves in the process of deformation of the sample remains constant. In other words, after nucleation of the microrelief the coating is deformed in such a manner that only the period of the microrelief changes but the number of the waves does not. Consequently, the evolution of the period of the relief

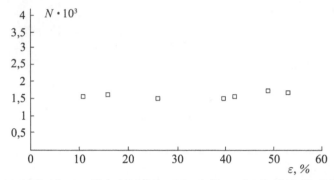

Fig. 9.15. Dependence of the number of the waves of the regular microrelief on the surface of a polyethylene terephthalate film sample with a thin aluminium coating, on the lateral compression strain of the film [57].

in compression of the polymer substrate is in fact identical with the phenomenon of the variation of the period in tensile drawing/compression of the accordion bellows.

In the quantitative aspect, the observed phenomenon can be expressed as follows

$$\lambda = \lambda_0(1 - \varepsilon) \tag{9.4}$$

The extrapolation of the dependence of the period of the microrelief on the compression strain of the polymer–substrate (Fig. 9.15 and equation (9.4)) on the ordinate gives the value of the period of the relief at the moment of the loss of stability by the system. The extrapolation of the same dependence to the abscissa intersects the latter at a point of ~100%. The physical meaning of this intersection is that if it would be possible to compress the coating by 100% (actually compress it to a point), the regular microrelief (RMR) period would naturally convert to 0.

The role of mechanical stress

Attention will be given to other factors which determine the formation of regular periodic structures in the deformation of the polymer films with a thin rigid coating. For this purpose, in [58] investigations were carried out into the effect of the tensile drawing rate on the parameter of the resultant microrelief. Figure 9.16 shows the dependence of the RMR period, formed in tensile drawing the polyethylene terephthalate samples, with a thin (4 nm) platinum coating, by 100% at a temperature of 90°C in the tensile drawing rate range from 1 to 1000 mm/min. It may clearly be seen that the

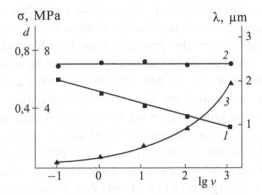

Fig. 9.16. Dependence of the period of the microrelief (1), the width of the deformed polyethylene terephthalate samples with a thin platinum coating (2) and the stress in the samples (3), on the tensile drawing rate at 90°C by 100% [58].

RMR period is linked with the tensile drawing rate of the polymer: as the tensile drawing rate increases, the period of the resultant RMR becomes smaller. It is therefore necessary to determine the reasons for the phenomenon observed in [58].

As shown previously, the RMR period is determined in particular by the degree of compression of the rigid coating on the polymer substrate. In [58] in independent experiments it was shown (Fig. 9.16, curve 2) that the width of the polyethylene terephthalate samples, deformed by 100% at different rates, is the same in all cases, i.e., the degree of compression of the coating does not depend on the tensile drawing rate of the polymer and, consequently, the observed dependence can be explained by the change of the degree of compression of the coating.

In these experiments, the tensile drawing rate affected mainly the level of the stress in the deformed sample because the other process parameters (temperature, specimen geometry, the thickness of the film in the coating, etc.) were maintained constant.

Actually, Fig. 9.16 shows that the tensile drawing rate results in a large increase of the stress in the polymer (curve 3). This stress increase is sufficiently large (by up to 15 times with the variation of the tensile drawing rate by five decimal orders of magnitude) and is clearly correlated with the variation of the RMR period: as the stress increases, the RMR period becomes smaller and vice versa. On the basis of these results it may be assumed that the detected effect is associated with the controlling role of the mechanical stress in the formation of the RMR. This conclusion is most probable but it can not follow unambiguously from these experiments. The point is that in the conditions of tensile drawing with a constant rate the mechanical behaviour of the polymer may be affected not only by its viscoelastic (relaxation) properties. In particular, it is well known [59] that the increase of the tensile drawing rate greatly changes the rate of heat removal from the deformed polymer. In this case, with increase of the tensile drawing rate the deformation conditions gradually change from isothermal to adiabatic. In turn, the adiabatic conditions in inelastic deformation of the polymer results in its auto-heating which in the case of polyethylene terephthalate is accompanied by a large number of difficult-to-control phenomena: crystallisation, destruction, the self-oscillatory mechanism of deformation, etc.

To avoid these complications and determine in the pure form the effect of the mechanical stresses on the process of formation of the RMR, the observed effect was studied under static conditions: i.e., in the stress relaxation conditions. In this case, it is not expected that the process of formation of the RMR will be accompanied by changes of some parameters responsible for the investigated process, for example, the temperature of the polymer or its phase state as a result of crystallisation.

To solve this problem, the following experiments were carried out in [58]. Samples of the polyethylene terephthalate film with a deposited thin platinum coating were tensile drawn at 90–100°C to 70% at a rate of 1 mm/min. After tensile drawing, the samples were released from the clamps after different periods of time (from 0 to 120 min) and the surface relief was studied in an electron microscope. The typical results of this study are presented in Fig. 9.17. It may clearly be seen that with increase of the relaxation time of the deformed polymer with a platinum coating in the clamps of the dynamometer at the deformation temperature the period of the resultant microrelief greatly increases, together with its regularity.

It may be assumed that the detected increase of the period of the microrelief with time takes place as a result of the reduction of its amplitude, i.e., as a result of the change of the magnitude of compression of the coating on a soft substratum. However, as in the previous case, in relaxation of the samples with the fixed dimensions there was no change in their surface area and, consequently, the degree of compression of the coating remained constant.

It should be mentioned that in the experimental conditions all the parameters of the system (geometrical dimensions, temperature) were

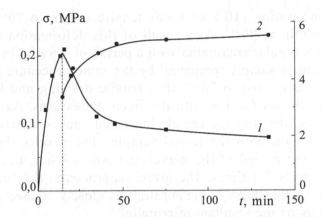

Fig. 9.17. Dependence of the stress in the polyethylene terephthalate sample in the process of tensile drawing (tensile drawing time is indicated by the vertical dashed line) and in the conditions of stress relaxation after arresting the tensile loading device (1) and the appropriate change of the period of the microrelief in the conditions of stress relaxation (2) after interrupting tensile drawing at 100°C at a rate of 1 mm/min, on the tensile drawing time [58].

constant. The only parameter which could change in the system is the stress in the deformed specimen with the fixed dimensions. Figure 9.17 (curve 1) shows the dependence of the stress in the polymer substrate on the tensile drawing time and subsequent relaxation of the investigated samples. It may be seen that after arresting the dynamometer the stress in the specimen decreases quite rapidly. The same figure (curve 2) shows the dependence of the period of the microrelief on its relaxation time in the fixed state after 70% tensile drawing at 100°C. Figure 9.17 shows that the polyethylene terephthalate samples with a platinum coating and held after tensile drawing for a specific period of time (0 to 120 min) with the fixed dimensions at 100°C are indeed characterised by the increase of the period of the microrelief and this growth accelerates with time. It should also be mentioned that the period increases quite markedly (it almost doubles). The main increase of the period takes place in the first 30 min relaxation and then slightly increases in the entire time period studied. Comparison of the curves 1 and 2 in Fig. 9.17 shows that the variation of the stress in the polymer substrate taking place with time determines the observed change in the period of the microrelief. The effect of the stress on the parameters of the resultant microrelief acts not only in the conditions of stress reduction in the substrate. The authors of [58] presented the results of the following experiments. A polyethylene terephthalate sample with a deposited platinum coating (10.5 nm) was tensile drawn to 70% at a rate of 1 mm/min at 90°C. As a result of this deformation the sample acquired a regular microrelief with a period of 4.6 μm. In the second experiment, a sample 'prepared' by the same procedure as the first sample was cooled to 70°C after tensile drawing and held in the same conditions for 1 h with the fixed dimensions. As a result of cooling the stress in the sample increased approximately 1.5 times in comparison with the initial sample. The results show that in this case the period of the microrelief was 3.3 μm, i.e., decreased approximately 1.4 times. The given experimental data indicate that the level of the stress in the substrate is closely connected with the parameters of the resultant microrelief.

It should be mentioned that in this case it is not important whether the modulus of the substrate increases or decreases. As indicated by the above described results, both cases can be encountered in the experiments, and the variation of the modules is always accompanied by a corresponding change of the RMR. The presented experimental data show that the level of the stress in the substrate is closely connected with the parameters of the resultant microrelief. Regardless of whether the stress in the polymer substrate increases or decreases, the period of the microrelief of the coating 'follows' these changes of the stress and, therefore, maintains the balance of forces (minimum possible stress in the given conditions) in the system, i.e., compensates the change of stress in the substrate as a result of the corresponding change in the microrelief.

The important role of the mechanical stress in the formation of the RMR is confirmed by the results of studying this process in the temperature range close to the crystallisation temperature of the polymer–substrate. As is known, the glassy amorphised polyethylene terephthalate rapidly solidifies on heating above 105–110°C. Figure 9.18 shows the tensile drawing curve of glassy polyethylene terephthalate at 115°C. It may be seen that in the first tensile drawing stage (up to 25%) the stress in the polymer is not high (up to 1–1.5 MPa) in the rubbery state. In these conditions, the RMR forms by the same mechanism as in the previously examined cases with the formation of the RMR with a relatively large period (approximately 3.5 μm). However, as indicated by Fig. 9.18, starting at 25% strain there is a rapid and large increase of the stress in the specimen (by at least 15 times). Evidently, this effect is associated with the crystallisation of polyethylene terephthalate. It is well-known [60]

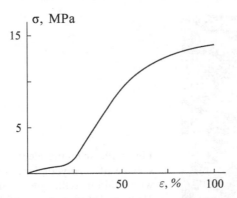

Fig. 9.18. Stress–strain curve of polyethylene terephthalate at 150°C at a rate of 10 mm/min.

that the rate of 'cold' crystallisation of polyethylene terephthalate is greatly increased if the polymer acquires molecular orientation.

Evidently, this effect is also illustrated by the tensile drawing curve shown in Fig. 9.18. In other words, at the initial moment the crystallisation rate of polyethylene terephthalate at 150°C is not high and has no significant effect on the modulus of the polymer so that the process of formation of the RMR takes place by the same mechanism as at lower temperatures. Further tensile drawing results in high-intensity orientation crystallisation so that the modulus of the polymer starts to increase rapidly. Orientation crystallisation has a specific effect on the process of formation of the RMR.

As indicated by Fig. 9.19, in this case, the rearrangement of the RMR in accordance with the increasing stress is quite unusual. The RMR, formed in the initial stages of tensile drawing and having therefore a large period, is not converted to the RMR with a new period, as demonstrated previously. It was shown as previously that the increase of stress results in a change (reduction) of the period of the RMR but the originally formed RMR does not disappear in this case. It may be be seen that in this case the new RMR with a period 4–5 times smaller than the previous one is superimposed on the latter and this results in the formation of a structure with two distinctive periods (3.5 and 0.5 μm).

It is possible that the observed phenomenon is associated with the above mentioned crystallisation of PET in deformation in the temperature range 110–120°C. As shown previously, the change of the modulus should be accompanied by the appropriate change of the period of the RMR. However, the rearrangement of this type is associated with the large deformation of the surface layer of the

Fig. 9.19. Scanning electron micrograph of the surface of a polyethylene terephthalate sample with a thin metallic coating after deformation by 100% at 115°C at a rate of 10 mm/min.

polymer so that the period of the RMR formed in these conditions and, consequently, its amplitude are large. In tensile drawing (Fig. 9.18) the modulus of the polymer rapidly increases as a result of its crystallisation. The RMR, formed at a high stress, has a considerably smaller period (and, consequently, the amplitude) and perturbs the polymer to a considerably smaller depth. Consequently, the RMR with a small period can form on the coating which already has the RMR with a considerably larger period. The complete relaxation of the originally formed RMR with a large period is not possible under these conditions so that this RMR appears 'fixed' as a result of the crystallisation process taking place.

Thus, the mechanical stress is one of the most important factors determining the parameters of the regular structures formed in the course of deformation of the polymer films with a thin rigid coating.

Role of the mechanical properties of the polymer-substrate
Detailed studies of the conditions of formation of RMR in [56, 58] show that the stress is not the only controlling factor. An important role in this process is played by the mechanical properties and, in particular, the modulus of the polymer-substrate. A direct confirmation of this assumption is provided by the following experimental data. A PET sample with a thin Pt coating was tensile drawn by 50% at 90°C. This temperature range is characterised by the formation of the RMR with a period of approximately 1 μm (Fig. 9.20 *a*). The sample, prepared by this procedure, was cooled to room temperature and then again tensile drawn. Because in the first tensile drawing the sample of the polymer was deformed to a low strain in the rubbery state (above its glass transition temperature) the sample did not acquire significant molecular orientation. Repeated tensile

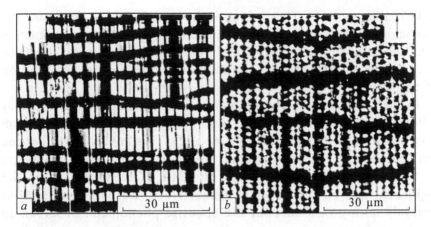

Fig. 9.20. Scanning electron micrograph of a polyethylene terephthalate sample, tensile drawn to 50% at 90°C (*a*); the same specimen after tensile drawing with the formation of a neck at room temperature (*b*).

drawing at room temperature took place at a considerably higher stress and the resulted in the formation of a neck. The section of the polymer deformed to the neck is shown in Fig. 9.20 *b*.

It may be seen that the period of the microrelief does not change, although tensile drawing of the polymer in these conditions caused changes in the fracture pattern of the metallic coating. Tensile drawing at room temperature to the formation of a neck takes place at a considerably higher stress in comparison with the first drawing cycle and, therefore, one could expect an appropriate change in the value of the period of the RMR since, as mentioned previously, both parameters (stress and period) are closely linked. However, the results show that in these deformation conditions when the polymer is in the glassy state the period of the microrelief does not change. The RMR, produced in the first tensile drawing, simply 'flows over' to the region of the neck fully reproducing its period. Evidently, the system, formed in this case is non-equilibrium because the higher stress at which it forms is not compensated by the appropriate change of the RMR. Nevertheless, it was stable because of the high modulus of the polymer-substrate, which was therefore capable of storing high internal stresses and resist the changes of the parameters of the RMR.

9.6. Mechanism of the formation of the regular microrelief

In the previous sections, we discussed the role of several factors which determine the conditions of self-organisation in the polymer–

rigid coating systems. These conditions are sufficient to determine the mechanism of the observed phenomena. Using the above experimental data, the authors of [50–53] carried out the theoretical analysis of the phenomena of formation of the regular microrelief in deformation of the polymer film with a thin rigid coating.

In the general form, the problem of the formation of the regular relief in the compression of the rigid coating on the soft substratum was investigated theoretically in [61, 62]. This analysis was carried out long before the experimental observation of the phenomenon of formation of different types of RMR in compression of the rigid coatings on the polymer films. Here, we carry out similar analysis taking into account the specific phenomena observed in the experiments.

Elastic substratum

We examine an example of analysis for the case in which the soft substratum is elastic (the case of cross-linked rubber). For the theoretical analysis of the formation of RMR the authors of [51, 52] used the model shown schematically in Fig. 9.21 and solved the problem of the loss of stability in compression of a rigid elastic plate with an ideal adhesion bond with a less rigid elastic half plane (substrate).

In the first approximation, no account was made of the shear stresses formed in bending of the coating at its interface with the substrate. Consequently, the compressive stresses on the axis Y (Fig. 9.21) can be regarded as constant. The coating will be treated as an elastic beam. The deformation of the elastic substrate in this approximation is described by the harmonic equation [63]

$$E\frac{\partial^2 u}{\partial x^2} + G\frac{\partial^2 u}{\partial y^2} = 0 \tag{9.5}$$

where E and G are the elasticity moduli of the substrate in tensile drawing and shear, u is the displacement from the equilibrium position. The axial stress in the substrate is determined by the Hooke's law

$$\sigma_x = E\frac{\partial u}{\partial x} \tag{9.6}$$

The bending of the coating in the elasticity theory is described by the conventional fourth order differential equation [64]:

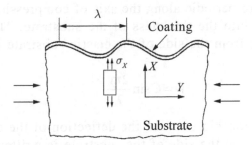

Fig. 9.21. Diagram of the polymer sample with a metallic coating used in [51] for calculating the wavelength of the microrelief in the case of the loss of stability of the coating on a soft substratum.

$$E_1 I w^{\langle IV \rangle} + N w^{\langle II \rangle} + q = 0 \qquad (9.7)$$

where E_1 is the elasticity modulus of the coating; w is the deflection of the coating in relation to the plane $Y = 0$; N is the applied compressive force; q is the returning force, acting on the coating on the axis X on the side of the half plane; I is the moment of inertia of the cross-section of the coating equal for the unit width of the coating to $h^3/12$, where h is the thickness of the coating. Differentiation is carried out with respect to the coordinate Y.

As it is well-known that in the loss of stability the small deflection of the beam is described by the sinusoidal function, the solution of equation (9.5) will be obtained by the method of separation of variables in the form

$$u = u_x \sin \frac{2\pi y}{\lambda} \qquad (9.8)$$

where u_x depends only on the variable x, and λ is the wavelength of deflection of the coating to be determined. Substituting equation (9.8) into (9.5) gives the following solution tending to 0 at $x \to \infty$:

$$u = C \exp\left(-\sqrt{\frac{G}{E}} \frac{2\pi}{\lambda} x\right) \sin \frac{2\pi y}{\lambda} \qquad (9.9)$$

here C is the integration constant. The axial stresses is determined by substituting equation (9.9) into equation (9.6):

$$\sigma_x = -C \sqrt{EF} \frac{2\pi}{\lambda} \exp\left(-\sqrt{\frac{G}{E}} \frac{2\pi}{\lambda} x\right) \sin \frac{2\pi y}{\lambda} \qquad (9.10)$$

The stresses are periodic along the axis of compression and decrease exponentially into the thickness of the substrate. The force acting on the coating from the side of the elastic substrate is $-\sigma_x$ at $x = 0$:

$$w = C \sin \frac{2\pi y}{\lambda} \tag{9.12}$$

Thus, the relationship between the deflection of the coating and the returning force on the side of the substrate is written in the form

$$q = \sqrt{EG} \, \frac{2\pi}{\lambda} w \tag{9.13}$$

The compressive stress is determined by substituting the equations (9.12) and (9.13) into equation (9.3):

$$N = E_1 I \left(\frac{4\pi^2}{\lambda^2} + \sqrt{EG} \, \frac{\lambda}{2\pi E_1} I \right) \tag{9.14}$$

Figure 9.22 shows the dependence of the compressive force N on the wavelength γ for the single values of the coefficients in front of $1/\lambda_2$ and λ. Function $N(\lambda)$ separates the figure into two sections. The area below the curve $N(\lambda)$ corresponds to the planar stable form of the coating. Similarly, the cross-hatched region corresponds to the loss of stability of the coating. The function $N(\lambda)$ is always greater than zero and has a minimum at a certain value of λ^*. This means that in loading below a certain value (critical load) the coating does not lose its stability and remains flat. The loss of stability takes place when the minimum of the function $N(\lambda)$ is reached; this minimum is determined from the condition of equality to 0 of the derivative $dN/d\lambda = 0$. Taking into account that the shear modulus is linked with the elasticity modulus by the relationship $G = E/2 \, (1+v)$ leads to

$$\lambda^* = 2\pi \sqrt[6]{\frac{1+v}{18}} h \sqrt[3]{\frac{E_1}{E}} \tag{9.15}$$

where v is the Poisson coefficient of the substrate. It may be seen that the equation (9.15) is completely identical with (9.2) obtained in [42–44] for the period of the relief formed in the film with a coating in the planar compression conditions.

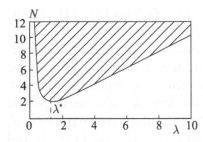

Fig. 9.22. Dependence of the compressive force n on wavelength λ for unit values of coefficients in front of $1/\lambda_2$ and λ.

The experimental results show that, firstly, in the conditions in which the irregular microrelief forms the coating should be bent with the formation of a wavy microwave with a period λ^*. Secondly, the wavelength of the period λ is determined only by the thickness of the coating and the ratio of the moduli of the coating and the substrate. The results can explain qualitatively the main special features of the formation of the regular microrelief during deformation of the 'rigid coating on the elastic polymer substrate' system and the experimental data published in [51].

Equation (9.15) was verified by experiments using, as the polymer-substrate, the cross-linked rubber and showed a 10% difference between the theoretical and experimental values. According to the authors, this is a sufficiently good agreement because they assumed that the modulus of the platinum in the thin layers and in the block is the same, although such an assumption is the subject of discussion.

Thus, it may be concluded that the mechanism of formation of the regular microrelief in deformation of the elastic polymer (rubber) with a rigid coating consists of the loss of stability of the coating as a result of the effect of compressive stresses, formed due to the Poisson compression of the polymer substrate.

The ductile substrate
The above analysis of the loss of stability phenomenon has been used to explained the mechanism of the phenomenon for the case of using an elastic (rubber) substrate. Similar studies of the processes of relief formation in deformation of the 'rigid coating on a soft substratum' polymer system using thermoplastic polymers (polyethylene terephthalate, polyvinyl chloride) [65] as a substrate indicate that the above analysis was not successful.

According to the elasticity theory [64], the loss of stability is a critical phenomenon. At stresses lower than critical the coating on the elastic substrate should remain flat. At the critical stress the coating should loose stability over the entire surface. The systems based on a rubber substrate lose stability in the same fashion [51, 53] and the RMR forms at a certain moment over the entire sample surface. The analysis, described in the previous section, was carried for this case.

In [66] it was shown that when using thermoplastic substrate, the mechanism of formation of the regular relief is completely different. Figure 9.23 shows the micrographs of the samples of a thermoplastic polymer (polyethylene terephthalate) with a platinum coating (a–d) and deformed by different strains. It may be seen that in the initial stages of tensile drawing (Fig. 9.23 a) the coating loses stability not over the entire area, as predicted by the theory, but only in individual areas in the vicinity of nucleation cracks. In this stage of tensile drawing a large part of the coating retains its smooth initial relief. With the development of deformation (Fig. 9.23 b), larger and larger areas of the coating are included in the process of the loss of stability and the coating acquires some relief. In this stage, the microrelief is not yet sufficiently regular. Many elements of the microrelief are not clear, the individual nucleation folds do not spread from one edge of the coating to the other, they are short and 'called off'. Only further deformation improves the resultant RMR which then occupies the entire area of the coating (Fig. 9.23 c). However, in this stage of formation the ERM is still incomplete and shows a large number of disruptions of regularity. It may be seen that many folds branch out, have a large number of 'dead ends' or 'dislocations' in the structure of the RMR. The relief becomes complete only at 100% strain and forms a single regular network on the surface of the polymer (Fig. 9.23 d). Thus, the experimental data show that the loss of stability of the rigid coating on the thermoplastic substrate is not of the critical nature and takes place by a completely different mechanism in comparison with the rubber substrate.

It was shown in [66] that the absolute values of the periods of the relief also do not correspond to the values determined using equation (9.15). The following analysis was made (Fig. 9.24) to explain the mechanism of the observed phenomenon in [66]. It is assumed that the loss of stability ('swelling') of the coating takes place at some point when the stress at this point reaches the yield stress. The increase of the amplitude (height) of swelling is evidently accompanied by a reduction of the stress in the vicinity of the resultant perturbation. If the amplitude is sufficiently high, then as a result of swelling of the coating at point A in Fig. 9.24 the stress at this point decreases to 0. It is also clear that the stress in the coating increases with increase of the distance from the point A.

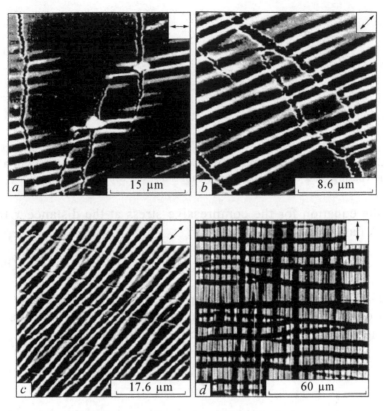

Fig. 9.23. Scanning electron micrographs of polyethylene terephthalate samples deposited with a thin (20 nm) platinum layer and tensile drawn subsequently at 90°C by 5 (*a*), 10 (*b*), 15 (*c*) and 100 (*d*) % [66].

At point B at a certain distance from the point A, the compressive stress again reaches its yield stress. In other words, at point B the coating 'forgets' about the defect. Consequently, the stress at this point again reaches the yield stress of the coating so that its swelling can again take place. Thus, in this case we are concerned with the situation of the 'local loss of stability' which can take place in a certain weak area (defect) of the surface of the coating. In other words, we analyse the situation which occurred in the actual system evidently containing a large number of uncontrollable defects (weak areas) in the coating.

Evidently, shear stress forms in the vicinity of the area of swelling of the coating on the polymer–coating interface. On the condition that the substrate behaves as a rigid–plastic body, the magnitude of the shear stress is restricted by the yield stress of the substrate material. Since the sum of the forces applied to the coating is equal

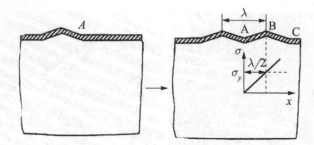

Fig. 9.24. Schematic representation of the process of the loss of stability of a thin metallic coating on a thermoplastic polymer substrate [66].

to 0, the equation for the compressive stress at the distance x from point A has the form:

$$F = \tau x \qquad (9.16)$$

here F is the compressive force in the coating, τ is the shear yield stress of the substrate. The compressive stress σ is obtained by dividing the force by the cross-section of the coating

$$\sigma = \frac{\tau x}{h} \qquad (9.17)$$

With increase of the distance from the point A the compressive stress increases in proportion to the distance to this point. The dependence of stress is shown schematically in Fig. 1.29. The figure shows that at some distance λ from the point A, the stress in the coating reaches the yield stress

$$\lambda = \frac{\sigma_y h}{\tau} \qquad (9.18)$$

here σ_y is the yield stress of the coating. Evidently, at distance λ from point A local swelling of the coating may again take place. Thus, equation (9.18) can be used to estimate the wavelength of the loss of stability (period of the relief). Since the yield stress of the substrate in shear τ is equal to half the yield stress in tension [67], the wavelength of the loss of stability for the hard elastic substrate is described by the equation

$$\lambda = \frac{2\sigma_y h}{\sigma_m} \qquad (9.19)$$

here σ_m is the yield stress of the substrate in tension.

In Fig. 9.25 the experimental data, obtained in the determination of the period of the RMR for the thermoplastic substrates (polyethylene terephthalate, polyvinyl chloride) are compared with the appropriate values calculated from equation (9.18). It may be seen that theory and experiment are in good agreement indicating that the assumptions made regarding the mechanism of the loss of stability of the metallic coating in compression of an inelastic (plastic) substrate were substantiated.

Thus, in [66] the results show the principal difference in the mechanisms of the loss of stability of the rigid coating on the elastic and plastic substrates. Actually, the results show a new type of the loss of stability of the rigid coating on the compliant base which takes place during deformation of the plastic substrate (PET, PVC) – hard coating system.

Nevertheless, in both cases, the physical reasons for the formation of RMR are the same. Here we are concerned with the phenomena associated with the uniaxial compression of anisodiametric solids where the controlling role is played by the macroscopic mechanical properties of the materials so that these phenomena are studied in certain sections of the solid state physics [68]. It is therefore not surprising that the relationships which were obtained for determining the wavelength of the resultant regular relief include various

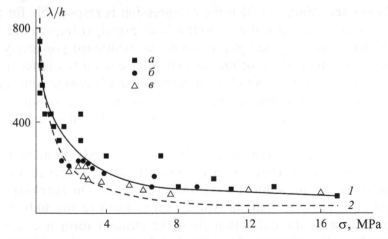

Fig. 9.25. Dependence of the period of the microrelief, reduced to the thickness of the coating, on the magnitude of the applied stress for the systems based on polyethylene terephthalate (a), plasticised polyvinyl chloride (b) and impact resistant polyvinyl chloride (c) tensile loaded at temperatures higher than the appropriate glass transition temperatures (1), and the theoretical curve calculated from equation (1.18) [66].

macroscopic characteristics of the materials, such as modulus, stress, yield stress, thickness of the coating (equations (9.2), (9.15), (9.19)).

The quantitative analysis of the loss of the mechanical stability of an elastic solid was carried out for the first time than 200 years ago by the outstanding physicist and mathematician Euler. He established [68] that in the uniaxial compression of the anisodiametric hard elastic solid, for example, a rod, the latter is suddenly bent on reaching some ultimate stress σ in such a manner that its shape becomes sinusoidal. The wavelength of this sinusoid is twice the distance L between the ends. The ultimate stress of this type of loss of stability in this case is equal to:

$$\sigma_c = 2\pi \ E \ J/L^2 \qquad (9.20)$$

here E is the elasticity modulus of the material of the rod, and J is the moment of inertia of its cross-section. This type of the loss of stability of the solid is well-known and is referred to as the Euler loss of stability. It may be seen that as the length of the elastic rod increases, the force required for the loss of stability decreases.

It may easily be shown that the formation of the regular microrelief in uniaxial tensile drawing of the polymer films with a rigid coating [56] (Fig. 9.26) is associated with the Poisson transverse compression of the polymer substrate during its uniaxial tensile drawing. The conclusion according to which the compression is responsible for the formation of the discussed microrelief is of principal importance for understanding this physical phenomenon. As mentioned previously, in the uniaxial compression of the anisodiametric solid the latter loses its stability on reaching the ultimate strength and becomes sinusoidal with a wavelength equal to the double length of the compressed solid. The classic case of the Euler loss of stability shown in Figs. 9.26 *a*, *b*.

However, if this asymmetric object is strongly bonded with the soft substratum (Fig. 9.26 *c*), the situation observed in the compression of the subject drastically changes. On reaching the critical compressive load the solid cannot acquire the half wave shape because in the deflection from the straight form it could be subjected to the effect of the returning force from the side of the substrate, with the force proportional to the magnitude of deflection. As a result of this interaction between the external applied force and the resultant internal resistance from the side of the substrate, the

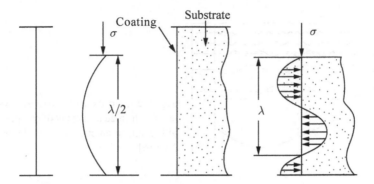

Fig. 9.26. Diagram of the loss of stability of the asymmetric solid in the free state (*a, b*) and on a soft substratum (*c, d*).

coating unavoidably becomes sinusoidal with the period of the wave equal to λ (Fig. 9.26 *d*).

The period of the relief is determined by the following circumstances. The deformation work in compression of the anisodiametric solid (the coating in the present case) evidently increases with increase of the number of completed bends (with the reduction of the period of the relief), Fig. 9.27. It is not accidental that in the absence of the substrate the anisodiametric solid acquires the shape of the half wave, i.e., a relief with the maximum period forms or, which is the same, with the minimum number of bends. However, the soft but very long substratum, 'secured' to the coating, makes corrections to this process. Evidently, as the period of the relief increases, the amplitude of the relief also increases, with other conditions being equal (Fig. 9.27). The increase of the amplitude of the relief indicates the drawing out of the part of the polymer, 'secured' to the substrate, over a relatively large distance from its initial flat surface. In other words, the increase of the period of the resultant relief, which is so 'advantageous' for the coating, is completely 'disadvantageous' for the soft polymer substratum. The period of the relief, observed in the actual situation, is determined from the condition of the minimum total balance of the forces (stresses) in the coating and the substrate.

The minimisation of the energy of the system of this type gives the value of the period of the relief in the case of the deformation of the plastic substrate:

$$\lambda = 2h\,\lambda_y/\sigma_m(1 - \varphi) \qquad (9.21)$$

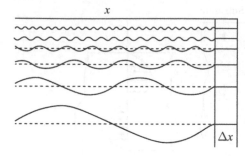

Fig. 9.27. Folds with different wavelengths, formed as a result of compression of the anisodiametric solid by the length Δx [56].

here h is the thickness of the coating, σ_y is the yield stress of the coating, σ_m is the yield stress of the substrate under tensile drawing, and φ is the degree of compression of the coating on the polymer-substrate surface.

9.7. Regular fragmentation of the coating

As shown previously, the formation of the RMR is associated with the compression of the coating on a soft polymer substratum. The driving force of this process is the compression of the polymer substratum associated with the reduction of its surface area determined in turn by the transfer of material from the surface to the bulk of the polymer. At the same time, a reversed process – increase of its interfacial area – also takes place during deformation of the polymer. In this case, the coating on the surface of the polymer is fragmented because its fracture elongation is considerably smaller than in the polymer-substrate.

In fact, Fig. 9.13 shows that in addition to the formation of the regular microrelief, quite regular fragmentation of the coating also takes place. As mentioned previously, to understand the mechanism of the observed phenomenon it is important to take into account the fact that the polymer film, subjected to uniaxial tensile drawing, is subjected simultaneously to two types of deformation. Evidently, the fragmentation of the coating is associated with the fracture component of its deformation. In particular, this is indicated by the fact (Fig. 9.13) that the cracks in the coating propagate almost always in the direction normal to the direction of uniaxial tensile drawing.

Thus, the regular fragmentation of the hard shell is also associated with the special features of the transfer of mechanical stress from the soft substratum of the rigid coating through the interface. Here, it should be mentioned that the nature of regular fragmentation of the coating depends in particular on the deformation mechanism of the substrate. It is well known that the polymer films can be deformed by at least two mechanisms: uniformly (affine), like the deformation of a piece of rubber, or non-uniformly, like the deformation of a glassy or crystalline polymer (by the development of a neck or crazes). In the latter case, the fragment of the initial non-deformed polymer and the fragment of the deformed polymer, transferred to the oriented state, exist simultaneously. Although the fragmentation of the rigid coating takes place regularly in both cases, the mechanisms of reaching this regularity differ.

In the case of uniform deformation of the polymer-substrate the regular fragmentation of the coating develops after a certain period of time. In the first stages of tensile drawing in this case the dimensions of the fragments of the coating are not identical. Evidently, this effect is associated with the results which show that in the initial stages of fracture (at small elongation of the polymer-substrate) the controlling contribution to the fragmentation of the coating comes from the surface microdefects characteristic for any real solid which induce fracture of the coating in the area of their localisation. The defects of this type are evidently distributed randomly in the coating and this also results in the irregular, random failure of the coating. Experimental evaluation shows that the size distribution of the fragments in the initial stages of tensile drawing is very wide. However, after this randomly formed size distribution of the coating fragments, the next stage is the start of the very interesting and unique process of further failure of each of the fragments formed. The point is that after the initial random breakdown of the coating into fragments, the process of tensile drawing of the substrate continues and, consequently, each resultant fragment remains under load.

The stress in each resultant fragment is distributed highly non-uniformly (Fig. 9.28). Evidently, this stress at the ends of the fragment is equal to 0. With increase of the distance from the ends, the stress in each fragment of the coating increases and reaches its maximum value exactly in the centre of the fragment. With further tensile drawing of the substrate the stress in each fragment increases and, in the end, reaches the strength value. In turn, the fracture stress is reached exactly in the centre of the fragment. These circumstances lead to the very interesting process of failure the coating by its separation into two equal parts.

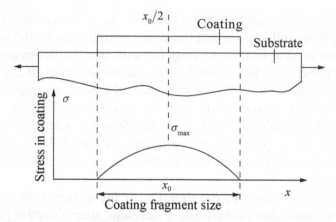

Fig. 9.28. Diagram of the distribution of the stresses in the fragment of the metallic coating on the surface of the deformed polymer substrate [69].

This process can be detected in direct microscopic studies (Fig. 9.29). The mechanism of 'separation' into two equal parts continues until the weak soft substratum can transfer the stresses higher than the strength of the coating to the coating fragments. Subsequently, as the dimensions of the fragments become too small for the substrate to transfer the fracture stress to them, the 'separation' process is completed. Because of these reasons, the dimensions of the fragments are equalised and a system of coating fragments with a very narrow size distribution forms on the surface of the substrate.

Thus, on the basis of these results it may be assumed that the fragmentation process continues until the stress in the substrate increases or until the size L of each fragment becomes so small that the weak soft substratum no longer transfer to the coating the stress sufficient for reaching its ultimate strength. In both cases, the fragmentation of the coating is interrupted, regardless of the continuing tensile drawing of the substrate and the fracture fragments will be subsequently simply pushed away from each other whilst retaining their dimensions (Fig. 9.29). Under these assumptions, the size of the coating fragment (L) is determined as follows [69, 70]

$$L = 2h\sigma^*/\sigma_0 \qquad (9.22)$$

where h is the thickness of the coating, σ^* is its tensile strength and σ_0 is the stress in the substrate.

In [56, 59, 70] the results show a distinctive relationship between the visually detected parameters of the microrelief, formed in

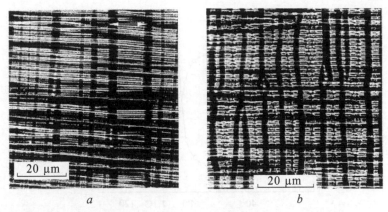

Fig. 9.29. Scanning electron micrograph of a polyethylene terephthalate sample with a thin metallic coating after deformation by 50% at 100°C (a); the same sample after additional tensile drawing by 50% at 80°C (b).

deformation of the polymers with the coatings, and the mechanical properties of the polymer-substrate. The relationship (9.22) shows that the size of the fragments of the coating depends in particular on the stress in the deformed polymer-substrate σ_0. This figure compares the temperature dependence of the average size of the fragments of a platinum coating, formed in tensile drawing of an amorphised polyethylene terephthalate film-substrate and the actual part of the dynamic complex modulus of polyethylene terephthalate. In may be seen that these independently determined characteristics are in distinctive correlation. In fact, while the polyethylene terephthalate is in the glassy state (T_g = 78°C), the size of the fragments is small because the stress at which the glassy polymer is deformed is relatively high. In transition through T_g the stress in the polymer rapidly decreases and this influences immediately the average size of the fragments of fracture (the size rapidly increases). Polyethylene terephthalate is a polymer capable of the so-called cold crystallisation, i.e., as soon as the mobility of the macromolecules become sufficiently high, the polymer rapidly crystallises. The highest crystallisation rate is observed at ~120°C. As a result of crystallisation of the stress in the polymer rapidly increases which has the controlling effect on the ovary size of the fragments of fracture of the coating. As indicated by Fig. 9.30, the parameters of the relief, formed in deformation of the polymer with the coating, are unambiguously linked with the properties (with the stress in this case) of the polymer-substrate. The results confirm the assumptions made when deriving equation (9.22).

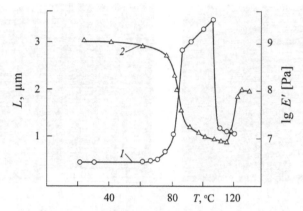

Fig. 9.30. Dependence of the average size of the fragments (L) of fracture of a platinum coating on the temperature of tensile drawing of the polyethylene terephthalate substrate at a rate of 0.1 mm/min by 100 % (1) and the temperature dependence of the actual part of the complex modulus (E') of polyethylene terephthalate (2) [56].

It should be mentioned that the structure of the equation (9.22) is identical with that of the previously derived equations describing the fragmentation of the coating during tensile drawing of the polymer-substrate. It should be mentioned that the following relationship was obtained for the wavelength of the RMR in the case of the thermoplastic polymer-substrate:

$$\lambda = 2h\sigma_y/\sigma_m \tag{9.23}$$

here h is the thickness of the coating, σ_y is the yield stress of the coating, σ_m is the yield stress of the substrate in tensile drawing.

Dividing equation (9.21) by (9.22) leads to

$$\lambda/L = 2\sigma_y/3\sigma \tag{*(9.24)}$$

Equation (9.23) indicates that the ratio of the period of the wave to the length of the fragments of fracture is generally independent of the mechanical properties of the plastic substrate, and is determined by the ratio of the yield stress of the coating to its strength. On the basis of this circumstance we can obtain another independent experimental evaluation of equation (9.22). In fact, in the microscopic studies the period λ and the average size of the fracture fragments L can be determined simultaneously and independently.

Figure 9.31 shows the dependence of the ratio λ/L on the thickness of the coating. As indicated by the figure, the λ/L ratio is in fact

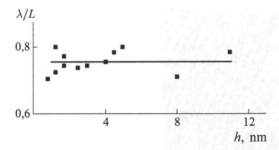

Fig. 9.31. Dependence of the ratio λ/L on the thickness of the coating h for the polyethylene terephthalate–platinum coating system (100% strain at 90°C at a rate of 10 mm/min) [65].

independent of the coating thickness. This result shows that the assumptions of the loss of stability of the rigid coating evidently correspond to the actual mechanism of the observed phenomenon.

There is another type of fragmentation of the coating in deformation of the polymer-substrate not associated with the mechanism of dividing the fragments into two equal parts [71]. This type of fragmentation is observed when the polymer-coating is deformed non-uniformly, for example as takes place in crazing of the polymer. In this case, immediately after the formation of the long zone of the plastically deformed polymer a very narrow size distribution of the coating fragments appears (Fig. 9.32).

This is accompanied by the formation of a system of the thinnest strips of the coating of almost the same size and parallel to each other and propagating from one edge of the deformed specimen to another. The reason for the spontaneous formation of such a unique structure is that, in this case, the set of the characteristic defects of the coating does not take part in the process of its fragmentation. As mentioned previously, the process of non-uniform deformation is characterised by the continuous coexistence of the fragment of the initial non-deformed polymer and the fragment of the deformed polymer, transferred to the oriented state. This means that there also coexist two parts of the coating: the part broken up into fragments in the deformed part of the polymer, and the whole part, without fracture, coating the non-deformed part of the polymer.

All the events associated with the fragmentation of the coating take place in a narrow moving zone, situated between the oriented and non-oriented parts of the deformed polymer. The zone always contains the edge of the fractured coating in which the stress is evidently equal to 0. It is important to note that this interface is very

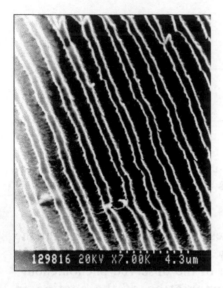

Fig. 9.32. Scanning electron micrograph of a polyethylene terephthalate sample with a thin metallic coating, after its deformation by the classic crazing mechanism [71].

sharp and thin in comparison with the non-oriented polymer–neck interface.

The mechanism of formation of the unique regular structural fragmentation of the coating is shown schematically in Fig. 9.33. With increase of the distance from this edge, the stress in the coating increases and rapidly reaches ultimate strength. At this moment, the next strip of the coating separates. In these conditions, the presence of the surface microdefects has almost no effect on the process of fragmentation of the coating because the surface of the polymer, which has not transferred to the oriented state, is virtually not deformed (the magnitude of elastic strain does not exceed several percent). The strips gradually separate from the coating when the ultimate strength of the material of the coating is reached and, consequently, the width of all fracture fragments is practically identical.

As shown in [69–71], regardless of whether the soft substratum and/or rigid coating is deformed elastically or not, the average size (L) of the fragment of fracture in the direction of the axis of tensile drawing is equal to:

$$L = 2h\frac{\sigma^*}{\sigma_0} \tag{9.25}$$

where h is the thickness of the coating, σ^* is its ultimate strength, and σ_0 is the stress in the substrate.

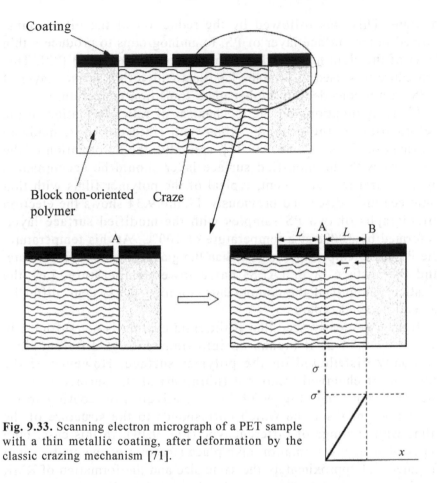

Fig. 9.33. Scanning electron micrograph of a PET sample with a thin metallic coating, after deformation by the classic crazing mechanism [71].

9.8. Surface structure formation in polymers with a chemically modified surface

The previously discussed data indicate that the polymer films with the hardness nanometric coatings are susceptible to the formation of regular periodic structures. For the regular surface structure formation of this type it is not necessary to coat the surface of the polymer with metallic or any other rigid coating. It is sufficient to modify the polymer (for example, by chemical modification) leading to the formation of a hard thin coating on the surface of the polymer.

Surface amination of polystyrene
This system was formed in [54, 55] where surface treatment (nitration)) of polymer films (PS) was carried out for certain periods

of time. This was followed by the reduction of the nitrogroups, formed in the surface layer of PS, to aminogroups to produce a thin layer of the rigid high modulus aminopolystyrene (APS) [72]. This procedure was used to produce PS films with a thin surface layer of APS with a considerably higher glass transition temperature.

If the assumptions on the mechanism of relief formation in the deformation of the polymer films with rigid coatings, made in previous sections [53, 56, 69–71], are valid, the deformation of the PS films with the modified surface layer should be accompanied by structural rearrangement, typical of the polymer films with thin rigid coatings, discussed previously. Figure 9.34 shows the electron micrographs of two PS samples with the modified surface layer, deformed by 100% at a temperature of 100°C. At this temperature, the PS has a temperature higher than the glass transition temperature, and the surface layer (APS) has a lower temperature, i.e., the elasticity moduli of the components, forming the system, differ by several decimal orders.

It may be clearly seen that if the rigid coating is not sufficiently thin (Fig. 9.34 *a*) it breaks up into fragments of different sizes randomly distributed on the polymer surface. However, if the duration of chemical treatment (nitration) of the surface layer of the PS is not long (Fig. 9.34 *b*), a relatively thin coating forms on the polymer surface which corresponds to the structure of the films with coatings of different type (Fig. 9.13). In this case, both types of structure formation take place (fracture of the coating into fragments of approximately the same size and the formation of RMR on these fragments). These results demonstrate the general nature of the phenomena taking place during deformation of the polymers with thin rigid coatings, irrespective of their nature.

Regular fragmentation of the surface of the polyamide fibre with a modified surface layer

Another example of surface modification of polymers was published in [73], in which the system was produced on the basis of an oriented polyamide fibre. The fibres based on the polycaproamide (PA-6) were treated with phenol formaldehyde oligomers of the rezol type [74]. These oligomers cause swelling of the polyamide and their subsequent rigidening by formaldehyde *in situ* greatly modifies the mechanical properties of PA-6 and their resistance to the effect of moisture. It is important to note that in the process of swelling the rezol resin penetrates only into a relatively thin surface layer of

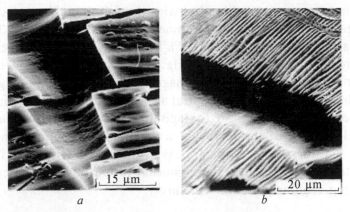

Fig. 9.34. Scanning electron micrographs of a PS sample with a modified surface layer, deformed at 120°C by 100%: *a*) processed with the nitration mixture at 60°C for 40 s, followed by complete restoration of the nitrogroups; b) scanning electron micrograph of a PS sample, treated with the nitration mixture at 40°C for 2 s followed by complete restoration of nitrogroups [55].

the fibre. Consequently, after hardening *in situ*, the fibre acquires a relatively thin hard shell made of the hardened rezol resin, strongly bonded with the core, consisting of pure PA-6.

It is well known that PA-6 dissolves easily and rapidly in acids, in particular, in formic acid. The pure non-modified fibre is completely dissolved in formic acid over several minutes. The shell produced from the hardened phenol formaldehyde resin, included in the surface layer of the fibre, does not dissolve in the formic acid. The interaction of the modified fibre with the formic acid takes place as follows. At the initial moment, the formic acid penetrates into the bulk of the fibre and causes extensive shrinkage (up to 50%). Naturally, shrinkage also causes the appropriate increase of the cross-section. Only then the molecular dissolution of PA-6 starts to take place.

If treatment is carried out on the modified polyamide fibre, the surface layer, insoluble in the acid, prevents shrinkage of the fibre. However, under the effect of internal stresses, this system, constructed on the principle of the rigid coating on a soft substratum, finds the exit from the existing situation. The process is accompanied by the formation of a crack on the polymer surface which grows along a surprisingly regular spiral (Fig. 9.35 *a*).

As a result, the surface layer of the fibre transforms to a spiral with a pitch of ~25 µm. The fracture surface shows that only the surface layer of the 'truncated' fibre has a spiral structure (Fig. 9.35

b). Further treatment of this fibre with the formic acid results in the complete dissolution of the non-modified core of the fibre. This is accompanied by the formation of a hollow cylinder with the walls penetrated by the spiral crack (Fig. 9.35 *c*). The resultant hollow cylinder can be stretched which makes it possible, firstly (Fig. 9.35 *d*), to obtain a strictly regular spiral and, secondly, detect the internal structure of the elements of the spiral. It may clearly be seen that the hard coating of the fibre has a multilayer structure. The data indicate that the polymers, regardless of their form (films are fibres), having thin rigid coatings, have unusual properties and are capable of forming regular periodic structures.

9.9. Polymer films with nanometric coatings – 'rigid coating on a soft substratum' systems

On the basis of the above results it may be concluded that the polymer films with a thin rigid coating have a number of characteristic properties. Firstly, compressive deformation of these films results in

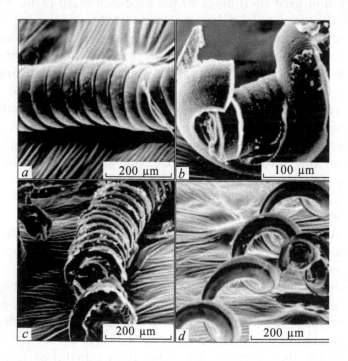

Fig. 9.35. Scanning electron micrographs of a polyamide fibre with a rigid shell of phenol formaldehyde resin and treated with formic acid, explanation in the text [73].

the formation of a regular microrelief with the parameter dependent on the properties of the coating and substrate materials, the thickness of the coating and the level of stress in the material. Secondly, in the tensile drawing conditions the rigid coatings on the surface of the polymer-substrate undergo regular fragmentation with the parameters of this process also dependent on the properties of the material of the coating and the substrate, the thickness of the coating and the special features of transfer of mechanical stress from the soft substratum to the hard coating. Finally, the resultant relationships (equations (9.15), (9.21) and (9.22)), linking the properties of the coating and the polymer substrate and verified by experiments, do not contain any restrictions for the nature of the coating and the substrate nor the scale of these elements. Taking these circumstances into account, it may be assumed that the polymer films without coatings are typical representatives of the group, refer to in [56] as the 'rigid coating on a soft substratum' systems (RCSS). It will be shown that the RCSS systems have in fact a number of characteristic properties and are widely found in the surrounding environment. Here, we shall formulate only the main features of the RCSS systems.

The RCSS system must satisfy three main conditions:

– the thickness of the coating should be negligible in comparison with the thickness of the soft substratum;

– the elasticity modulus of the coating material should be several orders of magnitude higher than the elasticity modulus (hardness) of the substrate material;

– the coating and the substrate should be characterised by optimum adhesion in the presence of which the mechanical stress may be efficiently transferred from the substrate to the coating.

If the above conditions are fulfilled, the previously mentioned surface structure formation (formation of a regular microrelief and/ or regular fragmentation of the coating) will take place regardless of the nature of the coating and substrate materials and irrespective of the scale of the RCSS system.

Conclusion

Thus, the deformation of the polymer is accompanied by the processes of mass transfer of the material from its surface of the bulk and vice versa. In fact, all the previously examined structural rearrangements are surface phenomena because they take place in the surface layers of the polymers with a thickness in the nanometric range. These processes are efficiently visualised by the deposition of thin rigid coatings on the surface of the polymer prior to deformation (shrinkage). This procedure makes possible not only to visualise the structural rearrangement, taking place in the polymer, but also characterise the fundamental properties of the RCSS systems [56].

References

1.   Vinogradov G.V., Malkin A.Ya., Polymer rheology, Encyclopedia of polymers, Moscow, Sov. entsiklopediya, 1977, V. 3, P. 340.
2.   Malkin A.Ya., Leonov A.I., Advances of polymer rheology, Moscow, Khimiya, 1970, P. 98.
3.   Malkin A.Ya., Highly elastic turbulence, Encyclopedia of rubbery polymers. Moscow, Sov. entsiklopediya, 1977, V. 3, P. 664.
4.   Satchell D.G., Mullin T., Proc. R. Soc. Lond. A, 2001, V. 457, P. 2567.
5.   Larson R., Rheol. Acta, 1992, V. 31, No. 3, P. 213.
6.   Karthaus O., Grasjo L., Maruyama N., Shimomura M., Thin Solid Films, 1998, V. 327–329, P. 829.
7.   Karthaus O., Grasjo L., Maruyama N., Shimomura M., Chaos, 1999, V. 9, No. 2, P. 308.
8.   Muller-Buschbaum P., Gutman J.S., Wolkenhauer M., Kraus J., Stamm M., Smilgies D., Petry W., Macromolecules, 2001, V. 34, No. 5, P. 1369.
9.   Gunther J., Stump P. I., Langmuir, 2001, V. 17, No. 21, P. 6530.
10.   Stump S.I., LeBonheur V., Walter K., Li L.S., Huggins K., Keser M., Amstutz F., Science, 1997, V. 276, P. 384.
11.   Gutmann J.S., Muller-Buschbaum P., Stamm M., Faraday Soc., 1999, V. 112, P. 285.
12.   Muller-Buschbaum P., Gutmann J., Stamm M., J. Macromol. Sci., B, 1999, V. 38, No. 5-6, P. 577.
13.   Morita H., Kawakatsu T., Doi M., Macromolecules, 2001, V. 34, No. 25, P. 8777.
14.   Yamamura M., Nakamura P., Kajiwara T., Kage H., Adachi K., Polymer, 2003, V. 44, P. 4699.
15.   Tanaka T., Sun P., Hirokawa Y., Katajama S., Kusera J., Hirose Y., Amija T., Nature, 1987, V. 325, No. 6107, P. 796.
16.   Hwa T., Kardar M., Phys. Rev. Lett., 1988, V. 61, P. 106.
17.   Dubrovsky S.A., Dokl. AN SSSR, 1988, V. 303, No. 5, P. 1163.
18.   Tanaka H., Sigehusi T., Am. Phys. Soc. Phys. Rev., 1994, V. 49, No. 1, P. R39.
19.   Matsuo S.E., Tanaka T., Nature, 1992, V. 358, P. 482.
20.   Tanaka T., Frontiers of Macromol. Sci. IUPAK 32 Int. Symp. On Macromolecules, Ed. by Saegusa T., Kyoto, Japan, 1989. P. 325.
21.   Matsuo S.E., Tanaka T., Nature, 1992, V. 358, P. 482.
22.   Hu Z., Chen Y., Wang C., Zheng Y., Li Y., Nature, 1998, V. 393, P. 149.
23.   Sekimoto K., Prost J., Julicher F., Europ. Phys J. E., 2004, V. 13, No. 3, P. 247.
24.   Suzuki A., Ishii T., J. Chem. Phys., 1999, V. 110, No. 4, P. 2289.
25.   Chan E.P., Karp J.M., Langer R.S., J. Polymer Sci. Part B, Polymer Physics, 2011, V. 49, P. 40–44.
26.   Schaffer E., Thurn-Albrecht T., Russel T.P., Steiner U., Nature, 2000, V. 403, P. 874.
27.   Schaffer E., Thurn-Albrecht T., Russel T.P., Steiner U., Europhys. Lett., 2001, V. 35, P. 518.

28. Steiner U., Schaffer E., The 2nd Int. Conf. on Scanning Probe Microscopy of Polymers, Weingarten, Germany, 2001, P. 28.

29. Landau L., Lifshitz E., Pitaevski L., Electrodynamics of continuous media, Oxford, Pergamon Press, 1984.

30. Herminghaus S., Phys. Rev. Lett., 1999, V. 83, P. 2359.

31. Chou S.Y., Zhuang L., J. Vac. Sci. Technol. B. 1999. V. 17, No. 6. P. 3197.

32. Karthaus O., Grasjo L., Maruyama N., Shimomura M., Thin Solid Films, 1998, V. 327–329, P. 829.

33. Shaffer E., Harkema S., Roedrink M., Blossey R., Steiner U., Macromolecules, 2003, V. 36, No. 5, P. 1645.

34. Schaffer E., Harkema S., Blossev R., Steiner U., Europhys. Lett., 2002, V. 60, P. 255-261.

35. Tobol'skii A., Properties and structure of polymers, ed. by G.L. Slonimsky and G.M. Bartenev, Moscow, Khimiya, 1964.

36. Gul' V.E., Kuleznev V.N., Structure and mechanical properties of polymers, Moscow, Khimiya, 1972.

37. Bartenev G.M., Frenkel' S.Ya., Polymer physics, Leningrad, Khimiya, 1990.

38. Volynskii A.L., Kechek'yan A.S., Grokhovskaya T.E., Lyulevich V.V., Bazhenov S.L., Ozerin A.N., Bakeev N.F., Vysokomolek. Soed. A, 2002, V. 44, No. 4. P. 615.

39. Volynskii A.L., Grokhovskaya T.E., Kechek'yan A., Bazhenov S.L., Bakeev N.F., Dokl. RAN, 2000, V. 374, No. 5, C. 644.

40. Volynskii A.L., Grokhovskaya T.E., Kechek'yan A.S., Bakeev N.F., Vysokomolek. Soed. A, 2003, V. 45, No. 3, P. 449.

41. Volynskii A.L., Grokhovskaya T.E., Lyulevich V.V., Yarysheva L.M., Bol'shakov A.V., Kechek'yan A.S., Bakeev N.F., Vysokomolek. Soed. A, 2004, V. 46, No. 2, P. 247.

42. Bowden N., Brittain S., Evans A.G., Hutchinson J.W., Whitesides G.M., Nature, 1998, V. 393, P. 146.

43. Bowden N., Huck W.T.S., Paul K.E., Whitesides G.M., Appl. Phys. Lett, 1999, V. 75, No. 17, P. 2557.

44. Huck W.T.S., Bowden N., Onck P., Pardoen T., Hutchinson J.W., Whitesides G.M., Langmuir, 2000, V. 16, No. 7, P. 3497.

45. Allen H.G., Analysis and design of structural sandwich panels, New York, Pergamon, 1969.

46. Dalnoki-Veress K., Nickel B.G., Dutcher J.R., Phys. Rev. Lett., 1999, V. 82, No. 7, P. 1486.

47. Volynskii A.L., Grokhovskaya T.E., Sembayeva R.Kh., Bazhenov S.L., Bakeev N.F., Dokl. RAN. 1998, V. 363, No. 4, P. 500.

48. Volynskii A.L., Grokhovskaya T.E., Sembayeva R.Kh., Yaminsky I.V., Bazhenov S.L., Bakeev N.F., Vysokomolek. Soed., 2001, V. 43, No. 2, P. 239.

49. Volynskii A.L., Grokhovskaya T.E., Sembayeva R.Kh., Bazhenov S.L., Vysokomolek. Soed. A, 2001, V. 43, No. 6, P. 1008.

50. Volynskii A.L., Chernov I.V., Bakeev N.F., Dokl. RAN, 1997, V. 355, No. 4, P. 491.

51. Bazhenov S.L., Chernov I.V., Volynskii A.L., Bakeev N.F., Dokl. RAN, 1997, V. 356, No. 1, P. 54.

52. Volynskii A.L., Bazhenov S.L., Lebedev O.V., Yaminskii I.V., Ozerin A.N., Bakeev N.F., Vysokomolek. Soed. A, 1997, V. 39, No. 11, P. 1805.

53. Volynskii A.L., Bazhenov S.L., Lebedeva O.V., Bakeev N.F., J. Mater. Sci., 2000, V. 35. P. 547–554.

54. Volynskii A.L., Kechek'yan A.S., Bakeev N.F., Vysokomolek. Soed. A, 2003, V. 45, No. 7, P. 1130-1134.
55. Volynskii A.L., Kechek'yan A.S., Bakeev N.F., Vysokomolek. Soed. A, 2005, V. 47, No. 3, P. 430-437.
56. Volynskii A.L., Bazhenov S.L., Bakeev N.F., Ros. Khim. Zh. (ZhVKhO im. D.I. Mendeleeva), 1998, V. 42, No. 3, P. 57.
57. He Jianping. Dissertation, Cand. chemical sciences. M.V. Lomonosov Moscow State University, 2001.
58. Volynskii A.L., Voronin E.E., Lebedev O.V., Bazhenov S.L., Ozerin A.N., Bakeev N.F., Vysokomolek. Soed. A, 1999, V. 41, No. 9, P. 1442.
59. Kechek'yan A., Andrianova G.P., Kargin V.A., Vysokomolek. Soed. A, 1970, V. 12, No. 11, P. 2424.
60. Bazhenov S.L., Kechek'yan A., Vysokomolek. Soed. A, 2002, V. 44, No. 4, P. 629.
61. Biot M.A., Quart. Appl. Math., 1959, V. 17, No. 1231, P. 722.
62. Biot M.A., J. Appl. Phys., 1954, V. 25, No. 11, P. 2133.
63. Manevitch L.I., Pavlenko A.V., Koblik S.G., Asymptotic method in the theory of elasticity of orthotropic solids, Kiev, Vysshaya shkola, 1982.
64. Landau L.D., Livshits E.M., The theory of elasticity, Moscow, Nauka, 1965.
65. Bazhenov S.L. Volynskii A.L., Lebedeva O.V., Voronina E.E., Bakeev N.F., Vysokomolek. Soed. A, 2001, V. 43, No. 5, P. 844–851.
66. Volynskii A.L., Voronina E.E., Lebedeva O.V., Yaminskii I.V., Bazhenov S.L., Bakeev N.F., Vysokomolek. Soed. A, 1999, V. 41, No. 10, P. 1627.
67. Kachanov L.M., Fundamentals of the theory of plasticity, Moscow, Nauka, 1969.
68. Feynman R., Leighton R., Sands M. The Feynman lectures on physics, V. 2, Physics of continua, Moscow, Mir, 1977.
69. Volynskii A.L., Voronina E.E., Lebedeva O.V., Yaminskii I.V., Bazhenov S.L., Bakeev N.F., Vysokomolek. Soed. A, 2000, V. 42, No. 2, P. 262.
70. Volynskii A.L., Bazhenov S.L., Lebedeva O.V., Ozerin A.N., Bakeev N.F., J. Appl. Polymer Science, 1999, V. 72, P. 1267–1275.
71. Bazhenov S.L., Volynskii A.L., Alexandrov V.M., Bakeev N.F., J. Polym. Sci. Part B: Polym. Phys., 2002, V. 40, P. 10.
72. Davankov V.A., Encyclopedia of polymers, Moscow, Sov. entsiklopediya, 1972, V. 1, P. 116.
73. Volynskii A.L., Bondarev V.V., Bakeev N.F., Vysokomolek. Soed. A, 2002, V. 44, No. 11, P. 2058.
74. Volynskii A.L., Volkov A.V., Bondarev V.V., Seleznev N.I., Arzhakov M.S., Bakeev N.F., Vysokomolek. Soed. A, 1990, V. 32, No. 11, P. 2323

10

Evaluation of the structural and mechanical properties of nanometric surface layers

The fundamental properties of substances refined to the nanodimensions are being studied intensively. As shown in recent years [1–7], the transfer of materials from micro- to nanodimensions results in qualitative changes in their physical, mechanical, physical-mechanical and other properties. Regardless of the exceptional importance of reports on the properties, in particular, the mechanical properties of matter in the 'nanostate', this problem is far from being solved. The literature contains only a small number of reports dealing with the construction of complicated and expensive equipment for evaluating the mechanical properties of solids of the micron dimensions in the uniaxial tensile drawing conditions [8–10]. There is almost no reliable information on the stress–strain properties of the material with the size of units or tens of nanometres. This is due mostly to the absence of reliable experimental methods. In fact, it is difficult to imagine how to test by the transitional methods the properties of the solid with the geometrical dimensions of tens or hundreds of angströms. Most information on the mechanical properties of the nanomatter is obtained mainly by indentation experimental methods [11–13]. These approaches have a number of shortcomings of the procedural nature: in particular, they cannot be used for the realistic evaluation of the most important characteristics of solids, such as breaking strain or fracture stress (strength). Therefore, it is very important to carry added investigations aimed at developing new approaches to examine the properties, in particular, the stress–strain properties of the solids, refined to the nanostate.

10.1. Physical fundamentals of the method for evaluating the stress–strain properties of surface layers and nanometric coatings deposited on polymer films

The previously discussed procedure (deposition of rigid nanometric coatings on the polymers followed by deformation) not only can be used to visualise the structural rearrangement of the deformed polymer (see chapter 1) or investigate the fundamental properties of the rigid coating on a soft substratum' systems (RCSS) (see chapter 9). This approach also provides an universal, simple and often efficient method of evaluating the stress–strain properties of matter refined to the nanostate.

It is useful to mention the main relationships obtained for describing the structural–mechanical behaviour of the polymer films with the nanometric rigid coatings.

Evaluation of the strength of surface layers and coatings
In fragmentation of the coating in uniaxial tensile drawing of the polymer-substrate, the average size (L) of the fracture fragment in the direction of the tensile drawing axis is determined using the relationship

$$L = 3h\sigma^*/\sigma_0 \qquad (10.1)$$

where h is the thickness of the coating, σ^* is its strength and σ_0 is the stress in the substrate. The width of the fracture fragments (L) is easily determined from the analysis of the micrographs of the surface of the deformed sample. The thickness of the coating is regulated in the stage of deposition of the coating on the polymer. The stress in the polymer substrate is determined reliably from the dynamometric curves of tensile drawing the sample. Thus, the only unknown quantity in the relationship (10.1) is the strength of the coating

$$\sigma^* = L\sigma_0/3h \qquad (10.2)$$

Evidently, the determined relationship is the basis of the direct original method of evaluating the strength of the solids in layers of any thickness, including in the nanometric range [14–16].

Evaluation of the yield stress of surface layers in coatings
As established in [17, 18], the wavelength of the loss of stability

λ (RMR period) is described by the relationship (see the previous section)

$$\lambda = 3\sigma_y h/\sigma_m \tag{10.3}$$

where σ_y is the yield stress of the coating, h is the thickness of the coating, σ_m is the yield stress of the substrate in tensile drawing. Thus, knowing the period of the microrelief (λ), which can be determined by analysis of the micrographs, the thickness of the coating h, which is regulated during deposition of the coating on the polymer, and also the stress in the polymer σ_0, which can be determined from dynamometric curves, it is possible to determine quite easily and simply the yield stress in the coating:

$$\sigma_y = \lambda\sigma_m/3h \tag{10.4}$$

Evaluation of the magnitude of plastic deformation in the surface layers and coatings

The determination of the strength of yield stress of the nanometric coatings, deposited on the polymer, does not exhaust the possibilities of the previously described approach. Analysis of microscopic patterns of the surface of the deformed polymer with the nanometric coating can also be used to determine the breaking strain of the deposited coating [19, 20]. The application of the fragmentation patterns of the coating in deformation of the polymer-substrate offers a simple procedure of determining the breaking strain of the coating material.

Attention will be given to an electron micrograph showing the typical fragmentation pattern of the metallic coating formed in the tensile drawing of the polymer-substrate (Fig. 10.1). It is evident that the sum of the widths of the metallic (or any other) fragments of the fractured coating in the direction of the tensile drawing axis is $L_{coat} = 1 + \varepsilon_{coat}$, where ε_{coat} is the irreversible (plastic) deformation in the coating. For the same specimen (or a section of the specimen) in which the value L_{coat} was measured, its total length is evidently equal to $L_{total} = 1 + \varepsilon$ (ε is the strain in the polymer-substrate, which is set by the tensile drawing machine). Consequently

$$\frac{L_{coat}}{L_{total}} = \frac{1+\varepsilon_{coat}}{1+\varepsilon} = \frac{\lambda_{coat}}{\lambda_{polym}}$$

where λ_{coat} and λ_{polym} is the draw ratio of the coating and the substrate, respectively. Consequently, the required value of the

$$L_{coat} = L_1 + ... + L_{10}$$

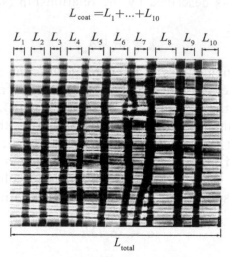

Fig. 10.1. Calculation of the plastic deformation in the coating in tensile drawing of the polymer-substrate.

plastic deformation of the coating is equal to

$$\lambda_{coat} = \frac{L_{coat}}{L_{total}} \lambda_{polym}$$

If the irreversible (plastic) deformation ε_{coat} is expressed in percent, then

$$\varepsilon_{coat} = \frac{L_{coat}}{L_{total}} \lambda_{polym} - 1 \tag{10.5}$$

The equation (10.5) shows that the plastic deformation of the coating can be determined by direct measurement on the micrographs of the sum of the widths of the fragment of the coating in the direction of tensile drawing L_{coat}, the total length of the sample L_{total}, for which L_{coat} was measured (Fig. 10.1), and also using the available values of the strain of the polymer-substrate λ_{polym}, given by the tensile drawing machine.

Thus, the expressions (10.2), (10.4) and (10.5), verified by experiments, can be used for easy and simple determination of the most important characteristics of the solids, such as strength, yield stress and breaking strain in layers of almost any thickness, deposited on the surface of the polymer-substrate. We shall show the possibilities of the previously described method for evaluating the structural–mechanical characteristics of solids of the nanometric size on the example of evaluating the stress–strain properties of chemically modified surface layers of polymers.

10.2. Modification of polymer surfaces

The efficiency of the methods for evaluating the stress–strain characteristics of the nanolayers of solids, developed in [14–20], will be shown on the example of chemically modified polymer surfaces. As shown previously, chemical treatment of the polymers is often accompanied by the formation of modified surface layers (Figs. 9.34, 9.35) whose properties differ drastically from those of bulk polymers. Nevertheless, so far it has not been possible to obtain specific mechanical characteristics of these layers since no efficient experimental method for this evaluation is as yet available.

It is important to mention that to evaluate the structural–mechanical characteristics of the surface layers of the polymers using the relationships (10.2), (10.4) and (10.5), it is important to ensure that the polymer system corresponds to the RCSS conditions. In other words, the surface layer should firstly be considerably thicker than the polymer-substrate and, secondly, the surface layer should be far harder than the polymer-substrate. In particular, these conditions are satisfied by the examples discussed below.

Surface treatment of polymers with gaseous fluorine
Generally speaking, obtaining information on the structure and properties of the modified surface layers of the polymers is an important scientific and applied problem. Chemical surface modification is one of the promising directions of the chemistry and technology of polymers. Conceptually, this approach is based on the chemical modification of surface layers of the polymers which results on the whole in new and often valuable physical–chemical and physical–mechanical properties of the material. Many aspects of this scientific direction have been studied extensively in recent years and generalised in a recently published monograph [21]. This monograph discusses in detail many fundamental and applied aspects of the surface modification of polymers. It is important to note that there is another problem (within the framework of the examined scientific direction) which has not yet been completely solved. It is the evaluation of the stress–strain properties of the layers formed on the surface of the polymer as a result of modification

additions, such as, for example, chemical treatment (modification), irradiation with cold (low-temperature treatment) plasma, ultraviolet radiation, etc. At the same time, the information of this type may be of controlling importance for understanding many phenomena of the surface modification of polymers and also for solving important applied problems. The absence of this information is associated mostly with the absence of general method for evaluating the stress–strain properties of the thin (to the nanometric level) polymer layers.

Consequently, the approach for evaluating the stress–strain properties of the surface polymer layers, developed in [14–20], is highly promising. In [22], investigations were carried out into the stress–strain properties of the surface layers of polyethylene terephthalate formed during treatment with gaseous fluoride by the procedure described in detail in [23, 24].

Figure 10.2 shows the scanning electron micrographs of a sample of the polyethylene terephthalate film treated with gaseous fluorine and subsequently deformed at room temperature with the formation of a neck. Figure 10.2 shows that the treatment of the surface of the polyethylene terephthalate films with fluorine is accompanied not only by the chemical modification of the surface layer, studied in detail and described in [21], but also by the formation of a layer in the form of a morphologically distinctive surface structure.

It may be seen that in the deformation of the polymer films treated with gaseous fluorine, the surface shows structure formation fully identical with that observed in the deformation of the polymer films with a thin metallic coating (chapter 9). The surface layer of the polymer breaks up into fragments of approximately the same size. Evidently, the detected phenomenon is associated with the formation on the surface of the polymer film of a thin cross-linked polymer layer, characterised by higher mechanical properties than the material of the polymer-substrate. The results lead to actual approaches to the evaluation of the stress–strain properties of the modified layer, formed on the surface of the polymers during its chemical treatment, in particular with gaseous fluorine.

The strength will be determined using equation (10.2). All the terms of the equation are known, with the exception of the conventional thickness of the fluorine-treated layer of the polymer. The thickness can be determined by atomic force microscopy (AFM). Figure 10.2 *b* shows the AFM image of a polyethylene terephthalate sample, shown in Fig. 10.2 *a*. The figure indicate that in deformation of the polymer-substrate the polymer breaks up into approximately

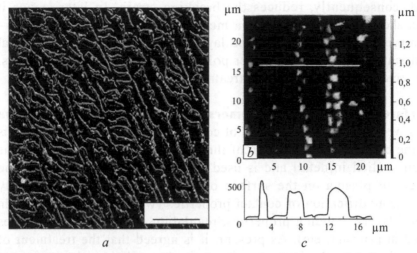

Fig. 10.2. Scanning electron micrographs of a polyethylene terephthalate sample, treated with gaseous fluorine for 30 min and subsequently tensile drawn at room temperature with the formation of a neck (*a*) and its AFM image (*b*), and also the appropriate cross section along the straight light line (*c*) [22].

identical fragments with a thickness of ~450 nm. This thickness is in good correlation with the thickness of the surface layers of the fluorine-treated polymers, determined by the methods [21, 25].

Substituting the thickness of the fluorine-treated layer of 450 nm to equation (10.2) shows that the strength of the fluorine-treated layer of the polymer on the surface of polyethylene terephthalate is 60.7 MPa during its deformation at room temperature and 3.8 MPa at 90°C. It may also be seen that the fluorine-treated polymer layer, regardless of its chemical modification, 'remembers' the glass transition temperature of polyethylene terephthalate because in transition through this temperature the strength of the fluorine-treated polyethylene terephthalate rapidly decreases, similar to any other polymer.

Equation (10.5) was used to determine the plastic deformation of the fluorine-treated layer on the polyethylene terephthalate surface. The results show that, regardless of some scatter of the data, the plastic deformation of the surface layer, produced in deformation of the polymer-substrate at room temperature, is equal to 15–30% and at 90°C it is 8–10% [2]. It may also be seen that long-term fluorine treatment of the polyethylene terephthalate greatly reduces its breaking strain. Evidently, this effect is associated with the processes of cross-linking and/or destruction of the macromolecules on the surface layer which complicates the orientation of the polymer and, consequently, reduces the breaking strain. In later stages it was shown [26] that the given method of evaluation of the stress–strain properties of the surface layers of the polymers is universal and suitable for studying other polymers, in particular, polyolefins, subjected to surface fluorine treatment.

Surface modification of polymers by treatment with cold plasma
It is well known that the effect of cold plasma on the polymers results in the chemical modification of their surface. This phenomenon has been studied in detail and is used widely in practice [27, 28]. The effect of plasma on the surface of the polymer is utilised mainly for the modification of contact properties (wetting, adhesion to thin metal layers, bonding properties, regulation of the adhesion of dyes used in printing, etc). As present, it is agreed that the treatment of the polymers with plasma results in the formation on the surface of modified nanolayers with a special chemical composition, and also special physical–chemical, electrical and other properties.

Evidently, the deformation–strength properties of the nanolayers of the polymers, modified by plasma treatment, should differ from the properties of the non-modified original polymer. However, the data of this type are not available in the literature. The main reason for this is, as mentioned previously, the absence of experimental evaluation methods. The approach developed and described previously for the evaluation of the stress–strain properties of the surface layers of the polymers provides facilities for the actual experimental evaluation. This approach can be used only if the treatment of the polymer with plasma results in the formation of a modified layer on the surface, with the breaking strain of the modified layer smaller than that of the polymer-substrate. Otherwise, there is no fragmentation of this layer which is the basis for the evaluation of its stress–strain properties using the relationship (10.2).

To verify this assumption, it is sufficient to deform the polymer sample treated with plasma and study its surface in an electron microscope. The results of these investigations are presented in Fig. 10.3 which shows the AFM images of the surface of a sample of a polyethylene terephthalate film treated in plasma for 1 min and subsequently tensile loaded at 90°C to 50% (Fig. 10.3 a), together with its three-dimensional reconstruction (Fig. 10.3 b) and the appropriate profile pattern (Fig. 10.3 c). It should be mentioned that the specimens of the same polymer, not treated with the plasma, had a smooth surface after deformation in the same conditions.

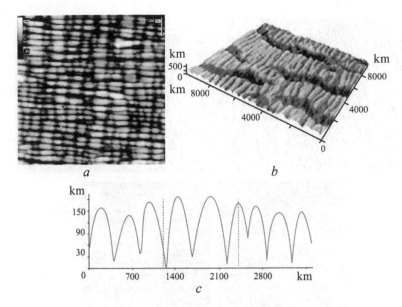

a   *b*

c

Fig. 10.3. AFM images of the polyethylene terephthalate, treated with plasma for 1 min and deformed at 90°C to 50% (*a*), its three-dimensional reconstruction (*b*) and the appropriate cross section (*c*).

Figure 10.3 shows that when the polymer films subjected to preliminary treatment in plasma arc deformed, their surface is characterised by the structure formation completely identical with that observed in deformation of any polymer film having the RCSS structure. The regular microrelief and fragmentation of the surface layer of the polymer are clearly visible. The detected phenomenon of surface structure formation in deformation of the polymers, treated with copper is of the general nature. This is indicated by the electron microscopic data, obtained for PET (*a*), PVC (*b*), HDPE (c), polyisoprene rubber (*d*), PVC containing a plastification agent (e), shown in Fig. 10.4. As indicated by the micrographs, this treatment results in the formation of the microrelief and fragmentation of the surface layer, regardless of the type of polymer.

Analysis of the electron micrograph shows that the size of the fracture fragments is almost completely independent of the treatment time of the polymer (polyethylene terephthalate) with plasma in the time period from 1 to 12 min and equals ~0.2 μm and ~0.44 μm in tensile drawing the polymer-substrate at 20°C (below T_g) and 90°C (above T_g), respectively. On the one hand, the plasma-modified layer 'remembers' the glass transition temperature of the initial polymer

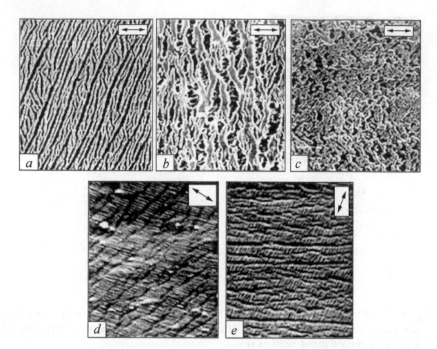

Fig. 10.4. Scanning electron micrograph of the structure of the surface layer of the polymers treated in plasma and deformed at 20°C: PET (*a*), PVC (*b*), HDPE (*c*), polyisoprene rubber (*d*), plasticised PVC (*e*). The direction of tensile drawing is indicated by the arrow [29].

(~75°C). On the other hand, the result appears to be unusual because the treatment time should be directly linked with the thickness of the modified surface layer. On the basis of these data it was concluded [29, 30] that the formation of the chemically modified layer takes place in the initial period of treatment of the polymer. Further treatment is not accompanied by any large increase of the thickness of the modified layer and the thickness remains almost completely constant. At the same time, the result is in agreement with the data obtained for other surface properties of the polymer in dependence on the plasma treatment time with plasma. Figure 10.5 shows the dependence of the contact wetting angle and the density of the surface charge on the duration of plasma treatment of polyethylene terephthalate [31]. It may be seen that this characteristic changes only in the first several tens of seconds of plasma treatment of the polymer and then remains constant.

Thus, the treatment of the polymers in the cold plasma results in the formation on the surface of a relatively rigid surface layer so that the layers acquire the properties of the RCSS. This circumstance may be utilised to evaluate the stress–strain properties of these layers

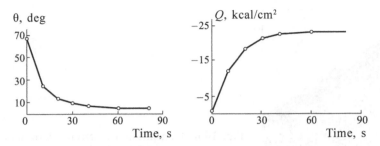

Fig. 10.5. Dependence of the contact wetting angle (a) and the density of the surface charge (b) on treatment time for a polyethylene terephthalate film, modified in a cathodic drop of a glow low frequency discharge (I = 100 mA) [31].

by the approach proposed in the study. To use equation (10.2), it is necessary to know the thickness of the plasma-modified surface layer. For this purpose, investigations were carried out on ultrafine slices of plasma-treated polyethylene terephthalate films by transmission electron microscopy. The typical experimental results are presented in Fig. 10.6. It may clearly be seen that the plasma treatment of the film surface results in the formation of a layer which is darker in comparison with the initial polymer. The thickness of the dark layer is almost completely independent of the duration of treatment of the polymer with plasma and reaches ~150 nm. It may be assumed that the dark layer on the micrographs is in fact the plasma-modified polymer. It should be mentioned that in the procedure used to produce the ultrafine slices the given slice was produced under an angle of ~40% in relation to the surface of the film [29, 30]. Therefore, the actual thickness of the modified layer should be equal to ~80–100 nm.

Additional information on the thickness of the plasma-modified surface layer of the polymer is provided by atomic force microscopy. Figure 10.7 shows the AFM image of the surface of a PET film treated with plasma for 4 min and subsequently deformed by 50%. All the special features, characteristic of surface structure formation of the RCSS systems, are clearly visible. At the same time, atomic force microscopy can be used to measure quite accurately the depth of the cracks, formed in the coatings a result of deformation of the polymer-substrate (Fig. 10.7 *b*). This depth is approximately 80 nm. It is important to note that this value changes only slightly in a wide range of the deformation of the polymer-substrate (Fig. 10.7 *c*). It may be assumed that the depth of the fracture cracks in the coating, measured by atomic force microscopy, corresponds to the thickness

Fig. 10.6. Transmission electron micrograph of an ultrafine slice of a polyethylene terephthalate sample, plasma treated for 1 min [29].

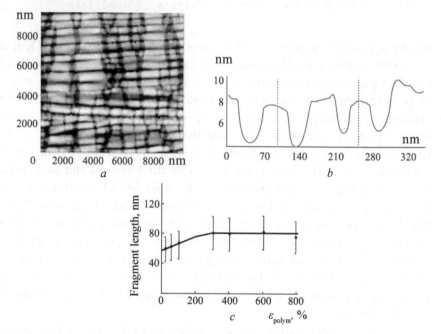

Fig. 10.7. Atomic force microscope image (a) and cross section (b) of a polyethylene terephthalate sample, plasma treated for 4 min and subsequently deformed to 50% at 90°C, and the dependence of the depth of the cracks in the coating on the deformation of the polymer-substrate (c) [29].

of the plasma modified polyethylene terephthalate layer. It is also important to note that there are also other estimates of the thickness of the plasma-modified layer on the polymer surface and they all give the values of the same order of magnitude (50–100 nm) [32–34].

Assuming that the thickness of the modified layer is $h = 80$ nm, the strength of the layer was determined using equation (10.2). The strength of the coating for the deformation of the polymer-substrate at 20°C, the stress in the substrate $\sigma_0 = 40$ MPa, and the average size of the fracture fragment of the coating $L = 0.2$ µm, is ~33.3 MPa. The same procedure was used to determine the strength of the layer formed in deformation of the plasma-treated polymer matrix at 90°C ($\sigma_0 = 2$ MPa and the average size of the fracture fragment of the coating $L = 0.44$ µm). In this case, the calculations using equation (10.2) give the strength of the plasma modified layer of ~3.7 MPa. As shown by the data, the strength of the modified polymer layer greatly changes in dependence on the deformation temperature. This is the first and still the only estimate of the strength of the surface layer of the plasma-modified polymer. It should be mentioned that the strength of the surface layer, formed in treatment of polyethylene terephthalate with cold plasma was close to the corresponding value of the strength of the surface fluorine-treated layer of polyethylene terephthalate (see the previous section).

It is justified to assume that in plasma treatment of the polymer the surface layer is characterised by a large number of chemical reactions, including cross-linking processes [27, 28]. Evidently, the cross-linking of this type causes that the breaking strain of the polymer in the plasma-modified layer to drop below the breaking strain of the initial polymer and, consequently, the layer is fragmented and it is then possible to determine its stress–strain properties. Nevertheless, this layer retains its polymer appearance and, consequently, remains sensitive to the transition between the glassy and highly elastic state.

In addition to the strength, in [29, 30] investigations were carried out to estimate plastic deformation of the plasma modified layer in deformation of the polymer-substrate. Figure 10.8 shows the dependence of the plastic deformation of the coating on the treatment time of polyethylene terephthalate with plasma after subsequent deformation at 90°C (curve 1) and 20°C (curve 2).

It may be seen that the coating formed on the polyethylene terephthalate surface is relative ductile. Its plastic deformation changes from ~20 to ~95% in deformation at 90°C and 20°C, respectively. Evidently, the detected effect is due to the fact that the deformation of the polymer-substrate at 20°C (below the glass transition temperature of polyethylene terephthalate) takes place at a stress an order of magnitude higher than at 90°C (higher than the glass transition temperature of polyethylene terephthalate). An additional factor, influencing the plastic deformation of the surface layer, is the strain of the polymer-substrate. In fact, the strain at 90°C was 50%, whereas deformation at 20°C took place with the

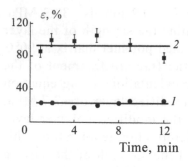

Fig. 10.8. The dependence of the plastic deformation of the rigid surface layer of polyethylene terephthalate, treated in plasma (ε) in deformation of the polymer-substrate at 90°C (1) and 20°C (2), on the treatment time with plasma [29].

formation of a neck in polyethylene terephthalate, with the natural draw ratio being 275%.

Thus, it was shown in this section that the applied method of direct evaluation of the fragmentation of the coatings taking place during deformation of the polymer-substrate is also suitable for describing and evaluating the stress–strain properties of the surface modified layers, formed in plasma treatment of the polymers. The results can be used to conclude for the first time that this layer forms at short (less than a minute) duration of treatment of the polymer with plasma and further plasma treatment has no significant effect on its thickness and mechanical properties. The approach, developed in [14–20], has been used for the quantitative evaluation of the mechanical parameters (fracture strength and plastic deformation) of the surface of the plasma-modified layers of polyethylene terephthalate and it has been shown that these characteristics are sensitive to the physical state of the polymer-substrate.

10.3. Evaluation of the stress–strain properties of coatings deposited on polymer surfaces

The deposition of the coatings on the polymer films should be treated as a radical modification of their surface properties. The two-layer materials of this type are used widely. In particular, the metallised polymer films are used in microelectronics [35, 36]. The aluminium coating, produced by vapour phase deposition and forming a very smooth surface, has the highest optical reflectivity [37]. The polymers with an organic silicon coating are efficient oxygen-isolating materials in the food and pharmaceutical industries [38], so that the volume of production of these materials now equals hundreds of thousands of tonnes. Evidently, the information on the stress–strain properties of the nanometric coatings, deposited on the polymer surface, is of considerable scientific and applied importance. It is also evident that the above approach to the evaluation of the stress–strain properties of the nanometric layers of the solids can be used efficiently for investigating the nanometric coatings on the polymers.

Prior to evaluating the stress–strain properties of the nanometric coatings, deposited on the polymers, it is useful to mention the main morphological special features of these materials subjected to tensile deformation.

Figure 10.9 shows the micrographs of two polyethylene terephthalate samples with a thin (10 nm) gold coating deformed at 90°C (a) and at room temperature (b). It is important to mention that the first photograph was taken from the sample deformed at a temperature higher than the glass transition temperature of polyethylene terephthalate ($T_g = 75°C$), and the second – at a lower temperature. The bright bands are the fragments of the fractured coating, the dark bands are cracks in the fractured coating.

Figure 10.9 shows that as a result of simple tensile drawing of the polymer-substrate at temperatures higher than the glass transition temperature, the coating on the surface of the polymer breaks up into a large number of regularly distributed islands of approximately the same size and a regular relief forms there at the same time. The surprising feature is the regularity of the spontaneously formed relief and its strict orientation in relation to the tensile drawing axis. The depressions and tips of the relief are always oriented strictly along (parallel to) the tensile drawing axis. In deformation of the polymer-substrate below T_g (Fig. 10.9 b) there is only regular fragmentation of the coating and the regular microrelief does not form.

Thus, simple tensile drawing of the polymer film with the thin hard coating results in the unique type of surface structure formation. It is important to note that both resultant structures (regular microrelief and the regular system of fragments of the fractured coating) are highly organised and periodic and can easily characterised by direct microscopic studies. Further studies showed that the formation of these structures is of the general nature and does not depend on the nature of the material of the substrate in the coating so that these systems can be referred to as the 'rigid coating on a soft substratum' (RCSS) [14–20].

It should be mentioned that the regular microrelief forms when the polymer is deformed at temperatures higher than its glass transition temperature (Fig. 10.9 a). Below T_g (in deformation of the polymer

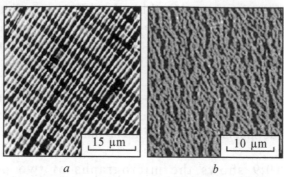

a   b

Fig. 10.9. Scanning electron micrographs of the polyethylene terephthalate specimens with a thin (10 nm) gold coating tensile drawn at 90°C by 50% (*a*), and at room temperature with the formation of a neck (*b*) [39].

with the formation of a neck) the polymer is evidently too hard to enable the thin coating to destabilise the surface of the polymer to such an extent as to allow the formation of RMR (Fig. 10.9 *b*). However, fragmentation of the coating is observed in both cases so that it appears feasible to use the approach developed in [14–20] for evaluating the strength and plasticity of the coatings by means of the relationships (10.2), (10.4) and (10.5).

10.4. Evaluation of the stress–strain properties of nanometric aluminium coatings

Prior to evaluating specifically the stress–strain properties of the nanometric aluminium layers, deposited on polymer substrates, we examine a number of data relating to the structure of such thin metal layers. Figure 10.10 shows the transmission electron micrographs of formvar films with aluminium layers of different thickness deposited on their surface by thermal vacuum sputtering. In the case of aluminium, even at the shortest sputtering time, corresponding to the coating thickness of ~1.8 nm (Fig. 10.10 *a*), a continuous metal layer forms on the polymer surface. Regardless of the continuity, this layer has a non-uniform thickness so that a mosaic pattern can be seen on the micrographs. Increasing the coating thickness to 4 nm (Fig. 10.10 *b*) does not change greatly the observed pattern. A further increase of the thickness of the aluminium coating (up to 16 nm) is accompanied by the formation of a large number of individual crystals (dark spots on the micrographs) with a size of 10–30 nm (Fig. 10.10 *c*, *d*). Formation of a continuous aluminium coating on the surface of the polymer creates suitable conditions for the formation of regular periodic structures on the surface during tensile drawing of the polymer-substrate and, consequently, for evaluating their stress–strain properties using the proposed procedure.

Figure 10.11 shows the typical electron micrographs of polyethylene terephthalate with an aluminium coating deposited on the surface after tensile drawing by 50% at 90°C. The use of different magnification makes it possible to evaluate both the regularity of the size of the fracture fragments (Fig. 10.11 a) and also the regularity of the resultant microrelief (Fig. 10.11 b). It may clearly be seen that in this case all the previously mentioned special features of surface structure formation, characteristic for the 'hard coating on the compliant base' silicon are indeed observed: fragmentation of the coating and the formation of the regular periodic relief.

The approach developed in [14–20] can be used to evaluate an important characteristic of the solids (coatings) of the nanometric dimensions like the yield stress. The data in Fig. 10.11 and the

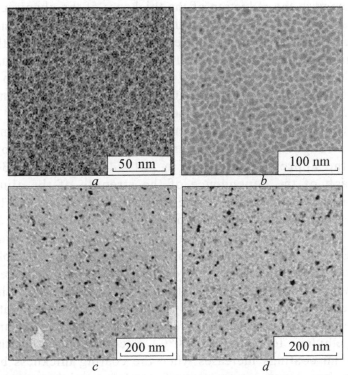

Fig. 10.10. Transmission electron micrographs of aluminium coatings with a thickness of 1.8 (a), 4 (b), 14 (c) and 16 nm (d) deposited by vacuum thermal sputtering on the formvar films.

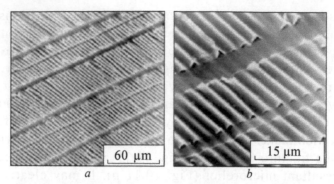

a   b

Fig. 10.11. Scanning electron micrographs of polyethylene terephthalate with an aluminium coating 10 nm thick after tensile drawing to 50% at 90°C.

equation (10.4) can be used for this evaluation. Figure 10.12 shows the results of calculating the yield stress for the aluminium coating using the equation (10.4). The data show that starting at a coating thickness of 10–12 nm the yield stress of the coating rapidly increases with the reduction of the thickness of the aluminium layer and reaches ~150 MPa. In the thickness range greater than 10–12 nm the yield stress tends to the value corresponding to the bulk aluminium (35 MPa) [40]. The results can be used to assume that the reduction of the thickness of the aluminium coatings to several nanometres leads to some structural transition, having the strongest effect on the properties of the metallic coating [41].

This conclusion is fully confirmed by the results of determination of the strength of the nanometric aluminium coatings, obtained using the approach described in [14–24]. Figure 10.13 shows the graph of the dependence of the strength of the aluminium coating on the thickness of the layer, calculated on the basis of the electron microscopic data using the relationship (10.2). Two facts should be considered. Firstly, the strength of the aluminium coating, calculated in deformation of the polyethylene terephthalate substrate at different temperatures, greatly differs. The strength of the coating, produced in deformation of polyethylene terephthalate at room temperature, is considerably higher than the appropriate value obtained at temperatures higher than the glass transition temperature of polyethylene terephthalate (90°C). Secondly, the strength of the aluminium coatings increases when the coating thickness becomes smaller than 4 nm, irrespective of the deformation temperature of the polymer-substrate. The increase of the strength of the aluminium

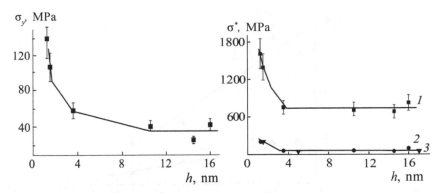

Fig. 10.12 (left). Dependence of the yield stress of the aluminium coating on its thickness in deformation of the polymer-substrate (polyethylene terephthalate) by 50% at 90°C [41].

Fig. 10.13. Dependence of the strength of the aluminium coating on its thickness. Deformation at 20°C (1) and at 90°C (2). The strength of bulk aluminium is 60–70 MPa (3) [40].

coating takes place symbatically to the increase of its yield stress (Figs. 10.12 and 10.13).

The approach described in [14–20] was also used to obtain the dependence of the plastic deformation of the aluminium coating on its thickness using equation (10.5) (Fig. 10.14). Regardless of the temperature of tensile drawing of the polymer-substrate, the elastic strain of the aluminium decreases with increase of the thickness of the metal layer. At the same time, it should be mentioned that the absolute values of the plastic deformation of the aluminium coating are considerably higher for the case of deformation of polyethylene terephthalate with a coating at 20°C in comparison with 90°C. In the first case, the plastic deformation may reach 140%, which is considerably higher than the available values of the plastic deformation of bulk aluminium (40–50%) [40]. Plastic deformation continuously increases with the decrease of the thickness of the deposited coating in the entire thickness range.

Thus, in the range of small thicknesses of the aluminium coatings there is a very unusual combination of their stress–strain properties: increase of the yield stress and ultimate strength is accompanied by a large increase of their plasticity. It should be mentioned that in crystalline solids the increase of the strength is usually accompanied by a reduction of plasticity [42]. This relationship is observed for both coarse-grained crystalline materials and nanostructured materials.

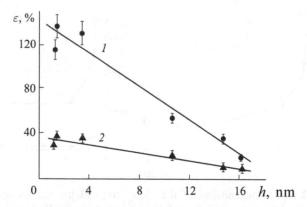

Fig. 10.14. Dependence of plastic deformation of aluminium on the thickness of the coating in deformation of the polymer-substrate at 20°C (1) and 90°C (2) by 50% [41].

The following questions should be answered. How to explain the difference in the strength of the aluminium coatings determined by experiments in deformation of the polyethylene terephthalate substrate at different temperatures (20°C and 90°C)? Why is the range of small thicknesses of the coatings characterised by a large increase of the strength and yield stress irrespective of the deformation temperature of the polymer-substrate? What is the cause of the increase of the plastic deformation of the coatings with the decrease of the thickness of the aluminium layer deposited on the polyethylene terephthalate films?

To explain the first of these features, it is necessary to examine in greater detail the differences in the deformation conditions of polyethylene terephthalate with the aluminium coating at the two previously mentioned temperatures (20°C and 90°C). Because the T_g temperature of polyethylene terephthalate is ~75°C, its deformation at 20°C takes place in the temperature range of the glassy state (with the formation of a neck) and at 90°C in the temperature range of the rubbery state (uniformly). This is not the only difference between these two processes: the deformation of the glassy polyethylene terephthalate takes place at a considerably higher (by an order of magnitude) stress in comparison with the rubbery state. It is also important to note that at 20°C the substrate is deformed by 275% (to form a neck) and at 90°C by only 50%.

The effect of this fact on the deformation behaviour of the Al coating was investigated. This was carried out on PET samples with an Al coating 9 nm thick, deformed with different strains in the range from 0 to 800% at 90°C (the effect of the draw ratio on the stress–strain behaviour of the aluminium coating at 20°C cannot be studied because the deformation of polyethylene terephthalate in these conditions takes place with the formation of a neck with a fixed natural draw ratio of 275%). It is important to note that the relationships (10.2) and (10.3) can be used to evaluate independently the strength and plastic deformation of the coating in deformation of the polymer-substrate. In the determination of the strength of bulk solids by the conventional method after reaching the fracture stress (growth of the main crack) the stress decreases in the material to zero and the experiment is completed. At the same time, when examining the processes of fracture of the coating on the substrate its fragmentation continues and after the cracks have penetrated through the cross-section of the coating and a family of individual fragments formed by further fracture of each of them. It is therefore possible to construct the true stress–strain curve of the coating deformed on the surface of the polymer-substrate.

This dependence is shown in Fig. 10.15. For comparison, the same figure shows the tensile drawing curve of bulk aluminium, obtained at the room temperature [40]. Figure 10.15 shows that the strength of the aluminium coating is many times higher than that of bulk aluminium. In addition to this, when the strain of the aluminium coating is increased to 40% its strength increases from 190 to ~800 MPa, whereas the stress in pure aluminium remains almost constant in the same strain range. The results show that the deformation of the aluminium coating on the polyethylene terephthalate substrate is characterised by the strong effect of strain hardening. This effect can be used to explain the difference in the strength of the aluminium coatings in deformation of the polyethylene terephthalate at 20 and 90°C, shown in Fig. 10.13. Actually, at 90°C the strain is only 50%, whereas at 20°C it is 275%. In addition, the deformation of the polymer-substrate takes place at a considerably higher (by an order of magnitude) stress in comparison with 20°C which increases, with other conditions being equal, the plastic deformation of the aluminium coating and, consequently, the strain hardening effect of the coating.

To explain the increase of the strength and yield stress of the aluminium in the range of the small coating thicknesses, we return to the electron microscopic structural data (Fig. 10.10). Depending on the thickness of the deposited layer, the structure of the aluminium coating undergoes extensive changes. The deposition of a relatively thick coating (10 nm or greater) results in the formation of a

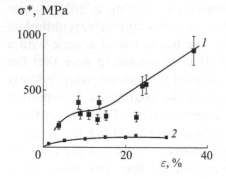

Fig. 10.15. Dependence of the strength of the aluminium coating 9 nm thick on strain at a temperature of 90°C (1) and the curve of tensile drawing of the bulk polymer at room temperature (2) [40].

distinctive crystalline structure (Fig. 10.10 *c*, *d*). The size of the individual crystallites in this structure is 10–30 nm. When thinner layers are sprayed, the structure greatly changes. It can be seen that in this case there is a continuous aluminium film with a mosaic structure (Fig. 10.10 *a*).

The electron diffraction data (Fig. 10.16) of the samples of the formvar films with the aluminium coatings with a thickness of 4 (*a*) and 16 nm (*b*) indicate that the aluminium coatings have different internal structures. The perfect crystalline structure of aluminium is observed only for the coating thicknesses greater than 5 nm. At small thicknesses the samples show only an amorphous halo indicating the amorphous state of the metal or, at least, a very low degree of its crystallinity. Evidently, the detected effect is associated with the fact that at thicknesses smaller than 5–10 nm aluminium cannot form an extended crystalline lattice with a long-range order.

Thus, the decrease of the thickness of the aluminium coatings below 10 nm leads to the amorphisation of the metal. Recently, methods for producing amorphous metals and their alloys have been developed [43]. One of these methods is high-speed ion-plasma sputtering of metals and alloys [44, 45]. The process can be conducted in such a manner that the metallic coatings have the nanometre thickness. Naturally, when the thickness of the deposited metallic films does not exceed tens of nanometres, as in [41], the structure of the metallic layers is amorphous because the crystalline phase with a long-range order can not form under these conditions.

It is important to stress that an intermediate case between the completely amorphous state of the metals and the crystalline state are the so-called amorphous–nanocrystalline composites consisting of an amorphous metallic matrix with nanosized crystallites in the form of individual inclusions. Attention to the structures was paid

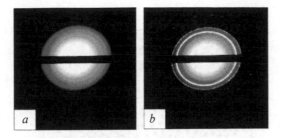

Fig. 10.16. Electron diffraction patterns of the aluminium coatings with a thickness of 4 (*a*) and 16 nm (*b*) [41].

for the first time in the middle of the 80s [46, 47] when examining the microstructure of the alloys produced in special conditions of rapid quenching from the melt [46], and also in annealing and partial crystallisation of the amorphous alloys [48]. On the basis of the results of microstructural studies, the authors of [46] proposed the model of the structure of the resultant state obtained in their work (Fig. 10.17) and reported the appearance of a new class of amorphous–crystalline materials with an exceptionally high mechanical strength. The examination of the mechanical properties of the resultant composite showed the unusually high values of microhardness (~21 GPa), more than twice the microhardness of the amorphous alloy of the same composition (~9 GPa). The fracture stress of the strip samples in uniaxial tension at room temperature was very high, 6–6.5 GPa. At the same time, it was shown that the materials of this type retain a high degree of plasticity. The theoretical studies of the model, presented in Fig. 10.17, showed how the unusually high mechanical characteristics (microhardness, yield stress and fracture stress) are combined with the high plasticity of this material [44]. It is assumed that the main factors of the large increase of these properties are the high mechanical properties of the nanocrystallites, determined by the small dimensions and presence of disclinations, and also by the extremely high concentration of these nanocrystallites in the matrix.

It should be mentioned that the structure of the amorphous–nanocrystalline composites, shown in Fig. 10.17, is externally very similar to the structure of the thin aluminium coating studied in [41] by electron microscopy (Fig. 10.10). This analogy and the previously cited literature sources can be used to make justified assumptions regarding the mechanism of the phenomena (increase of strength and yield stress irrespective of the deformation temperature of the

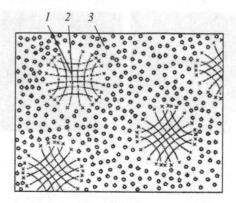

Fig. 10.17. The proposed structural model of the transitional amorphous–crystalline state, formed in cooling the melt with high rates: 1) the region of the crystallite with the variable parameter of the crystal lattice; 2) the region of the smooth transition from the crystalline to amorphous state; 3) thin amorphous interlayers [46].

polymer-substrate in the region of the small coating thicknesses and the increase of their plasticity with decreasing thickness of the metallic layer).

Thus, the special features of the stress–strain properties of the aluminium coatings, reported in [41], can be explained by their dimensional effects. The increase of strength and yield stress and also of plasticity are associated with the reduction of the thickness of the deposited coating and, consequently, with the effect of formation of the nanostructure.

Thus, the new direct microscopic method has been used for the first time for the evaluation of the stress–strain properties of aluminium in thin, nanometric layers, deposited on the polymer substrate. It is shown that these characteristics depend on the level of the stress in the substrate, the strain in the substrate and the coating thickness. The evolution of these parameters is associated with the strain hardening of the metal in tensile drawing and the effect of formation of the nanostructure of the crystalline materials in the small thickness range.

10.5. Evaluation of the stress–strain properties of nanometric coatings based on noble metals

Deposition of coatings by vacuum thermal evaporation of metal
In [39, 49] it was attempted to evaluate the stress–strain properties of gold coatings. It was shown that the method (proposed in [14–20])

of evaluating the structural and mechanical properties of nanometric layers, deposited on the polymers, has its limitations. Previously, we mentioned the fundamental properties of the RCSS systems, which enabled the relationships (10.2), (10.4) and (10.5) to be used for evaluating the fundamental mechanical characteristics of the deposited coatings. It appears that there is another condition essential for application of this approach. This condition, which appears to be evident, is the continuity of the deposited coating.

In this connection, it is useful to discuss briefly the method of determination of the thickness of the deposited coating. The method of determination of the thickness described in [14–20] was based on constructing a calibration graph of the dependence of the thickness of the coating on the sputtering time in the standard conditions. For this purpose, metal layers of different thickness were deposited on glass sheets, varying the sputtering time. Subsequently, a scratch was made with a wooden sharp edge in the glass of the coating and the resultant fracture was studied in an atomic force microscope. It should be mentioned that at a similar procedure using the polymer substrate instead of the glass substrate does not yield the desired result because the surface of the polymer is too compliant so that the sharp wood edge makes a very deep scratch and the thickness of the coating cannot be evaluated.

Typical results of this evaluation are presented in Fig. 10.18. The light section of the micrograph on the left is the gold layer, the dark part on the right – the surface of the glass. The fracture in the coating is efficiently recorded using the atomic force microscope (Fig. 10.18 a) and the corresponding cross section (Fig. 10.18 b) can be used to evaluate the thickness of the deposited coating with sufficiently high accuracy. This procedure was used to construct the calibration graph in the coating thickness – sputtering time coordinates which was subsequently used for the deposition of a layer of gold or platinum of the required thickness on polyethylene terephthalate films. Figure 10.19 shows the typical calibration graph of the dependence of the thickness of the metallic coating, deposited on the glass surface, on the sputtering time of the metal. Taking these factors into account, the small thickness of the coatings (at a sputtering time of <1 min) was determined by extrapolating the dependence of the thickness on the sputtering time (Fig. 10.19) to zero time. Thus, in the small thickness range (less than 10 nm) the thickness of the coating was not measured directly using the microscopic data and it was calculated assuming that the coating is

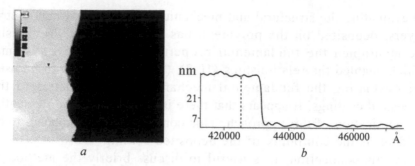

Fig. 10.18. AFM image of a scratch in a gold coating, deposited on a glass substrate by ion-plasma sputtering (*a*), and the appropriate cross section (*b*) [50].

continuous at all sputtering times and has a uniform thickness. In the case of aluminium coatings this assumption was justified and confirmed by direct electron microscopic data. Prior to using the approach for determining the stress–strain properties of the coatings in the thin layers, it is necessary to know the parameters of fracture of the coating as a result of tensile drawing the polymer-substrate. Therefore, in [39] investigations were carried out into the special features of fragmentation in the relief formation when deformation of polymers with gold coatings of the small nominal thickness (~8 nm). The results show that when tensile drawing the polyethylene terephthalate sample with a gold layer with a nominal thickness of 8 nm there is no fragmentation of the coating. It may be seen that the individual particles of gold separate as a result of the tensile drawing of the polymer-substrate without the formation of long fracture fragments.

Thus, taking into account the resultant microscopic data, it is not possible to evaluate quantitatively the fracture parameters of the gold coating in the small thickness range and, consequently, the proposed approach cannot be used for evaluating the mechanical properties of these objects.

At large thicknesses we obtain a single coating and the coating fragments during tensile drawing of the sample. However, according to the data obtained in the previous sections, the most distinctive and interesting anomalies in the structural–mechanical behaviour of the nanometric coating take place in the small thickness range, i.e., in the range where the continuous coating cannot be produced.

These difficulties were explained in direct microscopic studies of the process of deposition of noble metals on polymers by vacuum thermal sputtering [41]. Figure 10.20 shows transmission electron

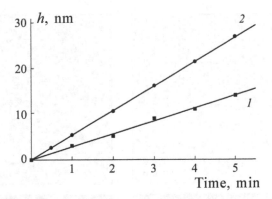

Fig. 10.19. Dependence of the thickness h of gold (1) and platinum (2) coatings on the duration of ion-plasma sputtering on a glass surface [50].

microscopic images of formvar films-substrates with gold layers with the effective thickness from 5 to 23 nm. Figure 10.20 shows that at a coating thickness of up to ~5 nm, no continuous metallic coating forms on the polymer surface. It may clearly be seen that the metal layer with a thickness of 5–8 nm, deposited on the polymer, is in fact a set of spherical nanoparticles uniformly distributed on the surface and not forming the continuous metallic coating (Fig. 10.20 *a, b*). The dimensions of the nanoparticles depend on the sputtering time (coating thickness). A further increase of the thickness of the gold coating on the polymer substrate results in the formation of a more complicated continuous set of metallic aggregates (Fig. 10.20 *c, d*). Thus, it is evident that in the small thickness range (up to 70 nm) no continuous coatings form on the surface of the polymer film which is the reason why in this thickness range there is no fragmentation of the coating during tensile drawing the polymer-substrate. Therefore, the very concept of the thickness of the coating at short sputtering times of the metal on the polymer surface becomes indeterminate. In this cases, we should talk about the effective nominal thickness of the coating because there is no continuous single coating on the surface of the polymer in the initial stages of sputtering the metal.

Naturally, the absence of fragmentation of this type prevents the approach developed in the studies to be used for evaluating the strength of the coatings because it assumes the presence of a continuous coating on the polymer matrix. This conclusion is also confirmed by the transmission electron microscopy data. Figure 10.21 shows the transmission micrograph of a polystyrene specimen with a very thin gold layer deposited on its surface. As indicated by the photograph, the amount of the deposited metal is small and,

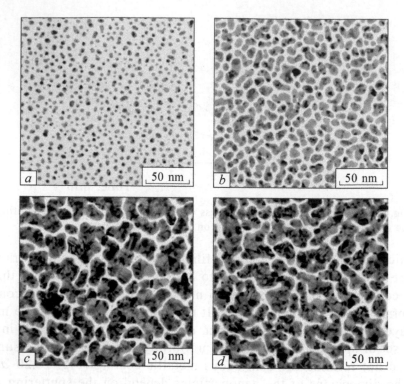

Fig. 10.20. Transmission electron micrographs of gold coatings with a thickness of 5 (*a*), 8 (*b*), 17 (*c*) and 23 (*d*) nm, deposited by vacuum thermal sputtering on formvar films [41].

therefore, the surface of the polymer contains a set of isolated nanoclusters and not a continuous coating [51]. It may clearly be seen that in deformation of the polymer-substrate the gold particles separate in relation to each other and there is no fragmentation of the coating in this case.

Thus, it is evident that the fragmentation of the coating takes place only if a specific percolation limit is reached and the metallic clusters on the polymer surface are spaced so closely that they begin to behave as an integral unit. Now it is clear why in the small thickness range there is no fragmentation of the gold coating deposited by vacuum thermal sputtering.

It is now necessary to clarify why there is no continuous metallic coating on the polymer surface at short sputtering times of the metal. This problem was the subject of detailed studies [52–54] in which it was shown that the noble metals are chemically inert substances. When falling of the polymer surface, it is more advantageous for them from the energy viewpoint to immediately form clusters

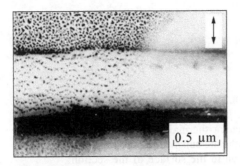

Fig. 10.21. Transmission electron micrograph of a deformed polystyrene film with gold particles deposited by vacuum evaporation [51]. The direction of tensile drawing is indicated by the arrow.

(islands) on the surface of the polymer substrate. This phenomenon can be regarded as a variety of dewetting (the process inverse to spreading of the liquid on the solid surface) and this results in the formation of a system of nanoclusters in the surface layer of the polymer.

This is also indicated by the results of studying the sections of polyethylene terephthalate films with a gold coating, deposited by vacuum thermal sputtering. Figure 10.22 shows a transmission electron micrograph of an ultrathin section of the surface layer of a polyethylene terephthalate film with a gold layer with a thickness of ~9 nm. It may be seen that the gold layer on the polymer surface is a non-monolithic openwork structure.

Thus, at the present time it is not possible to estimate the stress–strain properties of the gold coating deposited by vacuum thermal sputtering on the polyethylene terephthalate substrate. This is caused by the structural special features of the polymer films with the gold coating deposited by vacuum thermal sputtering. As a result of these structural features a porous openwork structure forms at the small effective coating thicknesses.

Evaluation of the stress–strain properties of coatings deposited by ion-plasma sputtering on the polymers

Prior to evaluating the properties of coatings of noble metals, deposited by ion plasma sputtering on the polymer surface, attention will be given to the direct microscopic data obtained for their structure. Figure 10.23 shows the transmission electron micrographs of gold layers with an effective thickness from 0.2 (*a*) to 29 nm (*f*). It may be seen that a continuous coating forms on the polymer surface starting only at an effective thickness of ~10 nm. The method of deposition of the noble metal on the polymer has only a slight effect on the structure of the resultant coating. This conclusion is

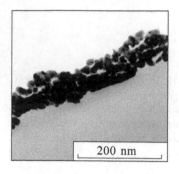

Fig. 10.22. Transmission electron micrograph of a gold layer with a thickness of 9 nm deposited by vacuum thermal sputtering on a polyethylene terephthalate substrate. The view in the tangential direction to the surface of the section.

confirmed by the data obtained in measuring the surface electrical conductivity of polymer films (polyethylene terephthalate) coated with thin gold layers (Fig. 10.24) [55]. The data shown in Fig. 10.24) confirm that up to the effective thickness of the metal layer of ~7 nm there is no percolation of the metal on the polymer surface, the metal is deposited by the island mechanism and there is no single continuous coating on the polymer surface. A single metallic coating forms on the polymer surface only when the metallic clusters reach the percolation threshold so that the electrical resistance of the coating decreases by ~12 orders of magnitude. It appears that the evaluation of the deformation–strength properties of the coatings using the procedure proposed in [14– 20] can not be carried out because no continuous coating forms on the surface of the polymer at short sputtering times. Previously it was shown (Fig. 10.21) that when individual metal clusters form on the surface of the polymer-substrate the tensile drawing of this system is not accompanied by the previously described surface structure formation, and the individual metal particles are simply pushed away from each other [51].

It was established unexpectedly (Fig. 10.25) that in deformation of PET films with a gold coating deposited by ion-vacuum sputtering with the thickness smaller than that at which a continuous film forms on the polymer surface (sputtering time from 3 s to 2 min, which corresponds to a nominal coating thickness of 0.2–5 nm, Fig. 10.25 a–c), the resultant patterns of surface structure formation are very similar to those observed in the deformation of the polymer-substrate with a continuous metallic coating. It is clearly seen that a system of asymmetric fragments of the metallic coating, oriented normal to the tensile drawing axis of the polymer, forms in all cases on the polymer surface.

The application of atomic force microscopy makes it possible to reveal and examine in detail the fragmentation pattern of the coating

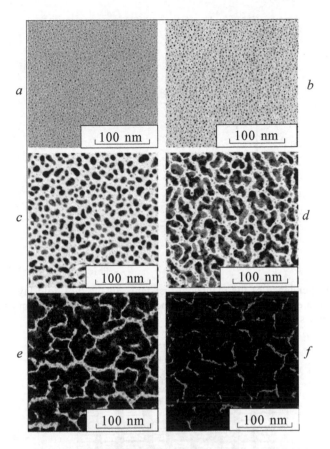

Fig. 10.23. TEM micrographs of formvar films coated with gold layers deposited by ion-plasma sputtering. The thickness of the deposited layer according to the calibration graph in Fig. 10.10: 0.2 (*a*), 0.4 (*b*), 3 (*c*), 5 (*d*), 16 (*e*) and 29 nm (*f*).

in the tensile drawing of the polymer-substrate. Typical results of this investigation are presented in Fig. 10.26. It is clearly seen that regardless of the absence of the single metallic coating on the surface of the deformed polymer, the surface layer of the polymer rapidly breaks up. These fragments are clearly visible on the surface of the polymer and their parameters can be easily measured. Atomic force microscopy can be used not only for the quantitative evaluation of the length (the size in the direction of tensile drawing) but also the depth of the depressions between the fracture fragments of the surface layer and also characterise the evolution of these parameters in the process of ion-plasma sputtering the polymer surface with a metallic coating [39].

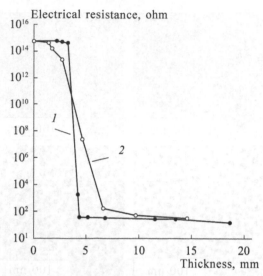

Fig. 10.24. Dependence of the electrical resistance of the thin gold layers on their thickness. The metal was deposited on the polymer (PET) by ion-plasma sputtering (1) and thermal sputtering (2) [55].

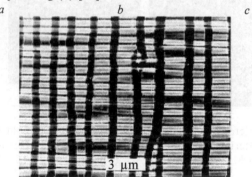

Fig. 10.25. Scanning electron micrographs of PET with a gold coating after uniaxial tensile drawing at 20°C; metal sputtering time: 30 (*a*), 15 (*b*) and 10 s (*c*) [39].

It is necessary to determine the mechanism of this phenomenon. To answer this question, it should be mentioned that in ion-plasma sputtering the deposition of the metal on the polymer surface is accompanied by its interaction with plasma. The effect of plasma on the polymers was examined in greater detail previously. In this section, the ion-plasma sputtering of the metal on the polymer results in initial (after several seconds, Fig. 10.5) in the formation of a hard, visibly cross-linked surface layer. Therefore, after its formation, a metal is deposited on this layer, and a three-layer polymer–plasma modified surface layer of the polymer–metallic coating system forms.

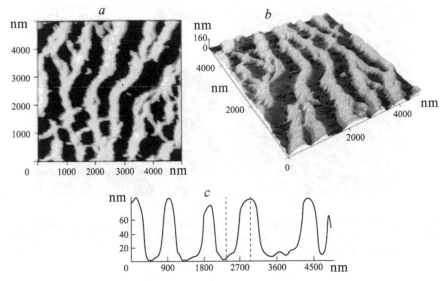

Fig. 10.26. AFM images of the polyethylene terephthalate film with a gold coating, tensile loaded at room temperature with the formation of a neck. The AFM image (*a*), the corresponding cross section (*c*), the three-dimensional reconstruction of the image (*b*). Sputtering time 5 s [39].

The presence of this two-layer coating can also be detected easily in cases in which a relatively thick metallic coating is deposited on the polymer. Figure 10.27 shows the electron micrograph of the surface of a low-density polyethylene sample with a deposit of a relatively thick gold layer (30 nm) after its deformation at room temperature with a rate of 1 mm/min (a neck forms in the polymer in these conditions). It is clearly seen that two groups of fracture fragments with greatly differing sizes form on the surface of the polymer. One part of the fragments has the size of 3–5 μm in the direction of the tensile drawing axis. These fragments separate without any order from the polymer substrate. The other part of the fragments has considerably smaller and more uniform dimensions (1–1.5 μm). These fragments do not separate from the surface of the polymer. At higher magnifications (Fig. 10.27 *b*) it is clearly seen that the specimens have a regular wavy microrelief and the main features of this microrelief were examined previously [8, 11–13]. Similar results were obtained when studying the films of polyethylene terephthalate with a thick (32 nm) platinum coating after tensile drawing at room temperature [39].

Thus, the method of evaluation of the stress–strain properties proposed in [14–20] has its limitations. The point is that the equation

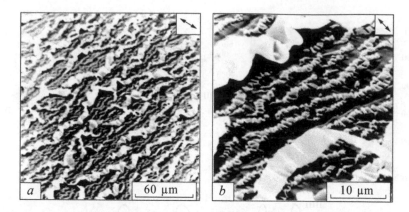

Fig. 10.27. Electron micrographs of the surface of a low-density polyethylene sample with a gold coating (25 nm) after deformation at 20°C with the formation of a neck: a) small magnification; b) high magnification.

(10.2) used for evaluating the strength of the coatings in this study gives adequate strength only if a two-layer polymer–metallic coating system forms which in [56] was referred to as the 'rigid coating on a soft substratum'. In this case, the main parameter which can be used to evaluate the strength is the size of the fracture fragments of the coating (L), which form during tensile drawing of the polymer-substrate. At the same time, as shown previously, the ion-plasma sputtering of the noble metals on the polymer results in the formation of a three-layer system (instead of the two-layer system): metal–plasma-modified layer–polymer-substrate. In addition, as indicated by the presented data and a large number of studies by other authors [52–55], the continuous coating on the polymer substrates in ion-plasma or thermal sputtering of noble metals on the polymer forms only after a long delay. In most cases, the initial portions of the sprayed metal condense on the polymer surface by the island mechanism. In other words, the first stages of sputtering the metal on the polymer surface are characterised by the formation of a system of discrete nanoclusters whose size, size distribution and also mutual distribution on the substrate depend on the sputtering rate, substrate temperature, the reactivity of the metallic atoms, etc. In this connection, the very concept of the thickness of the coating at short sputtering times of the metal on the surface of the polymer is a key factor for the method of evaluating the strength of the coating [14–20] becomes uncertain because h is included in the equations (10.2) and (10.4).

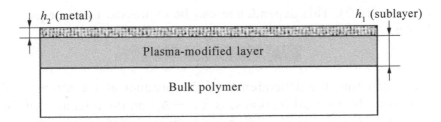

Fig. 10.28. Schematic representation of the polymer film with a layer of noble metal deposited on its surface by ion-plasma sputtering.

Therefore, it is important to derive quantitative relationships for the analysis of the stress–strain properties of the three-layer polymer–coating systems which form in particular in the ion-plasma sputtering of noble metals on the polymers. It will be attempted to develop an quantitative approach for analysis of this type of three-layer metal–plasma-modified surface layer of the polymer–polymer-substrate systems. The construction of such a system is shown schematically in Fig. 10.28.

As mentioned previously, one of the main special features of the investigated system is the assumption that the plasma-modified sublayer, formed in the initial moments of ion plasma sputtering of the metal, has a constant thickness, and during further sputtering only the layer of the pure metal situated on the sublayer changes its thickness. This assumption has been efficiently justified by the experiments [29, 30].

To analyse this situation, attention will be given to a thin coating, consisting of two layers (Fig. 10.28). Let the strength of the layer 1 (the so-called sublayer) with a thickness h_1 be equal to σ_1, and the strength of the layer 2 (the pure metal layer, situated on the sublayer, with a thickness h_2 be equal to σ_2. Then at fracture of the coating the latter will carry the load

$$F = \sigma_1 h_1 w + \sigma_2 h_2 w$$

where w is the width of the coating (sample). The effective strength of the coating in this case is determined by dividing the total applied force F by the cross-section of the coating $w\,(h_1 + h_2)$:

$$\sigma_A = \frac{\sigma_1 h_1 + \sigma_2 h_2}{h_1 + h_2}$$

where σ_c is the total strength of the coating determined using equation (10.2). This dependence can be expressed by the equation (10.6):

$$\sigma_c (h_1 + h_2) = \sigma_1 h_1 + \sigma_2 h_2$$

Constructing the dependence of the product of the strength of the coating by its total thickness $\sigma_c (h_1 + h_2)$ on the thickness of the metallic coating h_2, we should obtain from the angle of inclination of the straight line the strength of the bulk metal and from the section cut on the coordinate the product of the strength of the sublayer (σ_1) by its thickness (h_1). Naturally, we confine ourselves to h_2 values higher than 10 nm because at small thicknesses no single metallic coating forms on the polymer surface. If the assumption on the two-layer structure of the coating is valid, this dependence should be straightened in the coordinates of the equation (10.6).

The result of processing these experimental database is presented in Fig. 10.29. Firstly, it is the satisfactory straightening of the experimental data in the coordinates of equation (10.6). This fact indicates that the assumptions made regarding the two-layer structure of the coating are confirmed by experiments. Also, the straightening of the data in the coordinates of the equation (10.6) shows that for the case of ion-plasma sputtering the values of the strength of both gold layer and of the plasma-modified sublayer are constant. According to the results, in the range of small thicknesses (nanostructured formation effects) there is no increase of the strength of the gold coating in this case. In all likelihood, this effect is not manifested due to the fact that no single continuous coating forms in the small nominal thickness range and, therefore, the proposed method of determining the strength can not be applied here. The evaluation of the strength of the pure metals from the angle of inclination of the dependences 1 and 2 in Fig. 10.29 gives a value of 243 MPa for platinum and 160 MPa for gold. These values are in quantitative

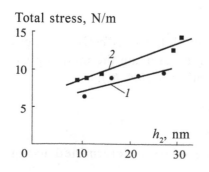

Total stress, N/m

h_2, nm

Fig. 10.29. Dependence of the strength of the two-layer coating, multiplied by its total thickness $\sigma_c (h_1 + h_2)$ on the thickness of the metallic coating h_2 for the platinum (1) and gold (2) coatings deposited by ion-plasma sputtering on the polyethylene terephthalate substrate.

agreement with the available values of the strength for bulk metals (240–350 MPa for platinum and 176–250 MPa for gold) [57].

The values determined by the dependences 1 and 2 on the ordinate were used to determine the value $\sigma_1 h_1$. Assuming that the thickness of the plasma modified layer (sublayer) is 80 nm, we obtain that its strength σ_1 is ~77 MPa for the case of ion-plasma sputtering of the gold coating and ~66 MPa for a platinum coating. As indicated by the results, the strength of the sublayer is similar in both cases.

In the previous section, relationship (10.2) was used to assess the strength of the plasma-modified layer in the experiments with plasma treatment of polyethylene terephthalate without sputtering the metal. In this case, the two-layer solid coating on the compliant base system forms and, therefore, it is fully justified to use the equation (10.2) for estimating the strength of the coating. In this case, the strength of the sublayer is ~33.3 MPa. It can be seen that in both cases the strength of the plasma-modified polyethylene terephthalate layer with a thickness of 80 nm is several tens of megapascals. The estimates are in agreement and this confirms the efficiency of the assumptions made when deriving the equation (10.6) and, in particular, the assumptions of the three-layer structure of the system formed in the ion-plasma sputtering of the noble metals on polyethylene terephthalate.

The observed differences in the determined strength values can be due to two reasons. Firstly, the combined effect of the plasma and atoms (clusters) of the noble metals on the polymer results in the formation of a sublayer containing a certain amount of the metal in its structure. Naturally, such a structure does not form in treatment of the polymer with the pure plasma. Possibly, it is the introduction of the atoms (clusters) of the metal in the initial stages of its sputtering to this plasma modified layer that leads to its hardening. Secondly, it is evident that this layer has no sharp boundary and, therefore, its schematic representation in Fig. 10.28 is idealised to a certain extent. The proposed new method for evaluating the strength of the coatings, deposited on the polymer substrate, can be used to analyse the stress–strain properties of the three-layer systems which form in particular in the deposition of nanometric layers of noble metals on the polymer films by ion-plasma sputtering.

10.6. The non-metallic coatings

Thus, the approach developed in [14–20] can be used to estimate directly the stress–strain properties of the modified polymer surfaces and the nanometric metallic films deposited on the polymer substrates. It may be expected that this approach will also provide the information of this type for coatings of other types.

In the previous sections, attention was given to the stress–strain properties of the metallic coatings deposited on the polymer by vacuum thermal sputtering. The results show that these properties depend on the nature of interaction between the polymer and the metal, the strain hardening of the metal, and also on the processes of its amorphisation in cases in which the thickness of the coating is in the nanometric range (several nanometres). At the same time, the polymer films with nanometric non-metallic coatings are used widely. In particular, the coatings based on silica (SiO_x), deposited by vacuum thermal sputtering, mostly on polyethylene terephthalate films, are employed extensively [58, 59]. It should be mentioned that a silicon oxide layer on the polymer surface represents a barrier to the penetration of oxygen through the film so that these systems can be used as packing materials in pharmaceutical and food industries [38].

Carbon coatings are also used in many applications. The thin carbon layers are regarded as promising materials for a wide range of biomedical applications: tissue regeneration, controlled supply of drugs, surface coatings for implants, etc. Ultrafine carbon films are also used for analysis and electronic applications [60]. In [61] it was also shown that the presence of carbon-containing structures has a beneficial effect on the biocompatibility of polymers with living cells. All these circumstances make the attempts to obtain information on the strength of these coatings, their resistance to various types of mechanical deformation, etc, very important.

It is important to note that both carbon and the SiO_x coatings, deposited by vacuum thermal sputtering on the polymers, are amorphous, in contrast to the coatings discussed in the previous section. At the same time, it is well known [1–7] that the transition of matter from micro- to nanodimensions results in the qualitative changes in physical, mechanical, physical-chemical and other properties. The changes of this type were described for the metallic coatings in the previous section. Changing the thickness of the coatings it is possible to change the geometrical dimensions of the solid in almost any range. It is therefore necessary to answer the question whether the geometrical dimensions of the amorphous solids (coatings) influence their structural–mechanical behaviour? To answer this question, in [39, 62] investigations were carried out into the structural–mechanical behaviour of polymer films with carbon coatings and coatings based on modified silica.

Analysis of the literature data indicates that at present there are no systematic unambiguous data on the strength of the coatings of amorphous carbon. The estimation of the mechanical properties is reduced mainly to the determination of the Young modulus and the hardness of the specimens by the nanoindentation method. In a number of studies, the authors determined the dependence of these characteristics on the depth of penetration of the plunger of the nanoindentor into the sample but not on the thickness of the deposited coating [63, 64]. Previously, it was mentioned that the main shortcoming of the nanoindentation method is the fact that it is not possible to determine the fundamental characteristics of the material, such as strength, modulus, fracture stress in the uniaxial tensile drawing conditions. The proposed approach can be used to estimate the deformation–strength properties of this type for any coatings, including carbon coatings. The consequences of tensile drawing of the polymer films with a carbon coating will now be discussed.

Figure 10.30 *a* shows a micrograph of the surface layer of polyethylene terephthalate with a carbon coating deformed by 50% at a temperature higher than the glass transition temperature of the polymer-substrate. It may be seen that deformation results in the fracture of the carbon coating with the formation of fragments in the form of strips oriented normal to the drawing axis, and in the formation of a regular microrelief, associated with the lateral contraction of the polymer during drawing. Figure 10.30 *b* shows a micrograph and an AFM image (Fig. 10.30 *c, d*) of polyethylene terephthalate with a carbon coating, deformed at room temperature. As indicated by the results, deformation of the sample at a temperature of 20°C is accompanied by the separation of the coating from the polymer matrix. At the same time, we can see the irregular fragmentation of the coating which indicates that the adhesion is sufficient for fragmentation and the proposed approach can be used to evaluate the strength of the coating.

Figure 10.31 shows the graph of the dependence of the strength of the carbon coating on the layer thickness. The data were obtained using equation (10.2). In particular, it should be noted that the absolute values of the strength of the amorphous carbon films were obtained for the first time. The strength varies in the range from 150 to 300 MPa, depending on the thickness of the deposited coating. It may also be seen that, as in the case of the aluminium coatings, the strength increases in the small thickness range (5–15 nm). For layers thicker than 20 nm the strength of the coating no longer depends on the coating thickness. The results cannot be explained by the nanostructure formation effects observed in the metals during amorphisation [44, 46] because the carbon coating is amorphous and its phase state does not depend on thickness. As regards the amorphous carbon coatings, there are no reliable data on their stress–strain properties. Information on these properties of the carbon coatings can usually be obtained by indentation methods. These methods are based on measuring forces in indentation of an indentor with a specific geometry into the surface of the investigated material (in the case of the coatings into the surface of the coating usually deposited on a hard substrate) [65].

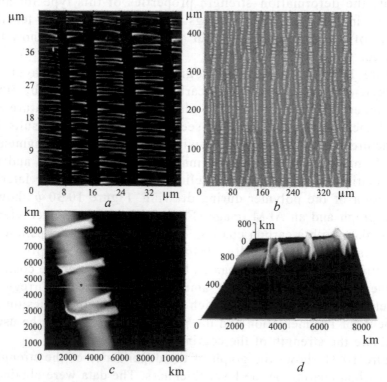

Fig. 10.30. Micrographs (*a*, *b*) of a polyethylene terephthalate film with a carbon coating, deformed in air by 50% at 90 and 20°C. The AFM image and its three-dimensional reconstruction (*c*, *d*) of the polyethylene terephthalate film with a 22 nm carbon coating, deformed at 20°C with the formation of a neck.

σ*, MPa

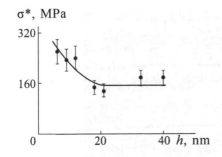

Fig. 10.31. Dependence of the strength of the carbon coating on the thickness of the sputtered layer in tensile drawing of the polyethylene terephthalate substrate by 50% at 90°C.

As indicated by Fig. 10.32, the deformation of the polyethylene terephthalate with the *m*-silica coating is accompanied by failure of the coating and leads, as in the case of the metallic coatings, to the formation of fracture fragments distributed in the direction normal to the drawing axis of the polymer. For the rubbery state of the polymer substrate (deformation of polyethylene terephthalate at 90°C), in addition to the disintegration of the coating into fragments, examination also showed the formation of a microrelief (Fig. 10.32 *b*) in the form of irregular folds, distributed in the direction of tensile drawing of the specimen. It should also be mentioned that the deformation of PET with the silica coating (coating thickness ~100 nm) results in the formation of very large folds and considerably smaller fracture fragments. By this feature the silica coating differs from the metallic coatings for which the size of the fracture fragments and the period of the microrelief are comparable in size.

Figure 10.33 shows the results of calculations of the strength using the proposed approach for the *m*-silica coating. It may be seen that the strength increases with the decrease of the thickness of the deposited layer (in the thickness range smaller than 150 nm). Assuming that the strength of the coating in the large thickness range corresponds to the strength of the coating in the bull, it may be assumed that the strength of the modified silica coating is approximately 20 MPa.

Thus, in [39, 62] the stress–strain properties of the m-silica coating were investigated for the first time. It is shown that the investigated coating is ductile but its fracture elongation is smaller than that of the polymer-substrate. Therefore, the coating can be investigated by the proposed structural–mechanical approach. The quantitative estimates of the mechanical properties of the *m*-silica were obtained for the first time and it was shown that, in contrast to the non-modified silica, this type of silica is a brittle coating and

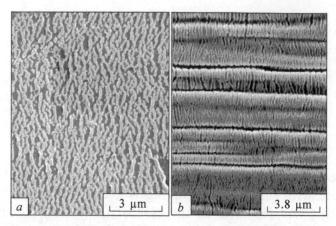

Fig. 10.32. Scanning electron micrographs of fracture of the modified silica coating with a thickness of 77 (*a*) and 100 (*b*) nm on a polyethylene terephthalate film after uniaxial tensile drawing at 20 and 90°C, respectively. The direction of drawing of the polymer is horizontal [63].

is a polymer having the typical polymer mechanical characteristics (low strength and high fracture elongation).

Thus, the inorganic amorphous coatings, deposited on the polymer substrates, show a large increase of the mechanical properties with a decrease of their thickness. Naturally, the growth of this type can not be explained by the analogy with the metallic (aluminium) coating (Fig. 10.13) because in principle they are not capable of crystallisation. It is possible that the detected effect is associated with the effect of the defectiveness of the solids, in particular coatings, on their strength, observed in the previous century by Griffith [66]. Griffith measured the strength of glass fibres of different diameter and found that the strength of the individual fibres increases with decrease of the diameter. In subsequent investigations, Griffith and other physicists observed the mechanism of the detected phenomenon. It may be seen that the strength is determined not only and not to such an extent by the strength of the interatomic or intermolecular bonds in the solids. The strength of the solid depends on its actual structure. In this case, the structure is represented by uncontrollable imperfections of various type, defects, secondary microscopic inclusions, microcracks, etc. In particular, these structural imperfections assemble around themselves (concentrate) stresses many times greater than the average applied stress. As a result of this effect, the cracks, leading to the failure of the solid, nucleate and propagate especially in the areas with the highest stress

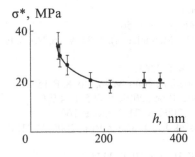

Fig. 10.33. Dependence of the strength of the silica coating on the thickness of the deposited coating at a tensile drawing temperature of 20°C.

concentration. The distribution of the imperfections of the structure in the volume and on the surface of the solid is random. Evidently, their total number depends on the dimensions of the solid: as the dimensions of the solid increase, the number of these defects in the solid also increases and the probability of failure of the solid on applying the same load becomes higher. This is why the strength of the solids increases with a decrease of their geometrical dimensions. It is very important to note that Griffith tested amorphous (not crystalline) glass fibres.

Conclusion

Consequently, it may be concluded that in this section we have described a new method of evaluating the stress–strain properties of the nanometric coatings deposited on polymer substrates. The method is based on the analysis of the parameter of the microrelief formed in deformation of the polymer films with a thin coating. Examples are given of the application of this approach to evaluating the stress–strain properties of the aluminium coatings and coatings based on noble metals, and the quantitative characteristics: elasticity modulus, fracture stress (strength), and plastic deformation of these materials in the nanometric layers, were obtained for the first time. It is shown that this method can be used universally for the evaluation of the stress–strain properties of the nanometric coatings, irrespective of their nature.

References

1.   Sergeev G.B., Nanochemistry, Moscow State University Press, 2003, 287 pp.
2.   Pomogailo A.D., Rosenberg A.S., Uflyand I.E., Metal nanoparticles in polymers, Moscow, Khimiya, 2000.
3.   Bronstein L.M., Sidorov S.I., Valetskii P.M., Usp. Khimii, 2004, V. 73, No. 5, P. 542-558.
4.   Suzdalev I.P., Nanotechnology. Physical chemistry of nanoclusters, nanostructures

and nanomaterials, Moscow, KomKniga, 2006.

5. Andrievsky R.A., Ros. Khim. Zh., 2002, V. 46, No. 5, P. 50.
6. Melikhov I.V., Ros. Khim. Zh. (ZhVKh im. D.I. Mendeleeva), 2002, V. 46, No. 5, P. 7.
7. Huang H., Spaepen F., Acta mater., 2000, V. 48, P. 3261.
8. Ivanchev S.S., Ozerin A.N., Vysokomolek. Soed. B, 2006, V. 48, No. 8, P. 1531.
9. Xiang Y., Chen X., Vlassak J.J., Mat. Res. Soc. Proc., 2002, V. 695, L. 4.9.1.
10. Xiang Y., Tsui T.Y., Vlassak J., J. Mater. Res., 2006, V. 21, No. 6, P. 1607.
11. Xiang Y., Chen X., Tsui T.Y., Jang J.-I., Vlassak J.J., J. Mater. Res., 2006, V. 21, P. 386.
12. Chen X., Vlassak J.J., J. Mater. Res., 2001, V. 16, No. 10, P. 2974.
13. Jen S.U., Wu T.C., Thin Solid Films, 2005, V. 492, P. 166.
14. Volynskii A.L., Voronina E.E., Lebedeva O.V., Yaminskii I.V., Bazhenov S.L., Bakeev N.F., Vysokomolek. Soed. A, 2000, V. 42, No. 2, P. 262.
15. Volynskii A.L., Bazhenov S.L., Lebedeva O.V., Ozerin A.N., Bakeev N.F., J. Appl. Polym. Sci., 1999, V. 72, No. 10, P. 1267.
16. Bazhenov S.L., Volynskii A.L., Alexandrov V.M., Bakeev N.F., J. Polym. Sci. Part B: Polym. Phys., 2002, V. 40, No. 1, P. 10.
17. Volynskii A.L., Bazhenov S.L., Lebedeva O.V., Bakeev N.F., J. Mater. Sci., 2000, V. 35, P. 547-554.
18. Volynskii A.L., Voronina E.E., Lebedeva O.V., Yaminskii I.V., Bazhenov S.L., Bakeev N.F., Vysokomolek. Soed. A, 2000, V. 42, No. 2, P. 262.
19. Volynskii A.L., Voronina E.E., Lebedeva O.V., Bazhenov S.L., Ozerin A.N., Bakeev N.F., Dokl. RAN, 1998, V. 360, No. 2, P. 205.
20. Volynskii A.L., Moses S., Dement'ev A.I., Panchuk D.A., Lebedeva O.V., Yarysheva O.V, Bakeev N.F., Vysokomolek. Soed. A, 2006, V. 48, No. 7, P. 1125.
21. Nazarov V.G., Surface modification of polymers, Moscow, MGUP, 2008.
22. Volynskii A.L., et al., Dokl. RAN, 2012, V. 442, No. 2, P. 203-205.
23. Nazarov V.G., Stolyarov V.P., Evlampieva L.A., Fokin A.V., Dokl. RAN, 1996, V. 350, No. 5, P. 639.
24. Nazarov V.G., J. Appl. Polym. Sci., 2005, V. 954, P. 897.
25. Nazarov V.G., Volynskii A.L., Yarysheva L.M. Stolyarov V.P., Bakeev N.F., Vysokomolek. Soed., 2012, V. 54, No. 9, P. 1355-1359.
26. Encyclopedia of low-temperature plasma, Ed. V.E. Fortov, Moscow, Fizmatlit, 2005, P. 822.
27. Rybkin V.V., Soros Educational Journal, 2000, V. 6, No. 3, P. 58.
28. Volynskii A.L., Panchuk D.A., Sadakbaeva Zh.K., Bol'shakova A.V., Yarysheva L.M., Bakeev N.F., Khimiya vysokikh energii, 2010, V. 44, No. 4, P. 369.
29. Panchuk D.A., Sadakbaeva Zh.K., Puklina E.A., Kechek'yan A.S., Bolshakova A.V., Abramchuk S.S., Yarysheva L.M., Volynskii A.L., Bakeev N.F., Vysokomolek. Soed., 2010, V. 52, No. 8, P. 794.
30. Drachev A.I., Gil'man A.B., Pak V.M., Kuznetsov A.A., Khimiya vysokikh energii, 2002, V. 36, No. 2, P. 143.
31. Klemberg-Sapieha E., Poitras J.D., Martinu L., Yamasaki N.L.S., Lantman C.W., J. Vac. Sci. Technol. A, 1997, V. 15, No. 3, P. 985.
32. Nazarov V.G., Vysokomolek. Soed. B, 1997, V. 39b, No. 4, P. 734-738.
33. Bergeron A., Klemberg-Sapieha J.E., Martinu L., Vac. Sci. Technol. A, 1998, V. 16, No. 6, P. 3227.
34. da Silva Sobrinho A.S., Schuhler N., Klemberg-Sapieha J.E., Wertheimer M.R., Adrews M., Gujrathi S.C., J. Vac. Sci. Technol. A, 1998, V. 16, No. 4, P. 2021.

35. Suo Z., Vlassak J., Wagner S., China particuology, 2005, V. 3, No. 6, P. 321.
36. Belyaev V.V., Opt. Zhurnal, 2005, V. 72, No. 9. P. 79.
37. Aluminium: properties and physical metallurgy, Ed. J. E. Hatch, translated from English. Moscow, Metallurgiya, 1989.
38. Panchuk D.A., Dissertation, Institute of Chemical Physics, Russian Academy of Sciences, 2010.
39. Metal constructions, Ed. E.I. Belen'. Moscow, Stroiizdat, 1986, P. 560.
40. Volynskii A.L., Panchuk D.A., Bol'shakova A.V., Yarysheva L.M., Bakeev N.F., Kolloidnyi zhurnal, 2011, V. 73, No. 5, C. 579-598.
41. Lyakishev N.P., Alymov M.I., Rossiiskie nanotekhnologii, 2006, V. 1, No.1, 2, P. 71.
42. Suzuki K., Fujimori X., Hashimoto K., Amorphous metals, ed. T. Masumoto, Moscow, Metallurgiya, 1987.
43. Gutkin M.Yu., Ovid'ko I.A., Defects and plasticity mechanisms in nanostructured and non-crystalline materialsm Moscow, Yanus, 2000.
44. Zolotukhin I.V., Barmin Yu.V., Stability and relaxation processes in metallic glasses, Moscow, Metallurgiya, 1991.
45. Glezer A.M., Molotilov B.V., Ovcharov V.P., Utevskaya O.L., Chicherin Yu.E., Fiz. Met. Metalloved., 1987, V. 64, No. 6, P. 1106.
46. Zhorin V.A., Fedorov V.B., Khakimov D.K., Galkina E.G., Tat'yanin E.V., Enikolopyan N.S., Dokl. AN SSSR, 1984, V. 275, No. 6, P. 1447.
47. Ping D.H., Xie T.S., Li D.X., Ye H.Q., Nanostruct. Mater., 1995, V. 5, No. 4, P. 457.
48. Panchuk D.A., et al., Vysokomolek. Soed. A, 2011, V. 53, No. 3, P. 372.
49. Volynskii A.L., et al., Vysokomolek. Soed., 2009, V. 51, No. 3, P. 436-446.
50. Felts J.T., J. Plast. Film. Sheet., 1993, V. 9, No. 139, P. 201.
51. Faupel F., Willecke R., Thran A., Mater. Sci. Eng. R., 1998, V. 22, No. 1, P. 1.
52. Strunskus T., Keine M., Willecke R., Thran A., Bechtolsheim C.V., Faupel F., Materials and Corrosion, 1998, V. 49, No. 3, P. 180.
53. Strunskus T., Zaporojtchenko V., Behnke K., Faupel F., Adv. Eng. Mater., 2000, V. 22, No. 8, P. 489.
54. Svorcik V., Slepitska P., Svorcikova J., Spirkova M., Zehentner J., Hnatowicz V., J. Appl. Pol. Sci., 2006, V. 99, No. 4, P. 1698.
55. Trent J.S., Palley I., Baer E., J. Mater. Sci., 1981, V. 16, P. 331.
56. Volynskii A.L., Bazhenov S.L., Bakeev N.F., Ros. Khim. Zh. (ZhVKh im. D.I. Mendeleeva), 1998, V. 42, No. 3, P. 57.
57. Handbook of chemistry and physics, Ed. in chief Charles D. Hodgman
58. Leterrier Y., Boogh L., Anderson J., Manson J.-A. E., J. Polym. Sci. Phys., 1997, V. 35, No. 9, P. 1449.
59. Leterrier Y., et al., J. Polym. Sci. Phys., 1997,V. 35, No. 9, P. 1463.
60. Svorcik V., Hubacek T., Slepicka P., Siegel J., Kolska Z., Blahova O., Mackova A., Hnatowicz V., Carbon, 2009, V. 47, No. 7, P. 1770.
61. Svorcik V., et alk. J. Mater. Sci.: Mater. Med., 2006, V. 17, No. 3, P. 229.
62. Panchuk D.A., et al., Vysokomolek. Soed. A, 2011, V. 53, P. 539–546.
63. Logothetidis S., Int. J. Modern Phys. B, 2000, V. 14, No. 2-3, P. 113.
64. Logothetidis S., et al., Diamond and Relat. Mater., 2000, V. 9, No. 3–6, P. 756.
65. Fedosov S.A., Peshek L., Advanced methods abroad, Moscow, The Faculty of Physics of the Moscow State University, 2004, P. 100.
66. Narisawa I., Strength of polymeric materials. Moscow, Khimiya, 1987.

11

Natural systems constructed on the basis of the 'rigid coating on a soft substratum' principle

In the previous chapters, it was shown that the deformation of the polymer films with the nanometric coatings is accompanied by a number of general phenomena. In [1], these systems are referred to as a 'rigid coating on a soft substratum' (RCSS). It should be mentioned that the deformation of the systems results in a special type of surface structure formation. Different types of deformation result in the formation of regular microreliefs on the surface of the films with the hard nanometric coating. The parameters of the resultant relief contain information on the properties of the material of the polymer substrate and the coating [2–5]. The deformation of this type is a simple and useful method for evaluating quantitatively the mechanical properties of the solids in cases in which the evaluation of this type is very difficult or generally not possible, for example, in the evaluation of the stress–strain characteristics of the nanometric coatings [6–8]. In fact, it is difficult to imagine the changes of the fundamental characteristics such as the yield stress, fracture strength or fracture strain of the solid with nanometric dimensions. As shown in chapter 10, this problem can be easily solved in the deformation of the polymer films with the nanometric coatings followed by analysis of the parameters of the resultant microrelief.

Figure 11.1 shows the electron micrograph of the surface of a PVC film with a thin (16 nm) platinum coating, tensile loaded to 50% at 90°C. It is clearly seen that two types of phenomena occurred in this case: fragmentation of the coating into strips of approximately the same size, and the formation of a surprisingly regular microrelief

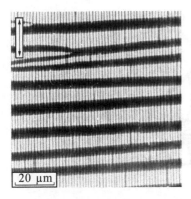

Fig. 11.1. Surface of a PVC sample coated with a thin layer (16 nm) of platinum, and subsequently tensile drawn to 50% at 90°C (the image was produced in a scanning electron microscope). Here and in later the double arrow indicates the direction of tensile drawing.

in these fragments. This relief is always strictly oriented along the tensile drawing axis of the polymer-substrate indicated by the arrow on the micrograph. The mechanism of both these phenomena was examined in detail previously (chapters 9, 10) and here we mention only the main conditions for its formation.

It should be mentioned that for this surprising microrelief to form, the system should correspond to the RCSS structure and the following conditions should be fulfilled: 1) the thickness of the coating should be negligibly small in comparison with the thickness of the substrate, and 2) the modulus of elasticity of the coating material should be several decimal orders of magnitude greater than that of the substrate. No further restrictions are imposed on the relationships used for evaluating the properties of the coating and the substrate.

11.1. Examples of 'rigid coating on a soft substratum' natural systems

The RCSS systems, corresponding to the two previously mentioned criteria, are widely found in nature. These systems include plant products, bodies of animals and many geological structures. In this chapter, we will try to use the considerations made in [1–8] regarding the properties of the RCSS systems for the analysis of several phenomena occurring in nature.

The simplest example of this type is shown in Fig. 11.2 which compares the electron micrograph of a polymer sample with a nanometric metallic coating (Fig. 11.2 *a*), subjected to planar tensile deformation, as described in chapter 1 (Fig. 1.3). It may be seen that these conditions result in the fragmentation of the coating

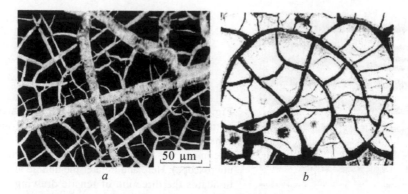

Fig. 11.2. Light micrograph of a polyethylene terephthalate sample with a thin aluminium coating, deformed to 35% in planar tensile drawing conditions (a) and crack networks on the surface of drying clay soil (b).

which has characteristic features. The main special feature of this type of fragmentation is the fact that each crack in the coating intersects some other crack under almost the right angle. Figure 11.2 *b* shows the photograph of a crust on the surface of drying clay soil formed during evaporation of moisture which rests on an almost incompressible substrate (underlayer saturated with moisture). With further evaporation, the resultant crust 'tries to compress itself, but the incompressible underlayer prevents this from taking place. Consequently, the hard crust is affected by internal stresses in planar tensile drawing. In other words, this crust is subjected to the same effect as the metallic coating shown in Fig. 11.2 *a*. It is therefore not surprising that the cast crust fragments in accordance with the same laws as the fragmentation of the metallic coating.

There are a large number of situations in which systems, constructed on the basis of the 'rigid coating on a soft substratum' principle are subjected in nature to planar tensile deformation. In the planar tension conditions, in which the crust exists on the surface of drying clay, hardening layers form on the surface of melts, for example, volcanic rocks. These conditions also result in the formation of a surface crust which breaks up into fragments because of the previously described reasons. In slow cooling of the melt the boundary between the hard layer and the not yet solidified liquid core moves into the bulk of the melt. The solid phase, coexisting continuously with the liquid phase, is constantly subjected to planar tensile deformation [9, 10]. In cases in which these process slows down, fragmentation takes place so regularly that it appears to be

a   *b*

Fig. 11.3. Columnar figures at Giant's Causeway in Northern Ireland: a) general vies; b) fragment of the structure [11].

made by humans. In particular, this mechanism forms the basis of the formation of a remarkable natural phenomenon – the so-called basalt fingers. Such objects are found widely in Northern Ireland and are referred to as the Giant's Causeway about (Fig. 11.3).

Here we will not discuss the special features of this type of fragmentation and we only mention that, firstly, the relationships governing the deformation of the RCSS systems are of general nature and do not depend on the scale of the elements forming them. Secondly, the above-mentioned circumstances indicate that the deformation of the polymer films with the nanometric coatings makes it possible to simulate and, consequently, investigate natural processes in cases in which the RCSS systems are deformed, i.e., the objects, constructed on the basis of the previously formulated principles, are deformed.

11.2. The Earth – the typical 'rigid coating on a soft substratum' system

In this connection, it is very attractive to analyse the structural–mechanical behaviour of a grandiose object such as our planet Earth.

Some current data on the internal structure of the Earth

At present, it is agreed that the Earth is the typical 'rigid coating on a soft substratum' system. According to the current views, the relatively thin (5–50 km) hard outer shell of the Earth (lithosphere) rests on the relatively compliant and thick (2900 km) shell – the

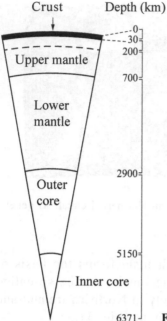

Crust   Depth (km)

Fig. 11.4. Internal structure of the Earth.

upper mantle (Fig. 11.4). The viscous, fluid matter of the Earth mantle is in the unstable state and continuously moves as a result of the thermal gradient [12]. In turn, this movement of the mantle matter (convection) generates mechanical stresses in the Earth's crust. These stresses are responsible for different geodynamic processes, taking place on the Earth (the continental drift, volcanic eruptions, earthquakes and relief formation processes). It may be seen that there is a distinctive and complete analogy between the films with the hard coating and the upper shells of the Earth. Thus, the structure of the upper shells of the Earth fully corresponds to the criteria of the 'rigid coating on a soft substratum' system.

Prior to analysing the relief formation on the Earth's surface, it is important to mention an important special feature of the structural-mechanical behaviour of the polymer films with the nanometric coatings. There are two mechanisms of formation of the specific microrelief on the surface of the polymer film with a rigid nanometric coating. Firstly, the initial non-oriented film with the coating can be deformed (this case is shown in Fig. 11.5 *a*). Secondly, the polymer film can be initially stretched and then coated with a hard coating, and this and be followed by stress relaxation in the stretched film so that the film can restore its initial dimensions. Figure 11.5 (*a, b*)

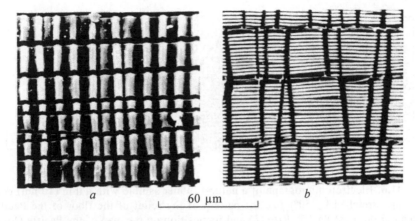

a |_____ 60 μm _____| *b*

Fig. 11.5. Scanning electron micrographs of a synthetic rubber sample coated with a thin gold layer (20 nm) and stretched subsequently by 100% (*a*), a natural rubber sample, pre-stretched by 100% and then coated with a thin layer of gold (20 nm) with mechanical stress relaxation (b) [1]. The tensile drawing axis is vertical in both cases.

compares the micrographs of synthetic rubber samples, produced by these two methods: direct stretching of the specimen with the coating and shrinkage of the specimen on which the coating was deposited after stretching.

It appears that in both cases a relief with these special features forms on the surface. The similarity of the structures also suggests identical mechanisms of the formation of these structures. However, there is a large difference between the data shown in Fig. 11.5 *a* and 11.5 *b*. The point is that the discussed microreliefs have different orientations in relation to the tensile drawing axis of the polymer. The microrelief, produced in shrinkage of the stretched polymer, is rotated by 90° in relation to the relief produced in direct tensile drawing. This important experimental fact is explained by assuming that the formation of the regular microrelief requires the stress leading to the compression of the hard coating which evidently is determined by the compression of the deformed polymer substrate. It is well known that in tensile drawing of the rubbery polymers with the latter not changing their volume there is extensive lateral contraction leading to the compression of the coating in the direction normal to the tensile drawing axis and also to the formation of the corresponding relief. In shrinkage of the stretched polymer the tensile and compressive stresses coincide. In other words, the directions of compression of the surface of the polymer in the two investigated cases are mutually perpendicular in relation to the tensile drawing axis of the polymer

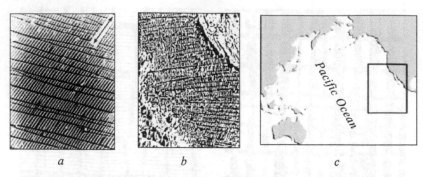

                a   b   c

Fig. 11.6. Electron micrograph of a natural rubber sample with a thin (10 nm) gold
coating, stretched by 50% (*a*), the pattern of the relief of the floor of the Pacific
Ocean in the East Pacific Rise (*b*) and its position on the map of the Pacific Ocean
(*c*) (indicated by the frame) [115].

and, consequently, the resultant microreliefs are perpendicular to each
other. It may be concluded that both the uniaxial tensile drawing
and uniaxial compression of the coating on the soft substratum
cause unavoidably the formation of a regular microrelief. The only
difference is the orientation of the microrelief in relation to the
tensile drawing axis (compression). The compression of the coating
on the incompressible substrate leads automatically to its stretching
in the normal direction and, conversely, stretching of the coating on
the soft substratum leads to the compression of the coating in the
normal direction. This circumstance is important for the simulation
of geodynamic processes.

   We compare the two reliefs – the relief of the polymer films with
a thin coating, stretched in one direction, and the actual relief of the
ocean floor. This can be carried out using Fig. 11.6 which compares
the electron micrograph of the surface of a stretched rubber film
with a gold coating and the pattern of the relief of a section of the
ocean floor in the region of the East Pacific Rise [13]. The relief of
the type shown in Fig. 11.6 *b* occupies huge areas of the ocean floor
measured in many thousands of kilometres (at least a third of the area
of the world oceans). It is noteworthy that the relief of the ocean
floor [14] greatly resembles the relief produced in stretching of the
polymer films with the hard coating [1–8]. This external resemblance
confirms the previously made assumptions regarding the generality of
the mechanism of the phenomena taking place in the 'rigid coating
on a soft substratum' systems of greatly different scales. It should
be mentioned that the origin of the system of the cracks (transform
faults) regularly distributed on the floor and parallel to each other

and of the folded relief on the surface of the fragments of the oceanic crust, situated between them, has not as yes being explained.

If the analogy between the examined objects is accurate, simple examination of the surface relief produces clear and very important information on the direction of compressive and tensile stresses, acting in the Earth crust. The stresses of this type have not as yet been assessed. This could be done if it is assumed that the Earth crust is an independent object, i.e., of the unthinkable dimensions but a single, real solid. This solid is spherical, has a non-constant chemical composition, defects, contains a temperature gradient and a large number of other complicating factors, but it is a single solid, resting on the relatively soft compliant substrate. The investigation of the Earth crust as a single object, capable of receiving and transferring mechanical stresses over huge distances (within the framework of oceans and, possibly, on the global scale) enables the approach developed in [1–8] to be used for evaluating the most important mechanical parameters of the Earth crust. Evidently, it is not in principle possible to determine these characteristics for the Earth crust as a single solid by some other method.

We mention briefly the current fundamental considerations regarding the tectonic processes taking place in the Earth crust and adjacent shells [16]. Approximately 200 million years ago the single Pangaea continent existing at the time broke up into several parts because of unknown reasons, with the parts starting to drift until the present day. Consequently, all the current continents and oceans formed as a result of this drift. The drift mechanism of the continents, with the discovery of this drift being one of the most outstanding achievements of science in the 20th century, may be described as follows. A system of gigantic cracks – mid-ocean ridges (MOR) – embracing the entire globe, formed and exists at the bottom of the ocean. The matter of the Earth mantle travels through these cracks, separating the edges leading to the expansion (spreading) of the ocean floor. As a result of this expansion, the continents 'travel' on the oceanic crust like on an escalator, gradually moving away from each other or coming closer together.

The drift mechanism of the continents has been formulated in the theory of lithospheric plates which can be found in, for example, the collection [16]. The tectonic pattern of the Atlantic Ocean is shown schematically in Fig. 11.7. The Mid-Atlantic Ridge separates the American and African plates. The matter of the Earth's mantle, moving through the MOR, separates the Atlantic shores so that

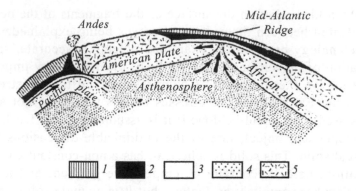

Fig. 11.7. Geodynamic circumstances in the region of the Mid-Atlantic Ridge from the position of the tectonics of lithospheric plates [16]: 1) water, 2–5) the lithosphere (2 – sediments, 3 – basalt oceanic core, 4 – upper mantle, 5 – continental crust).

America and Africa move away from each other with the speed of several centimetres per annum. The expansion of the floor of the Atlantic Ocean should be accompanied by the absorption of the oceanic floor in another area, because the dimensions of the Earth do not change. This area reduction actually takes place as a result of the shift of the Pacific Ocean plate below the American continent (subduction process) [12] (Fig. 11.7). It is evident that the movement of the Pacific Ocean and American plates towards each other results in their mutual compression. Since the uniaxial compression of the coating on the compliant substrate generates normal tensile stresses in the coating [1] (Fig. 11.5), it may be asserted that the oceanic crust in the Atlantic Ocean is compressed in the direction normal to the axis of the MOR and, correspondingly, stretched along the ridge. From the viewpoint of the investigated model it is very important to take into account that here we assume the existence of tensile stresses acting in particular in the direction normal to the axis of the MOR. This circumstance enables to explain qualitatively why the transform faults are always normal to the axis of the spreading (MOR). The point is that the fracture crack in the solid *always* grows in the direction normal to the axis of fracture (tensile) stress.

Simulation of relief formation processes in rock
The regular folding, characteristic of the relief of the oceanic floor, is one of the most frequently encountered morphological forms in the nature. It is therefore not surprising that considerable attention has been paid to the process of folding in the thickness of rock. The solution of this problem from the viewpoint of the loss of

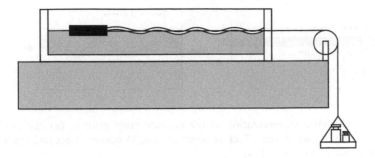

Fig. 11.8. Diagram of equipment for simulation of the process of folding in compression of a polymer film, floating on the surface of a fluid [22].

stability in the compression of the coating was evidently for the first time the subject of investigations carried out by Smolukhovskii at the beginning of the 20th century [17]. Later, this problem was further studied by Biot [18, 19] and Ramberg [20, 21]. In these investigations it was shown that the solid shell should bend under the effect of compressive stresses and gravitational forces. To confirm this relief formation mechanism, investigations were carried out to develop a method of simulation of processes of this type. The diagram of experimental equipment, used by Ramberg in these investigations, is shown in Fig. 11.8 [22]. Figure 11.8 shows that the load is applied to the elastic coating (rubber or gelatin plate) floating on a relatively dense but low viscosity fluid (mercury, saturated KI solution). Naturally, only the fault formation process can be simulated in these conditions and the process of cracking of the coating, which takes place in the nature, can not be simulated in these experiments (Figure 11.6). We believe that the application of mechanical stress directly to the hard coating does not at all reflect the actual situation. The only driving force of the tectonic relief formation of the Earth's surface is the convective movement of the partially molten upper mantle of the Earth. In particular, this motion transfers the stress (strain) required for the relief formation to the Earth's crust floating on it.

Taking into account the above considerations, the approach to investigating the RCSS systems, proposed in this study, can be regarded as a new efficient method of laboratory simulation of global tectonic processes responsible for the formation of the relief of the Earth's surface. This approach will be demonstrated by examining the processes of fold formation in the oceanic crust, shown in Fig. 11.6.

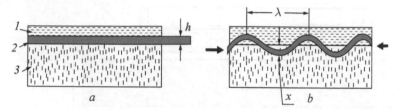

Fig. 11.9. Schematic representation of the oceanic crust prior to (a) and after (b) the loss of stability under the effect of compression: 1) water; 2) oceanic crust, and 3) the soft underlying layer [24].

11.3. Evaluation of the thickness of the Earth's crust

Thus, in order to simulate the oceanic crust [23], attention will be given to an elastic plate floating on a fluid with density ρ and coated with water (Fig. 11.9).

The increase of the density of the plate in the depth will be ignored. The density of the molten fluid will be assumed to be equal to the density of the plate. Under these assumptions, the displacement of some element of the plate from the equilibrium position will result in the formation of a returning force, determined by the hydrostatic pressure:

$$f = (\rho - \rho_w)\, gx \tag{11.1}$$

where ρ is the density of the plate and the melt, ρ_w is the density of water, g is the free fall acceleration, and x is the displacement from the hydrostatic equilibrium position. The bending of the Earth's crust with (11.1) taken into account is described by the equation [25]

$$EI\frac{d^4x}{dy^4} + N\frac{d^2x}{dy^2} + (\rho - \rho_w)gx = 0 \tag{11.2}$$

where E is the elasticity modulus of the oceanic crust, N is the compressive force, I is the moment of inertia of the cross-section of the crust equal at its unit width to $h^3/12$, where h is the thickness of the crust. The solution of the equation (11.2) is the periodic function

$$x = C\sin\frac{2\pi y}{\lambda} \tag{11.3}$$

where C is the bending amplitude, λ is the wavelength. The compressive stress is determined by substituting equation (11.3) to (11.2):

$$N = EI\left(\frac{4\pi^2}{\lambda^2} + \frac{(\rho-\rho_w)g\lambda^2}{4\pi^2 EI}\right)$$ (11.4)

The function $N(\lambda)$ is always greater than zero and has a minimum at some value λ^*. This means that at stresses smaller than some value (critical stress) the oceanic crust does not lose its stability and remains flat. The loss of stability takes place when the function $N(\lambda)$ reaches a minimum determined from the condition $dN/d\lambda = 0$. Taking into account that the elasticity modulus E and the speed of sound C are linked by the relationship $E = \rho C^2$, the wavelength of the loss of stability is determined by the equation

$$\lambda^* = \pi\sqrt[4]{\frac{4\rho c^2 h^3}{3(\rho-\rho_w)g}}$$ (11.5)

where c is the speed of sound in the crust. Equation (11.5) can be used to estimate the thickness of the crust layer which has lost its stability on the basis of the wavelength of the loss of stability in Fig. 11.9. Substituting the typical values of the wavelength of the length and stability (from the map of the relief of the floor, Fig. 11.6 b) into equation (11.5) $\lambda^* = 70$ km, $C = 6.5$ km/s, $\rho = 2600$ kg/ m^3 and $\rho_w = 1000$ kg/m^3, we obtain the estimate of the thickness of the layer $h \approx 5.0$ km. It may be seen that even such rough estimate gives a very rational value of the thickness of the oceanic crust which coincides quantitatively with the estimates of the thickness of the Earth oceanic crust obtained by studying the propagation of seismic waves formed during earthquakes [12]. The result validates the assumptions made when deriving equation (11.5). It should be mentioned that at present this is the only method of evaluating the thickness of the Earth crust based on the results of studying the relief of the Earth's surface and not on the measurements of the speed of propagation of the seismic waves.

11.4. Evaluation of the strength and longevity of the Earth's crust

We will try to use the approaches developed in [1–8] for evaluating the most important stress–strain properties of the oceanic crust, forming the floor of the Atlantic Ocean. The most characteristic special feature of the relief of the floor in this region of the world ocean is the presence of MOR, situated directly in the middle

Fig. 11.10. Chart of the relief of the Atlantic Ocean floor.

between Africa and America and repeating the contour of their shores
(Fig. 11.10). As in all other cases, the MOR has a system of the
regularly distributed parallel cracks (transform faults), intersecting
the MOR under almost the right angle.

In chapters 9 and 10 it was shown that the strength of the coating,
deposited on the polymer film, can be estimated by analysing the
pattern of relief formation under the effect of the mechanical stress
applied to the compliant polymer substrate. Measuring the distance
between the fragments of the coating, we can estimate the magnitude
of the fracture stress:

$$L = 4h(\sigma^*/\sigma_0) \qquad\qquad (11.6)$$

where L is the size of the fragment of the coating in the direction of
the tensile drawing axis; σ^* is the strength of the coating, σ_0 is the
stress in the substrate, and h is the thickness of the coating.

To evaluate the strength of the oceanic floor [26] in the vicinity of the Mid-Atlantic Ridge, we introduce the notation of the geodynamic characteristics in equation (11.6) $L = 4h \, (\sigma^*/\sigma_0)$. The same symbols will denote: L is the mean distance between the cracks (transform faults); h is the thickness of the coating (oceanic crust); σ^* is the fracture strength of the coating (oceanic crust), and σ_0 is the stress in the substrate (the relief forming stress). The value L can be measured directly on the chart of the oceanic floor (Fig. 11.10). To use the equation in practice, it is necessary to know the relief forming stress (σ_0) in the layer under the crust of the upper mantle of the Earth.

At present, it is generally assumed that the driving force for the spreading and, therefore, the continental drift, is the viscous convective flow of the matter of the upper mantle of the Earth (Fig. 11.7). The viscous flow process can be described efficiently by the universal Newton law which can be used to calculate the stress sustaining the flow of the liquid (this is also the stress in the viscous matter of the mantle σ_0):

$$\sigma_0 = \eta \, d\varepsilon \, /dt \qquad (11.7)$$

where η is the viscosity of the plastic matter of the Earth mantle, $d\varepsilon/dt$ is the strain rate of the matter of the Earth's mantle, caused by the stress σ_0. We estimate this stress, assuming that the viscosity of the matter of the Earth mantle, adjacent to the lithospheric plate, is $\eta = 10^{21}$ Pa·s [27]. The speed of continental drift is evidently equal to the speed of the viscous displacement of the matter of the Earth mantle, adjacent to the lithosphere.

Therefore, the flow speed of the matter of the upper mantle is assumed to be equal to the reliably determined spreading speed (the speed of continental drift) which can be estimated as approximately 10 cm/year. This speed, i.e., the strain rate in the equation (11.7), will be equal to: $d\varepsilon/dt = (\Delta l/l_0)/\Delta t$, where l_0 is the initial distance between the continents (~5000 km), Δl is the displacement per annum (10 cm) and Δt is the time (1 year). Substituting these values into equation (11.2) gives $\sigma_0 \approx 0.6$ MPa. This is also the stress sustaining the convective motion (flow) of the mantle matter or, which is the same, the relief forming stress in the Earth oceanic crust. It is unexpectedly small. Such a small stress in the actual oceanic crust is quite difficult to determine by experiments. Experiments are usually carried out in the oceanic crust to estimate the stress in the areas of stress concentration, for example, in the areas of bending of the crust – in the subduction zones. However, the relatively high local stresses can be caused by the very small background stress acting over a long period of time and determined using equation (11.7), possibly throughout the entire spreading period (~200 million years).

Nevertheless, the resultant value can be used with equation (11.7) to estimate the strength of the Earth oceanic crust if it is regarded as a single solid. For this purpose, we use the value of the thickness of the Earth's crust in the vicinity of the MOR $h = 10$ km [12], the average distance between the transform faults $L = 200$ km, which is determined from the chart (Fig. 11.8) and the tectonic stress $\sigma_0 \approx$ 0.6 MPa, calculated using the Newton law. Substituting the resultant values into equation (11.6) shows that the strength of the oceanic Earth crust in the area of transform faults is approximately 3 MPa. This is the first and still the only estimate of the strength of the Earth crust as an integral solid. It should be mentioned that the approach we use here gives the unexpectedly high strength estimate for the solids in very thin (nanometric) layers (chapter 10), whereas the resultant estimate for the solids of colossal dimensions is very low. Can the Earth oceanic crust have such low strength?

It should be mentioned that, generally speaking, the strength of the solids (although it is included in the appropriate handbooks) is not their constant as, for example, the melting point or specific heat capacity. In the current views of the strength of the solids (the kinetic theory of strength [28]) the fracture of the solids is interpreted as a thermal fluctuation process, determined by the thermal oscillatory motion of the molecules. In particular, the processes taking place on the molecular level result in the final analysis in the previously mentioned scale natural phenomena. The molecules of any solid are subjected to thermal motion, i.e., oscillate around the equilibrium position. The fracture or rupture of a chemical bond takes place under the effect of thermal fluctuations when oscillations of the atoms (molecules) with the giant amplitude form spontaneously. In other words, the fracture of the solid is a probabilistic process and in cases in which probabilistic laws operate the most important factor of this process – time – becomes effective.

The assumptions of the kinetic theory of strength of solids can be easily verified by simple experiments. Figure 11.11 shows the dependence of the fracture stress on the loading time for three materials of different nature: rock salt (I), polycrystalline aluminium (II) and caprone (III) [26]. It may be clearly seen that the strength of the solid, irrespective of its chemical nature, is actually not a material

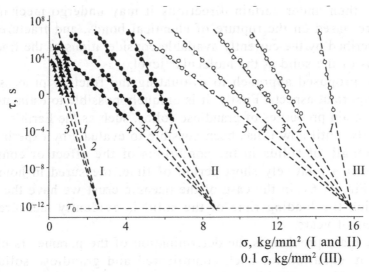

Fig. 11.11. Dependence of the longevity on the stress at different temperatures. 1) rock salt: 1 – 400; 2 – 500; 3 – 600°C; II) aluminium (polycrystalline): 1 – 18; 2 – 100; 3 – 200; 4 – 300°C; III – caprone (oriented fibres): 1 – –180; 2 – –120; 3 – –75; 4 – +20; 5 – +80°C [28].

constant and depends decidedly on the duration of loading, the magnitude of the load and temperature. The quantitative relationship between the strength of the solids, the duration of constant loading and the temperature is determined by the Zhurkov equation:

$$\tau = \tau_0 \exp[(U_0 - \gamma\sigma)/RT] \tag{11.8}$$

where τ is the time to fracture, U_0 is the activation energy of the fracture process (fracture of the chemical bond in the solid), γ is the activation volume which characterises the volume in which the elementary act of the fracture process takes place (fracture of the bonds), τ_0 is the frequency of thermal oscillations of the individual atoms in the solid in relation to the equilibrium position, σ is the acting constant stress, R is the universal gas constant, and T is absolute temperature. The experimental data, presented in Fig. 11.11, satisfy completely the Zhurkov equation and can be used to determine all the parameters of the equation.

Of course, the determination of the parameters of the Zhurkov equation for the Earth's crust is still an unsolved problem. In fact, the Earth' crust has a temperature gradient, a pressure gradient, a chemical composition gradient and a very large number of other complicating special features. However, if this object is the single solids, then under certain directions it may undergo mechanical fracture based on the rupture of chemical bonds, and fracture can be described by the currently available considerations of the fracture process of the solid at the molecular level.

The proposed approach to evaluating the fracture process has two important aspects. Firstly, it is uniquely possible to estimate the stress–strain properties of grandiose solids, such as the Earth's crust. Secondly, until now it has been possible to evaluate by experiments the strength of solids in the conditions of the effect of constant load over a relatively short period of time, measured in hours or days (Fig. 11.6). In the case of the oceanic crust we have the case in which the loading time of the solid is evidently measured in millions of years.

The main problem is the determination of the parameters of the Zhurkov equation for such complicated and grandiose solids as the oceanic crust. Table 11.1 shows the experimentally determined parameters of the Zhurkov equation for solids of different nature. It may be seen that parameter τ_0 changes only slightly for solids of different nature. This is not surprising because the frequency of thermal oscillations of the molecules in the solids should not depend strongly on their nature. The controlling parameter of the fracture process is U_0 which, according to the experimental estimates, is equal to several tens of kcal/mol, which corresponds to the fracture energy of different intermolecular bonds.

Since the physical meaning of the investigated parameters for complicated objects such as the Earth's crust is not clear at the moment, we use the average values of the parameters in the Zhurkov equation. To simplify calculations, for a rough estimate of the longevity (long-term strength) of the Earth crust we use the following average values: $\tau_0 = 10^{-12}$ s, $U_0 = 50$ kcal/mol, $\gamma = 1.0$ (kcal·mm²)/mol·kg, $R = 0.002$ kcal/(mol·K), $T = 400$ K and the fracture stress

Table 11.1. Parameters of the Zhurkov equation using the data published in [28]

	τ_0, s	U_0, kcal/mol	$\gamma, \dfrac{\text{kcal}\cdot\text{mm}^2}{\text{mol}\cdot\text{kg}}$
Metals	$10^{-12} \sim 10^{-13}$	25–170	0.7–0.6
Ionic crystals	$10^{-12} \sim 10^{-13}$	31–74	14–60
Covalent bonds	$10^{-12} \sim 10^{-13}$	91–113	–
Glasses	$10^{-12} \sim 10^{-13}$	45–90	–
Polymers	$10^{-12} \sim 10^{-13}$	25–53	0.14–0.9

σ = 0.2 kg/mm² which we determined using the equation (11.1). After substituting all these values into equation (11.4) we obtain that the time to fracture of the Earth's crust at such a low stress is approximately ~3·10⁷ (30 million) years. It may be seen that even with this greatly simplified estimate we obtain for the first time the completely geological times for the age of the important morphological forms of the Earth's crust such as the transform faults. It should be noted that if the age of the rocks can now be determined by reliable methods [12], at the moment there is no universal method for estimating the age of morphological forms on the Earth's surface and the above estimate is not only the first but also the only one.

After all, this is a very rough estimate because when using this approach it is necessary to take into account the specific data for the thickness of the crust, its temperature and other factors. The aim of this estimate is to demonstrate the possibilities of using this approach for the simulation of geodynamic processes.

Conclusions

Summing up, it may be concluded that the approach proposed in [6–9] is universal and can be used to simulate and estimate the deformation–strength properties of the solids irrespective of their size in cases in which the experimental determination is extremely difficult or completely impossible. The examined approach is already being used by scientists studying the evolution of the planets with time and for the analysis of their geodynamic circumstances although they have not as yet been visited by man. For example, analysis of the special features of the surface relief of Venus [2 29], using this approach, makes it possible to make several estimates regarding their structure and geological past.

References

1.   Volynskii A.L., Bazhenov S.L., Bakeev N.F., Ros. Khim. Zh. (ZhNKhO im. D.I. Mendeleeva), 1998. V. 42, No. 3. P. 57.
2.   Volynskii A.L., Bazhenov S.L., Leflooreva O.V., Bakeev N.F., J. Mater. Sci., 2000, V. 35, P. 547.
3.   Volynskii A.L., Voronina E.E., Leflooreva O.V., Bazhenov S.L., Ozerin A.N., Bakeev N.F., Vysokomolek. Soed. A, 1999, V. 41, No. 9, P. 1435.
4.   Volynskii A.L., Bazhenov S.L., Leflooreva O.V., Ozerin A.N., Bakeev N.F., J. Appl. Polym. Sci., 1999, V. 72, P. 1267.
5.   Bazhenov S.L., Volynskii A.L., Alexandrov V.M., Bakeev N.F., J. Polym. Sci. Part B: Polym. Phys., 2002, V. 40, P. 10.

6.   Panchuk D.A., Dissertation, Cand. Phys-Mat. Sciences, Moscow, Institute of Chemical Physics, RAS, 2010.

7.   Volynskii A.L. Panchuk D.A., Bolshakova A.V., Yarysheva L.M., Bakeev N.F., Kolloidnyi zhurnal, 2011, V. 73, No. 5, P. 579-598.

8.   Volynskii A.L., Panchuk D.A., Moiseeva S.V., Yarysheva L.M., Bakeev N.F., Rossiiskie Nanotekhnologii, 2008, V. 3, No. 1, 2, P. 97.

9.   Ryan M.P., Sammis C.G., Geol. Soc. Am. Bull., 1978, V. 89, No. 39, P. 1295.

10.   Koronovskii N.V., General geology, Moscow, MGU, 2002.

11.   Heezen B.C., Tharp M. World Ocean Floor, Office of Naval Research. US Navy, 1977.

12.   Volynski A.L., Bazhenov S.L., Geofisica internasional, 2001, V. 40, P. 87-95.

13.   Volynskii A.L., Bazhenov S.L., In Advances in Materials Science Research, V. 7, Ed. Maryann Wythers, 2011, Nova science Publishers Inc., Chapter 10.

14.   Smoluchowskii, Abh. Akad. Wiss. Krakau, Math., 1910, V. K1, P. 727.

15.   Lomize M.G., Plate tectonics, Encyclopedia, Moscow, Sovremenn. Estestvovanie, 2000, V. 9, P. 103.

16.   Biot M.A., J. Appl. Phys., 1954, V. 25, P. 2133.

17.   Biot M.A., Appl. Math., 1959, V. 17, P. 722.

18.   Ramberg H., Bull. Am. Assoc. Petrol. Geologists, 1963, V. 47, P. 484.

19.   Ramberg H., Tectonophysics, 1964, V. 9, P. 307.

20.   Walker D., V mire nauki, 1986, No. 12, P. 158.

21.   Kantha L.H., Geol. Mag., 1981, V. 118, No. 3, P. 251.

22.   Ramberg H., Stephansson O., Tectonophysics, 1964, V. 1, P. 101.

23.   Volynskii A.L., Priroda, 2007, No. 9, P. 10.

24.   Landau L.D., Lifshitz E.M., The theory of elasticity, Moscow, Nauka, 1965.

25.   Volynskii A.L. Yarysheva L.M., Moses S., Bakeev N.F., Ros. Khim. Zh. (ZhVKhO im. D.I. Mendeleeva), 2006, V. 50, No. 5, P. 126.

26.   Volynskii A.L. Bazhenov S.L., The European Physical J.E., Soft Matter, 2007, V. 24, No. 4, P. 317–324.

27.   Cathes L.M., The Viscosity of the Earth mantle, New York, Princeton University Press, 1975.

28.   Regel' V.R., Slutsker A.I., Tomashevski E.E., Kinetic nature of the strength of solidd, Moscow, Nauka, 1974.

29.   Koronovskii N.N., Dissertation, Cand. Geol.-Min. Sciences, Moscow, MGU, M.V. Lomonosov Moscow State University, 2003.

12

Perspectives for the practical application of surface phenomena in solid polymers

In the previous chapters, attention was given to the controlling effect of surface phenomena on the fundamental properties of solid polymers. The development and healing of interfacial surfaces in the polymers and polymer systems have been observed and that greatly different effects, such as mechanical [1–3] stresses and electrical [4] voltage, heat treatment of different type [5], surface chemical modifications [6], etc. Naturally, the general fundamental properties of the polymers like any other fundamental properties have a considerable potential for practical application. As shown in chapters 6 and 7, simple stretching of the polymer in the AAM results in the development in the polymer of a large interfacial surface and, consequently, almost any thermoplastic polymer is transformed to a non-specific adsorbent, capable of sorbing large quantities of low- and high-molecular substances from the surrounding space [7]. In addition, the polymer film, subjected to crazing in the AAM, automatically transforms to a porous separating membrane with the nanosized pore diameter [8], capable of efficiently separating both liquid and gas mixtures. The potential of this method for producing polymer sorbents and membranes is obvious because the formation of nanoporosity in the solid is a very difficult task which in practice is solved by time-consuming and labour-intensive procedure. These applied aspects of the crazing of the polymers in the AAM are evident and we will not discuss them in detail in this monograph.

Recently, the problems of formation of nanocomposites of different types have been discussed in the scientific literature and even in mass information media [9–12]. Considerable resources are used for the development of technological fundamentals of formation of these objects, and the advances in these areas are associated with the reliable development of greatly different branches of the industry and consumer aspects of our life.

One of the aspects of this direction is the justification of the scientific approaches to the development of nanocomposites with a polymer matrix. The number of publications in this area is increasing rapidly. We mention here only a number of applications in which the studies of this problem have already been reviewed [13–15]. We will discuss several approaches to the development of new types of nanocomposites with a polymer matrix which are based on polymer crazing in the AAM.

12.1. A new approach to the formation of nanocomposites with a polymer matrix

In the general form, the problem of formation of the nanocomposite with a polymer matrix consists of the addition to the polymer of a second (and in many cases of several) component of a different chemical nature. The problem of introduction of the modification additions to the polymers can be divided into two components: the addition of thermodynamically compatible and incompatible (with the polymer) additions. It is not difficult to add a compatible addition to the polymer. In fact, this addition is represented by a solvent for the given polymer. The additions of this type penetrate spontaneously into the polymer and are uniformly distributed in the volume of the polymer, forming a true solution. However, the compatible additions, added to the polymer, greatly reduce the most important characteristics of the polymer: hardness, strength, resistance to mechanical effects, etc. [16]. In other words, although such an addition may result in the required functional properties of the polymer, for example incombustibility or bactericidal activity, the polymers of this type cannot be used for producing any useful products, for example, films or fibres, with acceptable mechanical parameters.

The addition of incompatible components to the polymers also creates serious problems from the viewpoint of producing the final product with the optimum properties. Therefore, this type of addition

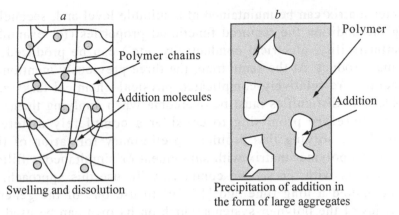

a

Polymer chains

Addition molecules

Swelling and dissolution

b

Polymer

Addition

Precipitation of addition in
the form of large aggregates

Fig. 12.1. Processes of adding compatible (a) and incompatible (b) additions to polymers.

is referred to as incompatible as it categorically 'does not want' to penetrate into the polymer. If it is attempted to ensure compatibility forcefully, it is necessary to work against surface forces. For example, grinding the addition together with the polymer in a mixer (extruder) it is usually not possible to produce a homogeneous mixture of the components with high mutual dispersion. In this case, the incompatible addition is precipitated into relatively long randomly distributed inclusions (fragments) in the bulk of the polymer, and we obtain an inhomogeneous, relatively coarse defective mixture of the polymer and the addition. In this case, as previously, we produce the material with poor mechanical properties which cannot be used to produce high-quality components. This problem is very important in producing the so-called volumeless product – films and fibres. The special features of the two examined methods of adding modification agents to the polymers are shown schematically in Fig. 12.1. Thus, the addition of modification compounds (both compatible and incompatible) of the polymer is associated with considerable difficulties. The real and only solution of the situation is the formation of the nanocomposite. In the general form, the formation of nanocomposite consists of the following stages. Firstly, it is necessary to reground at least one of the components of the composites to the nanodimensions (1 nm = 10^{-9} m). Secondly, it is necessary to mix the components of the system to produce a homogeneous mixture. Finally, the resultant system should be stabilised by some method to prevent spontaneous stratifying to the initial components because of their thermodynamic incompatibility. If it is possible to apply all these three procedures, it is hoped that, firstly, the properties of the polymer-matrix can be maintained at a suitable level and, secondly, using the addition the required functional properties (for example, incombustibility, electrical conductivity, etc.) can be produced in the final product. At the same time, the three previously mentioned procedures are relatively complicated physical–chemical tasks and considerable difficulties must be overcome when applying them.

It is therefore promising to consider a completely different approach for solving this problem: preliminary formation of the nanoporous polymer matrix with subsequent or simultaneous filling of the matrix with the second component. To use this approach, it was attempted a long time ago [17–19] to use one of the general properties of the polymer systems which on its own can be used to solve easily all the three previously formulated tasks – the crazing of the polymers.

The special features of this phenomenon were studied in previous chapters and here we only mention that the crazing phenomenon can be used to produce quite easily the nanosized porosity in the polymer. Since in this case the polymer acquires an excess of the interfacial surface [17–19], the structural transformation of this type in the polymer can be used to solve a very complicated physical-chemical problem because in this case it is necessary to counteract surface forces.

Secondly, the development of this type of porosity can take place only if the developing nanoporous structure is continuously filled with the surrounding liquid medium. In other words, when developing the nanoporous structure it is also necessary to solve the second previously formulated task of formation of the nanocomposite – supply of the second component to the polymer volume. The third task is to fix the addition in the volume of the polymer retaining its high nanosized dispersion – this task is easily also solved by crazing. This question will be examined later.

We should again mention the mechanism of structural rearrangement which accompanies deformation of the polymers in liquid media (chapter 6, Fig. 6.12). It was shown previously that the first stages of stretching the polymer in the AAM are accompanied by the formation of a certain number of crazes on the surface of the polymer, i.e., the zones with a unique fibrillar–porous structure.

The produced crazes grow during further deformation in the direction normal to the tensile loading axis of the polymer and maintain a constant and very small (fractions of a micron) width (the stage of growth of the crazes). This process continues until the growing crazes intersect the cross-section of the sample. This is followed by the start of the second stage of crazing of the polymer in the liquid medium – widening of the crazes when the crazes growing through the entire cross-section of the polymer increase in size in the direction of the tensile loading axis. Evidently, this is accompanied by the main transformation of the polymer to the oriented (fibrillised) nanoporous state.

The molecular mechanism of adding modification agents to polymers by crazing in liquid media
There is another stage in the crazing of the polymer in the AAM. Widening of the crazes is accompanied by a continuous increase of porosity and the specific surface of the polymer. This process cannot continue for a long period and the system 'finds' a method of getting rid of the excess surface area. When a large part of the polymer has been transferred to the oriented fibrillar state the porous structure starts to collapse [17–19] (chapter 7, Figs. 7.6–7.8). This stage is characterised by a large decrease of the cross-section of the deformed sample, accompanied by a decrease of its porosity, the average pore size and specific surface. In contrast to other crazing stages, this stage depends on the magnitude of the natural drawing of the polymer and also on the properties of the AAM and the geometry of the deformed sample. It should be mentioned that in collapse part of the liquid medium, trapped by the polymer in early deformation stages, is released into the surrounding space (syneresis phenomenon) [17–19].

Thus, the interaction of the deformed polymer with the surrounding AAM can be conventionally divided into two stages. In the first stage, the size of the pores and porosity continuously increase and the amount of the liquid, trapped in the nanoporous structure of the polymer, also increases. The second stage is characterised by the reverse process – collapse of the pores in the structure of the polymer and this is accompanied by a decrease of the size of the pores and porosity with partial release of the trapped liquid into the surrounding medium (syneresis). Evidently, these special features of crazing of the polymers in the AAM can be used for developing a new method of producing a wide range of nanocomposites with a polymer matrix.

Special attention will be given to the mechanism of interaction of the active liquid with the deformed polymer in cases in which the active liquid contains a modification addition. In [20, 21] the

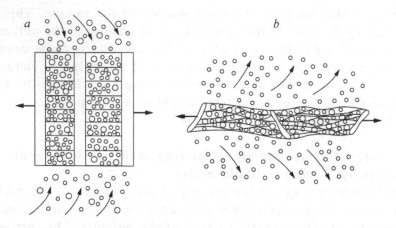

Fig. 12.2. Diagram of structural rearrangements, accompanying the crazing of a glassy polymer in a two-component AAM. The arrows indicate the direction of mass transfer of the low- molecular component in different stages of tensile loading [17].

modification addition was represented by organic dyes. The diagram of interaction of the deformed polymer with the solution of the modification addition in the AAM is shown in Fig. 12.2.

The initial stages of tensile loading are characterised by the growth of the crazes resulting in an increase of the total volume of the microvoids filled with the solution of the dye in the AAM (Fig. 12.2 *a*). For each polymer loading is carried out to the draw ratio at which the fibrillised material of the crazes starts to collapse and the porous structure changes to a compact one. As mentioned previously, this transition is accompanied by syneresis and release of part of the solvent from the volume of the crazes. However, the 'ejection' of the solvent into the surrounding space takes place through the microporous fibrillar structure of the polymer. The distances between the fibrils in such a structure decrease continuously as a result of their merger during drawing and become comparable with the large dye molecules. Consequently, ultrafiltration of the dye solvent takes place at the molecular level and, therefore, the dye molecules are trapped in the structure of the polymer because of purely geometrical (steric) reasons, and mostly the pure solvent is filtered into the surrounding space. Naturally, the amount of the dye trapped by the polymer increases with increasing strain and then remains constant because after transition of the polymer from the porous structure to a compact one there are no new portions of the solvent arriving in the polymer (Fig. 12.2 *b*) [17]. Naturally, these structural rearrangements also determine the possibility of adding different thermodynamically incompatible modification agents to the polymers, their uniform distribution in the volume and sustaining them in the structure of the fibre. It may be seen that the previously described first stage of production of the nanocomposite with the polymer matrix is also easily solved by crazing because the fixation of the modification addition in the structure of the polymer is based on the comparability of the molecular dimensions with the size of the pores in the polymer structure.

It is important to note that the addition of the low-molecular component, dissolved in the AAM, differs drastically from the addition of this component to the crazed polymer during its sorption from the solution. As shown in chapter 7 (see Figs. 7.20 and 7.21), the crazed polymer is capable of sorbing large quantities of the low-molecular component from the solution on the highly developed surface. The rule of equalisation of the Rehbinder polarities is fulfilled. This means that, for example, a large amount of an organic, relatively non-polar substance (dye) is sorbed from the polar solvent (water) on the surface of the organic polymer. However, if such a sorbed dye is transferred by the crazed polymer to the liquid (AAM) active as regards adsorption to the given polymer, the high-intensity desorption of the dye from the highly developed surface of the polymers starts to take place (chapter 7, Fig. 7.22). This phenomenon is caused by the fact that the adsorption from the solution is a competing process and the more active component (AAM) is capable of causing the desorption of the less active component (dye) from the polymer surface. All the processes of sorption to the nanoporous structure of the polymer and of subsequent desorption are determined by the relatively slow diffusion processes and are therefore quite time-consuming. The time to the establishment of equilibrium is measured in hours or even days [22].

The inclusion of the low-molecular component of the crazed structure of the polymer during its direct deformation in the AAM takes place by a completely different mechanism. In this case, the solution, for example a dye, is forced in the AAM into the developing porous structure of the polymer directly during drawing of the polymer. The duration of penetration of the dye solution to the volume of the polymer is determined by the rate of growth of the crazes which depends on the loading rate of the polymer and the processes of the viscous flow of the solution of the AAM to the areas of active deformation of the polymer. The saturation time of the porous structure of the polymer with the AAM solution

Dye content in the polymer·10^4, g/l

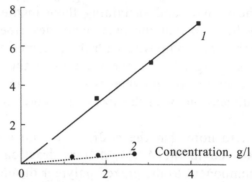

Fig. 12.3. Dependence of the content of rhodamine C in a polyethylene terephthalate film deformed in the AAM (*n*-propanol) by 400% on the rhodamine C concentration in the solution at its equilibrium sorption (2) and in direct drawing (1) on the solution concentration [18].

equals seconds or fractions of a second. It is very important to note that the content of the low-molecular component (dye) in the final material (after removing the AAM) is not determined by the equilibrium sorption process and is determined exclusively by the concentration of the addition added to the solution in the AAM in which deformation takes place.

Figure 12.3 shows the dependence of the content of the dye, added to the crazed structure of the polymer during its sorption, on the equilibrium concentration of the solution in which sorption takes place (2). It may be clearly seen that the adsorption of the dye in this case is extremely small because the AAM strongly competes with the dye during sorption. At the same time, the addition of the dye during direct drawing of the polymer in the dye solution in the AAM (curves 1) is not restricted by anything and is directly proportional to the concentration of the solution in which the polymer is deformed. This is understandable because the developing nanoporous structure of the crazes is continuously filled with the surrounding solution of a known concentration, and the subsequent collapse of the structure leads to the mechanical capture of the large addition molecules as a result of a large decrease of the pore diameter. As already mentioned, the pure solvent is released (syneresis) to the surrounding space in this case.

In [20, 21] the drawing of the polymer in the AAM resulted in the formation of coloured fibres and films of a number of polymers (PET, PVS, PVC, PC and a number of others). It should be mentioned that colouring of the synthetic fibre is a complicated, laborious and time-consuming process. It is well-known [23] that 'colouring is based on the spontaneous transfer of the dye from the solution or its dispersion to the fibre to the establishment of equilibrium. The colouring rate and the amount of the colouring substance, absorbed by the fibre, are determined by the laws of activated diffusion and sorption'.

In other words, firstly, colouring is a relatively slow process because it is determined by spontaneous processes of diffusion of the low-molecular component to the structure of the polymer. Secondly, to maintain the addition in the structure of the polymer, the addition and the polymer must contain active functional groups in their chemical structure capable of interacting with each other with the formation of relatively strong chemical bonds. Otherwise, the addition can leave the structure of the polymer by diffusion during, for example, treatment in water or water and soap solutions. This circumstance greatly complicates the addition of modifying components to the hydrophobic polymers which do not contain any active functional groups, such as polyolefins, polyesters, polyvinyl chloride. To overcome difficulties of this type, it is rational to use additional measures: 'the fibre is initially saturated with a slightly thickened dispersion of the suspensions of the dye, dried, and subsequently heated at 200–210°C (thermosol process) or subjected to the effect of trichloroethylene vapours ' [23]. Recently, the supercritical carbon dioxide has been used [24] as a loosening liquid during colouring of synthetic polymers. However, even the application of such active agent does not make it possible to organise colouring as a continuous process. In this case, the fibres are again treated in autoclaves at higher temperatures and pressures over a long period of time.

Thus, the processes of introduction of the modification agent to synthetic fibres are periodic, energy- and time-consuming. In addition to this, it is evident that in this case there are problems with the homogeneous distribution of the dye over the entire cross-section of the fibre. Therefore, attention will be given to a number of special features of crazing of the polymers in two-component liquid media.

The principal difference between the examined method of adding agents to the polymer and the traditional method [23] may be described as follows. The traditional method is based on the spontaneous penetration of the molecules of the modification agent to the structure of the polymer by diffusion. According to Rusanov's definition [25], crazing is accompanied by 'forced' injection of the solution of the modification agent in the AAM to the continuously developing nanoporous structure of the polymer. It is therefore not surprising that the amount of the agent added by this method may be considerably greater than the concentration of the same agent obtained in the equilibrium process of sorption from the same solvent [18].

We have examined the main special features of crazing of the polymers in liquid media from the viewpoint of development of nanosized porosity in them and filling with the second component. In fact, this justifies the general method of producing nanocomposites with the polymer matrix. In the most general form, the addition of the nanosized agent to the polymer (formation of the nanocomposite) may be aimed at solving at least two applied problems. Firstly, the addition can modify the properties of the polymer, for example, to ensure that the polymer is incombustible and electrically conducting. In this case, the addition should be strongly sealed in the polymer matrix and should not migrate from it under different types of influences of the type of abrasion and/or treatment with the solvent.

Secondly, there are applied tasks in which the final product (nanocomposite) should influence the surrounding medium, for example, suppress microorganisms or separate medical substances. In this case, the addition should be released from the polymer to the surrounding medium during long periods of time. The crazing of the polymer in the liquid media makes it possible to solve both these problems. We will examine these possibilities on specific examples.

Making polyester fibres incombustible
To make the polyester fibres incombustible, the fibres were deformed in an antipyrine solution (phosphonic acid amide) in laboratory equipment in the continuous mode and then subjected to heat setting by annealing at 120°C [26]. To produce the fibres, tests were carried out to determine the oxygen index and mechanical properties (strength), and the effect of standard treatment (washing, swelling, extraction, etc) on these parameters was studied. The results are formulated in Table 12.1. It may be seen clearly that simple tensile loading of the polyethylene terephthalate fibre in the antipyrine solution increases its oxygen index to the values greater than 40. This means that this material cannot be ignited and burn in air. At the same time, the strength of the fibres remains at the fully acceptable level and does not decrease by more than 25%. This effect is due to a large extent to the fact that in these experiments the strength was

Table 12.1. Effect of the method of treatment of polyester fibres, produced by crazing in a liquid medium containing antipyrine, on its strength and oxygen index

Treatment	Strength, gs/tex	Oxygen index
No treatment	32	43
Washing	30	37
Boiling in Na_2CO_3 solution	31	37
Swelling in surface active substance	30	31
Extraction by solvent (dichloroethane)	28	41

Comment: the strength of the fibre with no antipyrine 40 gs/tex, oxygen index 19; the same parameters after extraction by the solvent 38 and 19 gs/tex.

estimated by dividing the fracture force by the initial cross-section of the fibre. Since in this treatment a large amount of antipyrine is embedded in the polymer (up to 15%), the cross-section slightly increases and this influences the actual value of strength.

Should be mentioned that the resultant parameters are only slightly influenced by standard treatments and, in particular, multiple treatment (by up to 50 times) with soapy solutions (washing). Since the oxygen index of the comparison sample is 19, this means that this type of fibre can be easily ignited in air and burn in a stable manner in these conditions.

Production of electrically conducting polymer–polymer nanocomposites using crazing in liquid media

In this section, it is important to mention briefly a small number of attempts to produce polymer–polymer nanocomposites in which one of the components is an electrically conducting polymer. The production of the new types of electrically conducting polymers is the most important scientific and applied problem of the current physical chemistry of the polymers. Regardless of considerable successes recorded in recent years in this area, the practical application of the electrically conducting polymers is quite restricted because in most cases they are represented by infusible, insoluble powders which are difficult to process [27].

In [28–30], investigations were carried out to develop approaches to producing the nanocomposites in which the second polymer, added to the crazed polymer matrix, has high electrical conductivity. The procedure used in these studies was described previously.

The polymer films (high-density polyethylene and polyethylene terephthalate) were subjected to crazing during which appropriate monomers (acetylene and aniline) were added to the developing nanoporous structure. Subsequent polymerisation *in situ* of the added monomers and doping of the products resulted in the formation of a number of advanced nanocomposites which efficiently combine the excellent electrical conductivity, characteristic of polyacetylene and polyaniline, with the high mechanical parameters typical of polyethylene terephthalate and high-density polyethylene.

It has been shown that the addition of even small quantities of polyacetylene (8%) [28] to high-density polyethylene, and subsequent doping of the polymer mixture with iodine increase the bulk electrical conductivity by 14–16 orders of magnitude, i.e., to $(0.2-1.4) \cdot 10^2$ $Ohm^{-1} \cdot cm^{-1}$. Taking into account that the bulk specific electrical conductivity of pure polyacetylene is $\sim 10^2$ $Ohm^{-1} \cdot cm^{-1}$, this fact shows that the structure of the produced polymer mixture has a low percolation threshold of electrical conductivity. It is important to note that crazing produces the distinctive molecular orientation in the added electrically conducting polymer and this orientation greatly increases its electrical conductivity.

Successful attempt were also made to add a second electrically conducting polymer – polyaniline to the crazed matrices based on polyethylene and polyethylene terephthalate [29]. The main results of these investigations are presented in Tables 12.2 and 12.3. Table 12.2 compares the mechanical parameters of the pure polymer matrices, deformed by 100 and 200%, with the corresponding nanocomposites. It may be clearly seen that the most important characteristics such as the elasticity modulus and strength of the produced nanocomposites are similar to the limits of the appropriate characteristics for pure matrices and in some cases are even higher. At the same time, the produced nanocomposites show excellent electrically conducting characteristics which do not differ at all from those of the pure polyaniline (Table 12.3). Bulk electrical conductivity (σ_v) of both types of nanocomposites is very high and does not depend on the type of polymer matrix. At the same time, the data presented in Table 12.3 show that the surface electrical conductivity in the direction of the tensile loading axis ($\sigma_s^{\parallel}$) and in the normal direction ($\sigma_s^{\perp}$) greatly differ from each other. This circumstance can be used to influence the electrically conducting properties of the produced nanocomposites. Thus, the application of the crazed polymer matrices for polymerisation *in situ* of the electrically conducting polymers

Table 12.2. Mechanical characteristics of nanoporous matrices of polyethylene terephthalate and polyethylene and polymer–polymer nanocomposites with polyaniline based on them

Matrix or nanocomposite	λ, %	C, %	E, MPa	σ_f, MPa	ε_f %	E, MPa	σ_f, MPa	ε_f %
			Along drawing axis			Normal to drawing axis		
PET	100	–	690	48	190	660	34	380
PET–polyaniline	100	40.8	1210	51	7	1000	17	4
PE	100	–	310	64	150	360	29	550
PE–polyaniline	100	26.6	730	64	180	650	31	550
PE	200	–	320	88	80	280	28	680
PE–polyaniline	200	42.2	930	86	90	870	30	5

Table 12.3. Specific electrical conductivity of the polymer–polymer nanocomposites based on polyaniline and crazed matrices based on polyethylene terephthalate and polyethylene

Nanocomposite	Dopant	λ, %	C, %	σ_v, Ohm^{-1} · cm^{-1}	$\sigma_s^{\perp}$, Ohm^{-1} · cm^{-1}	$\sigma_s^{\parallel}$, Ohm^{-1} · cm^{-1}
PET–polyaniline	HBF$_4$	50	29.2	$1.6 \cdot 10^{-3}$	$3.7 \cdot 10^{-2}$	$5.4 \cdot 10^{-3}$
		100	40.8	$2.2 \cdot 10^{-2}$	$3.7 \cdot 10^{-1}$	$2.7 \cdot 10^{-2}$
		200	39.4	$7.2 \cdot 10^{-5}$	$1.0 \cdot 10^{-1}$	$4.7 \cdot 10^{-2}$
		300	12.8	–	$7.5 \cdot 10^{-4}$	$3.7 \cdot 10^{-3}$
PE–polyaniline	HCl	100	45.8	$4.8 \cdot 10^{-3}$	$2.9 \cdot 10^{-2}$	$3.6 \cdot 10^{-3}$
	HBF$_4$	200	42.2	$5.4 \cdot 10^{-2}$	$8.8 \cdot 10^{-1}$	$7.8 \cdot 10^{-1}$
	HCl	100	26.6	$2.9 \cdot 10^{-2}$	$5.4 \cdot 10^{-2}$	$3.0 \cdot 10^{-2}$
		200	42.2	$2.3 \cdot 10^{-2}$	$2.4 \cdot 10^{-1}$	$1.6 \cdot 10^{-1}$

efficiently combines in them the high mechanical parameters with optimum electrical conductivity. The experimental results show that the development of new types of electrically conducting polymer–polymer nanocomposites based on the crazed polymer matrices is highly promising for future applications.

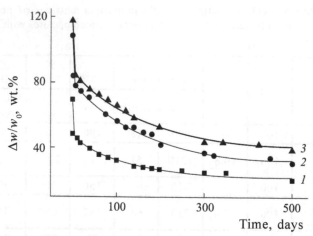

Fig. 12.4. Time dependence of the relative mass loss of polyethylene terephthalate samples containing glycerin. Samples were deformed in a liquid medium by 100 (1), 200 (2) and 300% (3) [34].

12.2. Production of polymer films and fibres capable of influencing the environment

The direct method of adding modification agents

Crazing can be used to produce new types of polymer nano-composites, capable of influencing the environment. This target is easily achieved if a viscous non-volatile liquid is added to the polymer [31, 32]. Figure 12.4 shows the weight loss curves of polyethylene terephthalate samples to which the non-volatile liquid (glycerin) was added by crazing. As indicated by Fig. 12.4 in this case the process of collapse of the structure of the crazes and the accompanying syneresis phenomenon stretch over many months or even years. Its rate depends in particular on the nature of the liquid medium and the polymer, the geometry of the sample and its deformation conditions. It is important to note that in the investigated case we are concerned with highly dispersed mixtures of the polymer and the incompatible low-molecular liquid. This incompatibility results in the phase separation in such systems. The slow rate of the process is explained by the fact that the diameter of the microvoids in the structure of the polymer is so small that it is comparable with the molecular dimensions of the added low-molecular liquid. Since the low-molecular liquid added to the volume of the crazes is incompatible with the polymer, the liquid is transported from the volume not by diffusion but by the viscous flow in very narrow pores. The departure of the liquid from the polymer is accompanied by its compression (contraction) and, consequently, the size of the pores gradually decreases. In turn, this reduction results in a progressive decrease of the rate of migration of the low-molecular component from the porous structure of the polymer.

As a result of the gradual decrease of the size of the pores part of the non-volatile low-molecular component remains in the polymer structure and does not leave it for an undefined long time. Actually, this means that the crazing of the polymer should be regarded as a special type of micro-encapsulation of the low-molecular liquids in the polymer films and fibres. It should be mentioned that the 'closing' of the liquid low-molecular component in the polymer matrix incompatible with this component is the most important scientific and applied task [33].

The phenomenon of the very slow release of the liquid non-volatile component from the crazed polymer can be used in cases in which the long-term effect, caused by some type of addition, for example, release of repellents, medicinal, bactericidal components or fragrances, is important. It should be mentioned that the polymer materials of this type can be produced in the form of fibres, films or fabric and, consequently, they are highly promising from the viewpoint of application in greatly different areas, for example in medicine, perfumery or textile industries.

Bactericidal films and fibers
As an example, consider the results of such studies as described recently in [35]. In this work, crazing was used to introduce bactericides katomine and altozan to PET fibres with the active antibacterial ingredients being alkyldimethylbenzylammonium chloride. The biocity of the produced fibres was investigated through microbiological testing. For this purpose, samples of fibres were placed in Petri dishes with bacteria cultures grown on an agar medium (*Staphylococcus aureus* and *Pseudomonas aeruginosa*). The diameter of the inhibition zone of bacterial growth was recorded in accordance with GOST 9.802-84. Typical results of the study are shown in Fig. 12.5. They show high bactericidal activity of the produced materials. It is particularly important to note that these bactericidal properties are preserved at the original level even after tenfold washing in soapy water solutions at 40–45°C.

In [36] crazing was used to add zinc salts from their solution to the PET fibres in the AAM. The resulting fibres had stable

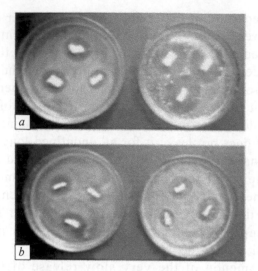

Fig. 12.5. Petri dishes with 15-day cultures *St. Aureus* (*a*) and *Ps. Aeruginosa* (*b*) in the presence of polyethylene terephthalate fibres modified by crazing and containing altozan (left) and katomine (right).

bactericidal properties unaffected for at least 10 soapy water treatment cycles.

Indirect method of adding modification agents

It is important to note that at present there are many bactericidal substances used widely in medicine, ecology and many other areas. However, the majority of these agents are not heavy non-volatile substances and, therefore, new polymer bactericidal forms cannot be produced using the previously described method of addition.

In chapter 8 it was shown that the nanoporous structure of the crazed polymer can be filled with a low-molecular component insoluble in the AAM. This filling method is based on adding some precursors to the nanoporous structure of the polymer. Subsequent chemical transformation *in situ* results in filling of the polymer matrix with almost any component of a different chemical nature. Using this method in [37–44] the authors produced a large number of types of nanocomposites based on crazed polymers, on one side, and metals, oxides and inorganic salts on the other side.

Following are the results of investigations carried out to develop methods of producing new types of metal–polymer nanocomposites of a synthetic polymer (polyethylene terephthalate), on the one side, and a noble metal (silver) on the other side [45]. Silver was selected due to the fact that additions of this metal to synthetic and natural polymers result in stable antibacterial and antiputrefactive properties of these polymers. As a result of this special characteristic of silver a 'shielding zone' free from microorganisms forms around the silver-containing materials. In most cases, this effect is implemented by simple surface treatment of the fibres or fabric (mostly cellulose and its derivatives) with silver salt solutions. As a result of the presence of the active functional groups, the silver ions are sorbed in the surface layer of these textile materials and give them the required bactericidal properties. However, this effect is unstable and lasts only a short period because the silver salts are easily dissolved in thermal aqueous-soapy treatments used in the service of textile materials.

The crazing method can be used to add modification components to the volume of the polymer and by this the method differs greatly from the surface treatment method. It is also very important to note the fact that to modify the polymers by the crazing method it is not necessary to have active functional groups in the modified polymer. The modification agent is not fixed in the volume by chemical interaction of the active functional groups of the polymers in the addition and is fixed by mechanical capture by the structure of the deformed polymer as a result of the comparability of the nanopores and molecular dimensions of the added agent [18].

Consequently, the first task is to select the AAM capable of not only crazing the modified polymer but also of carrying the required modification additions. Therefore, to produce the nanocomposites and add metallic silver to the PET films or fibres, they were deformed in a water–alcohol solution of $AgNO_3$. The presence of isopropyl alcohol ensured the deformation of polyethylene terephthalate by the crazing mechanism so that the polymer contained a porous nanostructure filled with the modification solution, containing $AgNO_3$. The samples were then dried to remove the liquid medium and produce the polyethylene terephthalate film (fibre), containing the required precursor (AgO_3) for subsequent formation of metallic silver in the polymer structure.

The concentration of the low-molecular compounds, added from the solution to the crazed polymer, is determined in the general case by the porosity of the polymer and by the content of the modification agent in the solution. The porosity of polyethylene terephthalate was close to 50%, and the increase of the $AgNO_3$ concentration in the solution from 0.05 to 2.5% increased the silver content in the nanocomposite from a trace amount to 2.5 wt.%.

It is well-known that the decomposition of silver nitrate to metallic silver may take place both in the thermal annealing conditions and under the effect of light. The reaction in the free volume takes place in accordance with the scheme:

$$2AgNO_3 = 2Ag\downarrow + 2NO_2\uparrow + O2\uparrow$$

The duration of ultraviolet radiation was selected in advance on the basis of the thickness of the colour layer of the film. In ultraviolet radiation, the milky white film of polyethylene terephthalate, containing $AgNO_3$, gradually became brown which is evidently associated with the release of metallic silver in its structure.

Confirmation that the process of ultraviolet irradiation of the film with $AgNO_3$ is accompanied by a chemical reaction with the formation of pure silver is obtained by x-ray diffraction analysis. The appearance of the maximum in the range $2\theta = 32.8$ on the diffraction pattern relates to the crystal phase of silver.

Previously [18], the structure of the crazed polymers was studied by scanning electron microscopy but only the transmission electron data can be used to evaluate the fine structure of the produced nanocomposite. In [45] structural–morphological studies were carried out by transmission electron microscopy using ultrafine sections.

Figure 12.6 *a* shows the micrograph of the ultrafine section of a polyethylene terephthalate nanocomposite containing 2.5 wt.% of silver. As shown by the micrographs, the silver particles are almost spherical. Figure 12.6 *b* shows the electron diffraction pattern of the same section of the polyethylene terephthalate sample containing 2.5 wt.% of silver. It may be seen that the diffraction pattern has the form of a single diffuse diffraction ring. The absence of maximum of higher orders is associated with the small size of the silver particles (approximately 5 nm) in which silver can not produce a lattice with a distinctive long-range order.

Thus, using the polyethylene terephthalate films as examples, it was shown that the crazing method can be used to add large quantities of metallic silver to the volume. The produced nanocomposites are characterised by a high level of mutual dispersion of the components and have a stable bactericidal effect [46].

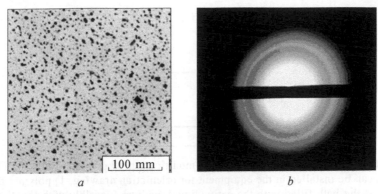

a   *b*

Fig. 12.6. Micrograph of the ultrafine section of the polyethylene terephthalate–silver film (2.5 wt.%) (*a*) and the microdiffraction pattern from the ultrafine section of the polyethylene terephthalate film with silver (*b*) [45].

12.3. Technological aspects of polymer modification by crazing

Thus, the crazing of polymers in the AAM is an universal method of adding modification agents to the polymers. An important distinguishing special feature of the production of composites based on crazed polymer matrices is the fact that, as already mentioned, crazing of the polymer is one of the types of inelastic plastic deformation of the polymer. This deformation is the basis of the method used widely in the production of polymer fibres and films – orientation drawing. Naturally, as a result of the considerable importance of this process, there are efficient facilities available for the orientation drawing of the polymers. At present time, the industry uses a very large number of high-speed machines for orientation drawing of the films and fibres in the continuous mode. Modification of the polymers by crazing in liquid media consists of the stage of orientation drawing of the polymer films and fibres with the addition of the modification agent. For this purpose, it is necessary to ensure orientation of the polymer in the solution of the modification agents in the AAM by including in the traditional technological procedure a relatively simple device shown schematically in Fig. 12.7 [19].

Because of the above reasons, many successful attempts have been made [48, 49] to carry out crazing of the polymer in the continuous mode. In other words, it is possible to produce nanocomposites in the form of polymer films and fibres having both the high mechanical parameters and different valuable properties (electrical conductivity, incombustibility, electrostatic properties, etc.) resulting from the

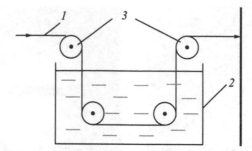

Fig. 12.7. Schematic of the device for modification of polymer fibres by crazing which can be installed in the equipment for orientation drawing: 1) polymer film of fibre; 2) the bath filled with the active liquid medium (possibly with the dissolved addition); 6) feed and receiving rollers [47].

addition of target agents. In particular, it is important to note that these composites are produced using the currently available equipment, in the continuous mode, with high speeds, characteristic of the advanced production of polymer films and fibres.

12.4. Methods for increasing the efficiency of crazing

Thus, the most important stages of crazing of the polymers in the liquid medium from the applied viewpoint are the formation and growth of crazes and also the collapse of the porous structure of the polymer. The first stage is associated with the formation and further development of porosity. This stage determines not only the general porosity of the resultant polymer but also the maximum possible amount of the target addition which can be 'supplied' to the volume of the polymer. Actually, the amount of the addition depends on the volume fraction of the resultant pores because the developing porosity is continuously filled with the solution of the addition in the AAM. The second stage – the collapse of the porous structure – ensures the 'capture' of the addition molecules in the porous structure of the polymer and its strong 'closing' in it (Fig. 12.2).

As shown previously, the initial stages of deformation of the polymer in the AAM result in the development of crazes of the polymers which are continuously filled with the surrounding liquid component. However, this is followed by the collapse of the developed porous structure of the polymer and the direction of mass transfer of the liquid changes to the opposite. This is followed by the syneresis of the liquid from the polymer structure and, evidently, in this stage there are no new portions of the surrounding liquid transferred to the polymer.

The amount of the modification agent, dissolved in the AAM, depends in particular on the time when the previously mentioned collapse of the porous structure starts after deformation. In other words, if the collapse takes place in relatively early stages of deformation, the general porosity, obtained by the polymer at this moment, will be small and, consequently, the amount of the addition will not be large. However, if the collapse starts at relatively high strains, the porosity of the polymer continues to increase for a relatively long time and the amount of the addition which fills the porous structure of the polymer will be larger.

Evidently, the larger the amount of the target addition which can be added to the polymer the higher the efficiency of modification of the polymer. Thus, the efficiency of crazing of the polymer in the liquid medium, which depends on the porosity which can be obtained in this process and, correspondingly, on the amount of the addition, is determined by two competing processes: development of porosity and the collapse of the resultant porous structure.

It is therefore necessary to decide whether the above processes can be influenced to some extent and, consequently, whether the efficiency of crazing as a universal method of modification of the polymer can also be affected. For this purpose, it is necessary to investigate the main factors influencing the start of the collapse of the porous structure of the polymer deformed in the AAM. This characteristic can be easily recorded by measuring the thickness of the film deformed in the AAM because, as mentioned previously, the collapse of the porous structure is accompanied by the decrease of the thickness (contraction) of the deformed polymer, visible by the naked eye.

Figure 12.8 shows the dependence of the relative change of the thickness of the polyethylene terephthalate sample on the draw ratio in the AAM for the samples with different initial thickness [50]. It may clearly be seen that the thickness of the sample starts to decrease at different rates as a result of the collapse for the films with different initial thickness. As the thickness of the initial film increases, the time to the start of the collapse increases with other conditions being equal, and vice versa. Evidently, as the initial film becomes thicker, the porosity which can be produced in the film increases because the start of the collapse indicates that no further portions of the AAM are supplied to the polymer. In turn, this means

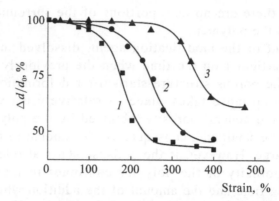

Fig. 12.8. Dependence of the relative change of the thickness of the polyethylene terephthalate sample on the tensile strain in the AAM. Thickness of the samples 50 (1), 100 (2) and 300 (3) μm [50].

that as the thickness of the initial film increases, the efficiency of the crazing becomes higher because, with other conditions being equal, the porosity produced in the polymer and filled with the AAM or with the solution of the target addition becomes greater. As the volume of the surrounding solution 'supplied' to the polymer increases, the amount of the addition found in the resultant product also increases.

It is important to discuss the mechanism of this phenomenon. Direct microscopic studies, carried out in [18, 19], can be used to explain the mechanism of these phenomena. In these studies it was shown that a considerably larger number of crazes develops and grows in a thicker film. This circumstance has very important consequences. The point is that the collapse of the porous structure of the craze can take place when the walls of the craze move away from each other to a relatively large distance or, which is the same, when the length of the fibrillar aggregates, connecting their walls, is sufficiently large. The increase of the length of the fibrils makes the ratio of the length to the diameter so large that these aggregates of the macromolecules become quite highly flexible. In particular, the flexibility, i.e., the capacity to change the form, enables these aggregates to 'catch up' with each other, come into contact, coalesce and ensures the general collapse of the porous structure.

Evidently, as the number of the crazes decreases the length (at the same tensile strain) of the fibrillar aggregates of the macromolecules increases and the time to the start of the collapse of the porous structures becomes shorter. In other words, as the crazes become wider, they are close to collapse, and vice versa. The reasons for this phenomenon may be described as follows. Loading of the polymer in the AAM is accompanied by the nucleation of a certain number of crazes which start to grow in the direction normal to the applied stress axis (chapter 6).

As shown in [18], the process of nucleation continues until sufficiently high stresses are maintained in the sample. The stress starts to decrease when some of the crazes growth through the entire cross-section of the deformed polymer. Since the growth rate of the crazes is almost completely independent of the thickness of the sample, it is evident that the growth time is considerably longer in the specimens with the larger cross-section. This means that in a thicker sample, with other conditions being equal, the crazes intersect the cross-section in a time considerably longer than in a thin sample. Consequently, the thicker sample is subjected to higher stress for a considerably longer time than the thin sample. As the time during which the polymer is subjected to high stresses increases, the duration of nucleation of the new crazes will take place longer and the number of the new crazes in the resultant polymer will be greater.

One of the factors influencing both the number of the nucleated crazes and the efficiency of crazing are the geometrical special features of the polymer sample. Some other factors which determine the number of the nucleated crazes in the polymer were investigated previously (chapter 6). One of these factors is the duration of the effect of mechanical stress. An even stronger effect on the processes of nucleation of the crazes is exerted by the absolute value of the applied stress. Another most important factor, which determines the processes of nucleation of the crazes and the efficiency of crazing, is the capability of the active liquid to lower the surface tension of the polymer (chapter 6).

Thus, to increase the number of crazes, crazing (tensile loading of the polymer in the AAM) should be carried out at the maximum stress. This aim is achieved by regulating the tensile loading rate of the polymer. Actually, the number of the crazes formed in stretching of the polymer films and fibres in the physically active liquid media increases with increasing tensile loading rate and this is associated with the increase of the deformation stress. Depending on the loading rate of the polymer or, which is the same, the stress in the polymer, the number of resultant crazes may vary from hundreds or thousands per millimetre of the length of the sample [51] to one craze for the entire sample [52].

However, high tensile loading rates cannot be used in all cases for stretching the polymers in the medium to carry out modification. The experimental results show that the maximum possible drawing rate is limited by the flow speed of the liquid to the active deformation zone [53]. For any polymer–liquid medium pair there is a critical tensile loading rate; above this rate the crazes do not form and the polymer is deformed through a neck, as is the case in stretching in air. This phenomenon is explained by the fact that to ensure effective crazing, the AAM must be supplied in sufficient quantities to the areas of active deformation of the polymer (to the tips of the growing crazes). Naturally, as the viscosity of the physically active liquid increases, the tensile loading rate resulting in the critical drawing conditions under which crazing appears decreases.

All the above-mentioned factors are determined by the characteristic properties of the polymer–liquid medium system: the geometry of the sample, mechanical stress, surface activity of the liquid medium. At the same time, as shown previously, the number of the resultant crazes is associated with the defectiveness of the surface of the stressed polymer.

Evidently, the number of defects can be influenced by introducing artificial nuclei in the surface layer of the deformed polymer and consequently affecting the number of resultant crazes. Therefore, a number of attempts have been made to produce the nucleation surface defectiveness which may influence the number of crazes formed in the polymer during its deformation in the AAM. This effect will be described by the deformation of a polyethylene terephthalate fibre in the AAM. Figure 12.9 shows the micrographs of a polyester fibre, stretched to different strains in the AAM. The initial fibre diameter is approximately 25 μm. In this case, the collapse started almost immediately because the 'loose' crazed structure does not form. It appears that the fibre is deformed with the formation of a neck so that tensile loading causes a change only in the ratio of the deformed and non-deformed parts of the fibre. Nevertheless, the process develops in this case by the crazing mechanism. This is indicated unambiguously by optical microscopy data (Fig. 12.9 *d*). It may be seen clearly that the collapsed parts of the fibre are distinctively coloured if a contrasting dye is added to the AAM.

One of the methods which can be used to regulate the number of nucleated crazes is the so-called method of preliminary nucleation

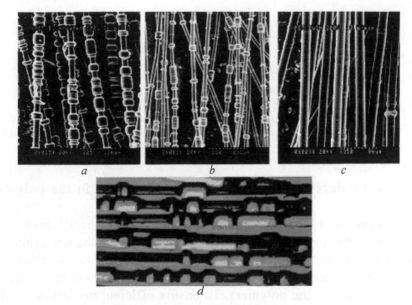

Fig. 12.9. Scanning electron micrographs of polyethylene terephthalate fibres stretched in the AAM (*n*-propanol) by 100 (*a*), 150 (*b*) and 300 (*c*) %; light microscopy of polyethylene terephthalate fibres stretched by 150% in the AAM (*n*-propanol) in which the contrasting dye (rhodamine C) was dissolved (*d*).

of crazes [54]. This method is based on the well-known phenomenon of 'dry crazing'. The point is that when holding the polymer at a sufficiently high constant load (slightly below the yield stress in air) for an hour or longer, the surface of the polymer is covered with a huge number of the so-called dry crazes. The structure of these crazes is identical with that of the crazes formed in liquid media. However, since the growth rate of the crazes in air is extremely low, these crazes grow only to a small depth into the polymer (of the order of several microns) and form a set of nucleation microdefects on its surface. Subsequent deformation of this sample in the AAM is accompanied by the nucleation and growth of a very large number of crazes. For example, in [54] it was shown that the method of preliminary nucleation doubles the number of crazes in polyethylene terephthalate (from 450 to 800 mm^{-1}).

The above method is in fact highly efficient from the viewpoint of regulating the number of nucleated crazes. However, it is completely unsuitable from the technological viewpoint. In fact, it is difficult to imagine how the currently available continuous high-speed processes of formation and stretching of the polymer fibres can be combined with the stage of holding the polymer in the stress state for several

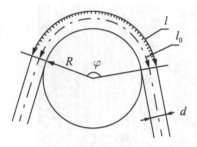

Fig. 12.10. Schematic representation of a separate section for nucleated additional crazes on the surface of the polymer [19].

hours to produce the required number of defects in the polymer surface.

Therefore, in [49] it was attempted to develop a special device for nucleating efficiently large numbers of crazes under the conditions of continuous stretching of polymer fibres and films. In this method, the crazes are nucleated in the low strain range (up to and in the range of the yield stress of the polymer). To ensure efficient nucleation of the crazes, prior to the stage of main stretching of the polymer the latter should be subjected to a low strain in the AAM. This is achieved by pulling the fibre around a circular template with a small diameter (Fig. 12.10). When bending the fibre in the AAM its external surface in relation to the template is subjected to tensile strain, and the internal surface to compressive strain. It is assumed that a fibre with the diameter d repeats the surface of the template with radius R at the distance of the angle φ and, therefore, the external surface of the fibre is subjected to tensile strain $\varepsilon = (l - l_0)/l_0$, where $l_0 = (R + d/2)\varphi$; $l = (R + d) \varphi$. Consequently, the strain to be determined is

$$\varepsilon = (l - l_0)/l_0 = d/(2R + d) \qquad (12.1)$$

The crazes nucleate in the conditions in which the polymer is subjected to the strain corresponding to the yield stress of the polymer. For example, the yield stress of polyethylene terephthalate is approximately 5–7%, and in polypropylene it is approximately 40%. Knowing the yield stress and the fibre diameter, equation (12.1) can be used to select quite easily the diameter of the template R at which the external surface of the fibre is deformed to the value equal to the yield stress. The required number of crazes can be produced on the polymer surface by multiple bending of the fibre in the AAM using the device shown schematically in Fig. 12.11.

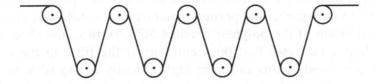

Fig. 12.11. Diagram of the device for nucleated crazes in the conditions of continuous drawing of the polymer fibre (film) [19].

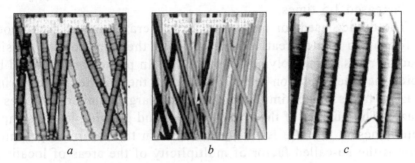

a   b   c

Fig. 12.12. Electron micrographs of polyethylene terephthalate fibres stretched by 50% in the AAM: a) the fibre stretched in the AAM using the device shown schematically in Fig. 12.7; b, c) prior to stretching the fibre was treated in the device shown schematically in Fig. 12.11 [49].

We determine how the preliminary nucleation of the crazes influences the deformation of the polymer in the AAM. Therefore, the polyethylene terephthalate fibre was stretched in the continuous mode in the AAM by 50% using the device shown in Fig. 12.7. Figure 12.12 a shows the external appearance of the polyethylene terephthalate fibres deformed under these conditions. It may be seen clearly that, regardless of the relatively low draw ratio, relatively large constrictions are detected in the fibre. These constrictions are determined by the collapse of the porous structure of the polymer as a result of large opening of the individual crazes. This effect is caused first of all by the small number of the nucleated crazes so that they open to a relatively large value. Evidently, the collapse of the porous structure under these conditions reduced the porosity and, consequently, the efficiency of crazing.

At the same time, the fibre treated in the device shown in Fig. 12.2 b prior to stretching does not contain at first sight crazes because there are no areas with a smaller diameter. However, under large magnification (Fig. 12.12 c) it may be seen that the fibre contains a very large number of the crazes which, however, have not grown to the collapsed stage so that there are no changes in the fibre diameter. In particular, because of the large number of crazes there is no collapse of the porous structure of the polymer, although the total strain of the polymer reaches 50%. In turn, the absence of the collapse indicates that the treatment of the fibre in the device shown previously helps to retain high porosity of the fibre to high strains. According to these experimental data, the porosity of the polymer increases 1.5 times, with other conditions being equal, and, consequently, the amount of the addition added to the polymer can be increased 1.5 times.

The device used for nucleation of the crazes in the continuous deformation mode greatly increases also the range of the tensile loading rates of the polymer in the AAM. In particular, it should be noted that the transition of the polymer to the oriented state during crazing takes place simultaneously in a large number of zones – crazes. The number of the crazes formed and growing in the sample determines in turn the level of the stress in the deformed polymer. This is the so-called factor of multiplicity of the areas of localized plastic deformation [55]. This factor can be described as follows. If the volume is transferred to the oriented state in the localized zone plastic deformation (crazes), the rate of this transition depends on the number of transition areas (crazes). This means that at a fixed strain rate of the sample the rate of transformation of the polymer to the oriented state in each craze decreases with increase of the number of these localized transition zones to the oriented state (crazes).

Previously, it was mentioned that the efficiency of the effect of the AAM depends in particular on the tensile loading rate of the polymer. At sufficiently high tensile stresses the crazing can be completely suppressed because for the crazing to take place the liquid must penetrate efficiently and in a sufficient quantity to the areas of active deformation of the polymer. At high strain rates the liquid is not capable of this and crazing is suppressed. As mentioned previously, the increase of the number of crazes lowers the actual rate of transformation of the polymer to the oriented state and, consequently, widens the tensile loading rate range of the polymer in the medium in which the polymer is deformed by the crazing mechanism

Thus, detailed studies of the crazing mechanism of the polymers in liquid media leads to an efficient universal method of modification of the polymers and to producing an almost unlimited range of the new types of polymer materials. The above-described analysis of the experimental data shows that crazing can be used as the universal method of adding modification agents to the polymers. This method is based on completely different mechanisms of 'supply' and retaining the modification agent in the polymer structure. The modification agent is not supplied by diffusion and is delivered by a much faster method of matter transfer by the viscous flow. To maintain (fix) the addition in the structure of the fibre, it is not necessary to have the low-molecular component of the active functional groups in the polymer capable of interacting with each other. Fixation is carried out by mechanical capture of the low-molecular component as a result of the comparability of its molecular dimensions with the size of the pores. The latter circumstance widens almost without bounds the range of additions that can be added. For example, by crazing the polymer in the AAM it is possible to colour efficiently hydrophobic polymers such as poplypropylene with water-soluble vat dyes.

We will sum up briefly the main advantages of the crazing in comparison with the traditional methods of adding the modification agents to the polymer films and fibres.

1. The modification agent is not added to the volume of the polymer by diffusion and is added instead by the viscous flow of the solution of the addition through the system of the microscopic pores characteristic of the structure of the craze. Since the transport rate by the viscous flow is considerably higher than in diffusion, the rate of the process of modification by crazing is sufficiently high to create a continuous, high-speed process of adding the low molecular agents by stretching the polymer in the appropriate liquid media. In fact, this makes it possible to combine the process of orientation stretching of the polymer with the process of adding modification agents. It is very important to note that to implement this process it is not necessary to develop new equipment and the currently available equipment for stretching polymer films and fibres can be used after slight modification.

2. Since the fixation of the addition in the fibre structure does not require any strong into molecular interaction between the polymer and the added agent, the range of the agents that can be added widens without borders. In particular, since the process is performed at low (room) temperatures, there are also no restrictions when adding thermally unstable additions and this also increases the range of the additions that can be used.

3. It is also important to note that any combination of the modification additions (for example, antipyrine and the dye, etc.) can be added to the polymer fibres in a single technological operation.

12.5. Producing the transverse microrelief in polymer fibres and films

The possibilities of practical application of crazing are not exhausted by the development of new types of polymer sorbents, membranes or nanocomposites. This phenomenon helps to solve another important applied problem. Crazing can be used to produce the transverse relief in the polymer films and fibres [18].

Generally, the formation of the relief is not a difficult problem in the case of fibres and films produced from the melt. This can be carried out using a die with the appropriate profile. However, in this case, the fibre has a longitudinal relief. At the same time, the natural fibres that are most valuable from the viewpoint of user properties have the distinctive transverse relief. As shown previously, in the first tensile loading stage the nucleated crazes propagate in the polymer in the direction normal to the axis of tensile stress. In particular, this circumstance can be used to produce fibres with a distinctive transverse relief. In this case, it is sufficient to nucleate and slightly grow the crazes by stretching the polymer in the AAM. If subsequent drawing is carried out in air with the formation of a neck, a transverse relief forms on the surface of the polymer. Figures 12.13 *a, b* show typical examples of the relief of the natural wool fibres. It is possible that the valuable user properties of these fibres are determined in particular by this specific relief. Using crazing, the transverse relief can be easily produced in the continuous mode (Fig. 12.13 *c, d*).

Finally, it is necessary to mention another practical application of polymer crazing. It is the formation of low-molecular nanoporous systems. As shown previously, crazing results in the formation of a system of mutually connected nanosized pores in the polymer. If this porous structure is filled with the low-molecular component capable of hardening *in situ*, a strong three-dimensional network of the second component will form in the crazed structure of the polymer. The removal of the polymer matrix helps to produce this network in the free state.

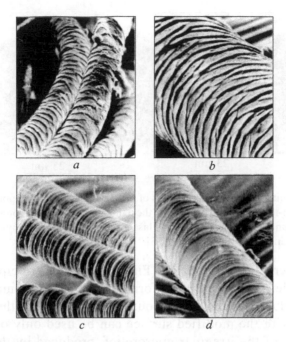

Fig. 12.13. Scanning electron micrographs of the natural wool fibres of sheep (*a*), dogs (*b*) and synthetic polyethylene terephthalate fibres, produced by the crazing mechanism (*c, d*) [18].

This approach to forming low-molecular nanosystems will be explained on a specific example. In [56–58] the crazed polymer films were filled with precursors which were used to produce oxides of silicon [56, 57] and titanium [57] by chemical transformations *in situ* (Fig. 12.14). The organic component was then fully removed by burn-out of the produced nanocomposite at a high temperature (~700°C). This treatment produced inorganic highly dispersed systems based on SiO_2 and TiO_2.

These inorganic materials have high porosity (up to 50%) and a large specific surface (hundreds of m^2/g) and can be used, for example, as a sorbents, carriers for catalysts, etc.

12.6. Practical application of polymer films with a regular microrelief

The modification of the polymer surfaces to produce a regular microrelief on them is highly promising for application in practice. We have already discussed the possibilities of producing a microrelief

a

Fig. 12.14. Scanning electron micrographs of brittle cleavage of nanoporous particles, produced by high-temperature treatment of the nanocomposites based on polypropylene $-SiO_2$: *a*) inorganic plate produced by classic crazing of polypropylene; *b*) using delocalized crazing of polypropylene [56].

on the polymer fibres by crazing (Fig. 12.13). The principal value of this microrelief is that it can be formed normal to the main axes of the fibre produced in the continuous mode. However, the polymers of this type with the modified surface can be used only on a limited scale because of the irregular microrelief, produced by this method.

At the same time, the demand for polymer films with a regular microrelief is very large. These materials can be used as templates for the ordered distribution of the nanoparticles on the hard surface [59], as substrates for orientation of liquid crystals [60, 61], flexible substrates for electronics [62], extensible electrodes and electrical connectors [63, 64], etc.

It is important to know that many properties of the solids are determined not only by the chemical composition but also by the micro/nanorelief resulting in unique properties, such as super hydrophobicity, self-cleaning, etc., so that these materials can be used in greatly differing areas of science and technology [65–69].

Another important application is the production of polymer transparent films with a regular microrelief for developing a new generation of liquid crystal (LC) displays. The display of this type based on a polymer is strong, light, flexible and unbreakable so that it is more suitable for application in practice. Figure 12.5 shows a photograph providing information on the properties of these displays. This photograph shows clearly the advantages of the flexible display and, undoubtedly, these displays will compete efficiently in future with the traditional displays with a glass base.

In general, the liquid crystal displays is a multilayer device constructed of a large number of optically transparent layers.

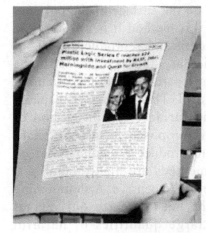

Fig. 12.15. A photograph of a flexible polymer-based display [70].

Blue Green Red

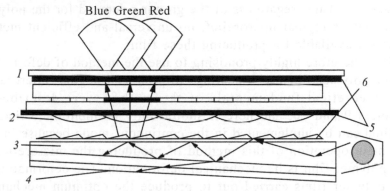

Fig. 12.16. Structure of a diffractive LC display with substrates with a microrelief of two types: 1) diffuser, 2) cylindrical microlenses, 3) light guide, 4) reflector, 5) polarisers, 6) grating [71].

The liquid crystal display contains elements for controlling the light fluxes, such as diffraction gratings, elements for the uniform distribution of illumination of the screen, elements for homogenising the light flux, and many others. These devices can be studied in greater detail in, for example, [71].

Figure 12.16 shows schematically the structure of the liquid crystal display of this type. In this monograph we cannot discuss the design features of such a complicated device as the liquid crystal display, but it is evident that its functioning requires transparent polymer films with a regular microrelief. It is therefore understandable why the effort of many researchers and engineers is directed to the development of methods of producing a regular microrelief on the polymer components.

At present, the periodic microrelief on the surface of the polymer is produced either by rubbing (but in this method it is not possible to ensure the required periodicity, depth and reproducibility), or by other expensive methods. In particular, study [72] describes a method of formation of a saw-like microrelief using a diamond cutting machine with programmed control. This machine is capable of producing millions of scans on the surface with the period of the fraction of a micron. In [73] it is proposed to use the sharp edge of the atomic force microscope to produce a regular relief on the surface of a polymer film. Both methods are highly laborious, require special equipment and are therefore expensive. A shortcoming of the chemical etching method is the need to carry out several photolithography operations using large quantities of harmful substances. Thus, regardless of the growing demand for the polymer films with a regular microrelief, no universal and efficient method is as yet available for producing these films.

It is therefore highly promising to use the method of deformation of the polymer films with coatings to produce a regular microrelief. The theoretical fundamentals of the method were described in chapters 9 and 10, and here it should only be mentioned that the method can be implemented in the continuous mode because in fact the main operation of this method of producing the microrelief on the polymer film is its mechanical deformation. The deformation of the polymer films carried out to produce the optimum mechanical properties (orientation drawing) is a well-developed method [74] and high productivity equipment has been constructed for this method.

The production of the regular microrelief by deformation of the polymer film with a thin hard coating does not present any serious problems [75]. In this case, it is sufficient to simply stretch the film so that a regular and very effective microrelief forms spontaneously on the surface of the film. Figure 12.17 shows this phenomenon. It may clearly be seen that the resultant microrelief is highly regular. This is indicated in particular by the presence of discrete equatorial refractions on the small angle diffraction diagram (Fig. 12.17 c). In other words, the film with the hard coating transforms during simple tensile loading to a sufficiently perfect diffraction grating.

The process of formation of the regular microrelief, with the example shown in Fig. 12.17, was implemented by deformation and shrinkage of the film with coatings [76, 77]. An additional advantage of this method is the fact that the initial material can be represented by the industrially produced biaxially oriented films (polyethylene

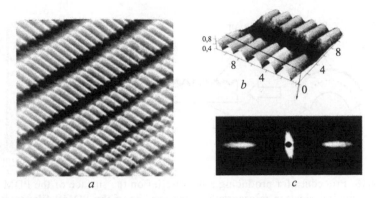

Fig. 12.17. Scanning electron micrograph (*a*), the AFM image (*b*) and the small angle diffraction pattern (*c*) of polyvinyl chloride samples with a thin (10 nm) platinum coating after 100% tensile loading at 90°C.

terephthalate, polypropylene) with a thin aluminium coating. In the continuous mode these films can be subjected to drawing, shrinkage or isometric annealing. In each of these cases a regular microrelief forms on the film surface, and the period and orientation of the microrelief can be easily regulated. The only serious shortcoming of the method is the need to remove the metal (aluminium) from the polymer surface after completing drawing (shrinkage) of the initial film. Dissolution of the aluminium coating in corrosive media (alkali, acid) results in the formation ecologically harmful aluminium solutions which must be neutralised.

A shortcoming of this method was overcome in [78] in which the initial product was represented by polymer films without metallic coatings treated with cold plasma. As shown previously (chapter 10), this treatment results in the formation on the polymer surface of a thin layer of a hard, visibly cross-linked but optically transparent surface layer. As a result of subsequent drawing the product becomes completely transparent, has a regular microrelief and does not require treatment with solvents.

It should be mentioned that the method described previously has been used quite widely for producing a microrelief on the surface of a rubbery polymer polydimethylsiloxane (PDMS) [79, 80]. This polymer can be easily oxidised on the surface in treatment with gases or plasma [81, 82] leading to the formation of a hard surface layer.

As an example of this type we consider the data in [83] where a bend plate (PDMS film) was treated on one side with a mixture of sulphuric and nitric acids. This type of treatment resulted in the formation of a hard layer of the oxidised PDMS on the PDMS

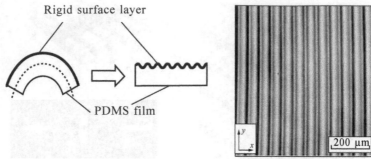

Fig. 12.18. Procedure for producing a microrelief on the surface of the PDMS film (*a*) and scanning electron micrograph of the surface of the PDMS film treated by the procedure *a* (*b*) [83]. Explanation in the text.

surface. Subsequent straightening of the PDMS films results in compression of the hard surface layer (Fig. 12.18 *a*). As a result of this compression the coating on the PDMS surface loses its stability and acquires a distinctive regular microrelief in complete agreement with the theoretical considerations presented in chapter 10 (Fig. 12.18 *b*). It is important to note that changing the duration of treatment of the PDMS surface with the oxidation agent, it is possible to regulate in a wide range the thickness of the hard layer formed on the polymer surface and, consequently, the period of the resultant microrelief.

The above results indicate that there is a universal method for producing the polymer films and fibres with a regular and, most importantly, regulated microrelief on the nanometric scale.

In conclusion, it should be mentioned that the deposition of the hard coating on the polymer films has many possible applications in practice, in addition to the formation of the regular microrelief on the polymer films. Firstly, the procedure can be used to visualise and study structural rearrangements of the deformed polymer (chapter 1) [84–88]. The information, obtained using this method, may be used to characterise in particular the mass transfer processes from the surface to the bulk and back in different types of deformation of the polymer which previously was not possible.

Secondly, the deposition of the coating on the polymers makes it possible to evaluate the properties of solids in cases in which this evaluation is difficult or impossible. Studies of the properties of the polymer films with the coatings allow the experimental evaluation of the mechanical parameters of the thin polymer layers [89–91] and other solids, deposited on polymer substrates [75, 92, 93]. In both cases we are concerned with the objects in the nanometric range. The evaluation of the structural–mechanical properties of such thin layers of the solids is of considerable scientific and practical importance but at the present time it is associated with considerable difficulties so that it is very important to develop approaches to solving this problem.

Finally, this approach can be used to evaluate the mechanical characteristics of solids of grandiose dimensions, such as the earth crust [94–97] which cannot be carried out by the currently available methods.

Conclusions

Thus, the set of the phenomena taking place at the interfacial boundaries of the solid polymers is of considerable applied potential. One of the aspects of this scientific direction is the practical application of polymer crazing in liquid media as an universal method of producing a new class of nanocomposites based on polymers. Firstly, the crazing of the polymers can be regarded as a process of self-dispersion of the polymer to the nanosized aggregates of the oriented macromolecules, separated in space by the microcavities of the same size. Secondly, crazing is also a method of supplying almost any low-molecular compounds to the nanoporous structure of the polymer. Finally, the collapse of the nanoporous structure of the crazed polymer in later stages of its deformation solves the problem of fixing the addition in the polymer.

All the three essential conditions of producing the nanocomposites are solved in deformation of the polymer in the AAM, and the polymer itself can be produced in a continuous high-productivity process. The experimental results show that it is possible to produce a wide range of the nanocomposites with a polymer matrix such as porous polymer sorbents, polymer separating membranes, new types of polymer–polymer nanomixtures, incombustible and electrically conducting polymer nanocomposites, metal-polymers, and a number of other materials.

Investigation of the processes of structure formation at the phase boundaries of the polymers with a hard thin coating (RCSS system) has a number of applied aspects. Firstly, this approach can be used to visualise the structural rearrangement of the deformed polymers and obtain new data on their deformation mechanism. Secondly, the deformation of the polymers with a thin hard coatings makes it possible to develop a new continuous process of producing the polymer films with a regular microrelief essential for producing light, unbreakable, flexible LC displays. Thirdly, the investigation of the processes of relief formation on the surfaces of the deformed polymer films with the coatings enables us to evaluate the deformation–strength properties of the solids of the nanometre range which is a very important applied in scientific task. Fourthly, this approach can be used to evaluate the solids of grandiose dimensions, such as the earth crust, which at present is not possible by any other method.

References

1.   Lazurkin S., Dissertation, Moscow, S.I. Vavilov Institute of Physical Problems.
2.   Bartenev G.M., Frenkel' S.Ya., Polymer physics, Leningrad, Khimiya, 1990.
3.   Kuleznev V.N., Shershnev V.A., Chemistry and physics of polymers, Moscow, KolosS, 2007.
4.   Electrical properties of polymers, Ed. V.I. Sazhin, Leningrad, Khimiya, 1970.
5.   Narisawa I., Strength of polymer materials, Moscow, Khimiya, 1987.
6.   Nazarov V.G., The surface modification of polymers, Moscow, MGUP, 2008.
7.   Volynskii A.L., Lopatina L.I., Arzhakov M.S., Bakeev N.F., Vysokomolek. Soed. A, 1987, V. 29, No. 4, P. 623.
8.   Volynskii A.L., Kozlova O.V., Bakeev N.F., Vysokomolek. Soed. A, 1986, V. 28, No. 10, P. 2230.
9.   Sergeyev G.B. Nanochemistry, Moscow, Moscow State University Press, 2003, 287 pp.
10.   Suzdalev I.P., Nanotechnology. Physical chemistry of nanoclusters, nanostructures and nanomaterials, Moscow, KomKniga, 2006.
11.   Andrievsky R.A., Ros. Khim. Zh. (ZhVKhO im. D.I. Mendeleeva), 2002, V. 46, No. 5, P. 50.
12.   Melikhov I.V., Ros. Khim. Zh. (ZhVKhO im. D.I. Mendeleeva), 2002, V. 46, No. 5, P. 7.
13.   Pomogailo A.D., Rosenberg A.S., Uflyand I.E., Metal nanoparticles in polymers, Moscow, Khimiya, 2000.
14.   Bronstein L.M., Sidorov S.I., Valetskii P.M., Usp. Khimii, 2004, V. 73, No. 5, P. 542-558.
15.   Ivanchev S.S., Ozerin A.N., Vysokomolek. Soed. B, 2006, V 48, No. 8, P. 1531.
16.   Kozlov P.V., Papkov S.P., Physico-chemical fundamentals of plasticizing polymers, Moscow, Khimiya, 1982.
17.   Volynskii A.L., Bakeev N.F., Highly dispersed oriented state of polymers, Moscow, Khimiya, 1985.
18.   Volynskii A.L., Bakeev N.F., Solvent crazing of polymers, Amsterdam, New York, Elsevier, 1996, P. 410.
19.   Volynskii A.L., Mikushev A.E., Yarysheva L.M., Bakeev N.F., Ros. Khim. Zh. (ZhVKhO im. D.I. Mendeleeva), 2005, V. 50, No. 6, P. 118.
20.   Volynskii A.L., Loginov V.S., Plate N.A., Bakeev N.F., Vysokomolek. Soed. A, 1980, V. 22, No. 12, P. 2727.
21.   Volynskii A.L., Loginov V.S., Plate N.A.,, Bakeev N.F., Vysokomolek. Soed. A, 1981, V. 23, No. 4, P. 805.

22. Loginov V.S., Dissertation, Cand. chemical sciences, M.V. Lomonosov Moscow State University, 1978.
23. Mel'nikov B.N., Dyeing of fiber, Encyclopedia of polymers, Moscow, Sov. entsiklopediya, 1972, V. 1, P. 1135.
24. Aiqin Hou, Kongliang Xie, Jinjin Dai J., App. Polym. Sci., 2004, V. 92, P. 2008-2012.
25. Rusanov A.I., 3rd International conference 'Chemistry of highly organised substances and the scientific basis of nanotechnology,' Abstracts of reports, St. Petersburg, 2001, 16 pp.
26. Volynskii A.L. Bakeev N.F., Vysokomolek. Soed., 2011, V. 53, No.7, P. 1203-1216.
27. Conducting polymers: Special applications, ed. by Alcacer L. Dordrecht, Reidel, 1987, P. 65.
28. Electronic properties of conjugated polymers, ed. by Kuzmany H., Berlin, Springer, 1987, P. 128.
29. Yarysheva L.M., et al., Vysokomolek. Soed. A, 1994, V. 36, No. 2, P. 363.
30. Saifullina S.A., Yarysheva L.M., Volkov A.V., Volynskii A.L., Bakeev N.F., Vysokomolek. Soed. A, 1996, V. 38, No. 7, P. 1172.
31. Saifullina S.A., Yarysheva L.M., Volynskii A.L., Bakeev N.F., Vysokomolek. Soed. A, 1997, V. 39, No. 3, P. 456.
32. Volynskii A.L., Arzhakov M.S., Karachevtseva I.S., Bakeev N.F., Vysokomolek. Soed. A. 1994. V. 36, No. 1. P. 93.
33. Volynskii A.L., Yarysheva L.M., Karachevtseva I.S., Bakeev N.F., Vysokomolek. Soed. A, 1995, V. 37, No. 10, P. 1699.
34. Kalinina O., Kumacheva E. Macromolecules. 2001. V. 34, No. 18, P. 6380.
35. Volynskii A.L., Bakeev N.F., Structural self-organization of amorphous polymers, Moscow, Fizmatlit, 2005.
36. Vinidiktova N.S., Borisevich I.V., Pinchuk L.S., Sytsko V.E., Ignatovskaya L.V., Khim. Volokna, 2006, No. 5, P. 34.
37. Weichold O., Goel P., Lehmann K., Moller M.J. App. Polymer Sci., 2009, V. 112, P. 2634-2640.
38. Volynskii A.L., Moskvina M.A., Volkov A.V., Zanegin V.D., Bakeev N.F., Vysokomolek. Soed., 1990, V. 32, No. 12, P. 933.
39. Stakhanova S.V., Nikonorova N.I., Zanegin V.D., Lukovkin G.M., Volynski A.L., Bakeev N.F., Vysokomolek. Soed. A, 1992, V. 34, No. 2, P. 133.
40. Stakhanova S.V., Nikonorova N.I., Lukovkin G.M., Volynskii A.L., Bakeev N.F., Vysokomolek. Soed. B, 1992, V. 33, No. 7, P. 28.
41. Stakhanova S.V., Trofimchuk E.S., Nikonorova N.I., Rebrov A.V., Ozerin A.N., Volynsky A.L., Bakeev N.F., Vysokomolek. Soed. A, V. 1997, V. 39, No. 2, P. 318.
42. Stakhanova S.V., Nikonorova N.I., Volynskii A.L., Bakeev N.F., Vysokomolek. Soed. A, 1997, V. 39, No. 2, P. 312.
43. Nikonorova N.I., Stakhanova S.V., Volynskii A.L., Bakeev N.F., Vysokomolek. Soed. A, 1997, V. 39, No. 8, P. 1311.
44. Nikonorova N.I., Stakhanova S.V., Chmutin I.A., Trofimchuk E.S., Chernavskii P.A., Volynsky A.L., Ponomarenko A.T., Bakeev N.F., Vysokomolek. Soed. B, 1998, V. 40, No. 3, P. 487.
45. Nikonorova N.I., Trofimchuk E.S., Semenova E.V., Volynskii A.L., Bakeev N.F., Vysokomolek. Soed. A, 2000, V. 42, No. 8, P. 1298.
46. Volynskii A.L., et al., Kolloidnyi zhurnal, 2010, V. 72, No. 4, P. 458-464.
47. Shelyakov O.V., et al., Method of production of polymer components on the basis of polyethylene terephthalate with antibacterial properties, Russian patent No.

2394948 on the application No. 2008126468, July 1, 2008, Publ. 20.07.2010, Bull. No. 20.

48. Volynskii A.L., et al., Khim. Volokna, 2006, No. 2, pp 46-50.
49. Guthrie R.T., Hirshman J.L., Littman S., Suckman E.J., Ravenscraft P.H., US Patent No. 4001 367, 1977.
50. Bakeev N.P., Lukovkin G.M., Marcus I., Mikouchev A.E., Shitov A.N., Vanissum E.B., Volynskii A.L., US Patent No. 5516 473 1996.
51. Volynskii A.L., Yarysheva L.M., Arzhakova O.V., Bakeev N.F., Vysokomolek. Soed. A, 1989, V. 31, No. 12, P. 2673.
52. Yarysheva L.M., Chernov I.V., Kabal'nova L.Yu., Volynskii A.L., Bakeev N.F., Kozlov P.V., Vysokomolek. Soed. A, 1989, V. 31, No. 7, P. 1544.
53. Sinevich E.A., Bakeev N.F., Vysokomolek. Soed. B, 1980, V. 22, No. 7, P. 485.
54. Volynskii A.L., Aleskerov A.G., Kechek'yan A.S., Zavarova T.B., Skorobogatova A.E., Arzhakov S.A., Bakeev N.F., Vysokomolek. Soed. B, 1977, V. 19, No. 5, P. 301.
55. Mironov A.A., Dissertation, Cand. chemical sciences, M.V. Lomonosov Moscow State University, 1994.
56. Volynskii A.L., Lukovkin G.M., Yarysheva L.M., Pazukhina L.Yu., Kozlov P.V., Bakeev N.F., Vysokomolek. Soed. A, 1982, V. 24, No. 11, P. 2357.
57. Trofimchuk E.S., et al., Rossiiskie nanotekhnologii, 2012, V. 7, No. 7-8, P. 18.
58. Volynskii A.L. Bakeev N.F., Nikonorova N.I., Trofimchuk E.S., Nesterov E.A., Muzafarov A.M., Olenin A.M., RF patent No.2320688 on the application No. 2006145417 on 21 December 2006, registered March 27 2008.
59. Volkov A.V., Polyansky V.V., Moskvina M.A., Dement'ev A.I., Volynskii A.L., Bakeev N.F., Dokl. RAN, 2012. V. 445, No. 3, P. 297.
60. Schweikart A., Fery A., Microchim. Acta, 2009, V. 165, P. 249-263.
61. Ohzono T., Monobe H.J., Colloid Interface Sci., 2012, V. 368, P. 18.
62. Voronina E.E., Yaminskii I.V., Volynskii A.L., Bakeev N.F., Dokl. RAN, 1999, V. 365, No. 2, P. 203-209.
63. Khang D.-Y., Jiang H., Huang Y., Rogers J.A., Science, 2006, V. 311, P. 208-212.
64. Watanabe M., Shirai H., Hirai T., J. Appl. Phys., 2002, V. 92, P. 4631-4637.
65. Lacour S.P., Jones J., Suo Z., Wagner S., IEEE Electron Device Lett., 2004, V. 25, P. 179-181.
66. Teixeira A.I., Abrams G.A., Bertics P.J., Murphy C.J., Nealey P.F., J. Cell Sci., 2003, V. 116, P. 1881-1892.
67. Chan E.P., Smith E.J., Hayward R.C., Crosby A.J., Adv. Mater., 2008, V. 20, P. 711-716.
68. Jiang X.Y., Takayama S., Qian X.P., Ostuni E., Wu H.K., Bowden N., LeDuc P., Ingber D.E., Whitesides G.M., Langmuir, 2002, V. 18, P. 3273-3280.
69. Uttayarat P., Toworfe G.K., Dietrich F., Lelkes P.I., Composto R.J., Biomed J. Mater. Res. A, 2005, V. 75A, P. 668–680.
70. Koch K., Bhushan B., Jung Y.C., Barthlott W., Soft Matter., 2009, V. 5, P. 1386-1393.
71. Jin Jang, Materials Today, 2006, V. 9, No. 4, P. 46–52.
72. Belyaev V.V., Opt. Zh., 2005, Vol 72, No. 9, P. 79.
73. Yamada F., Hellermark C., Taira Y., Proc. 1st Int. Display Manufacturing Conf., Seoul, 2000, P. 261.
74. Wen B., Mahajan M., Rosenblatt C., Appl. Phys. Lett., 2000, V. 76, P. 1240.
75. Slutsker A.I., Encyclopedia of polymers, Moscow, Sov. entsiklopediya, 1974, V. 2, P. 545.

76.   Volynskii A.L., Bazhenov S.L., Bakeev N.F., Ros. Khim. Zh. (ZhVKhO im. D.I. Mendeleeva), 1998, V. 42, No, 3, P. 57.

77.   Volynskii A.L., Bakeev N.F., Yarysheva L.M., Grokhovskaya T.E., Kulebyakina A.I., Bolshakov A.V., Olenin A.V., Method of creating microrelief on the surface of polymer films, Russian patent No. 2319556 of April 04, 2006, Bull. No. 8, 20.03.2008.

78.   Volynskii A.L., Bakeev N.F., Yarysheva L.M., Grokhovskaya T.E., Kulebyakina A.I., Bolshakova A.V., Olenin A.V., Method of creating microrelief on the surface of polymer films, Russian patent No. 2319557 of April 04, 2006, Bull. No. 8, 20.03.2008.

79.   Volynskii A.L. et al., Method of creating microrelief on the surface of polymer films (options), patent of the Russian Federation No. 2411258 dated 19 January 2011, Bull. No. 21, 27.07.2010.

80.   Hendricks T.R., Wang W., Lee I. Soft Matter., 2010, No. 6, P. 3701-3706.

81.   Hendricks T.R., Lee I., Nano Lett., 2007, V. 7, No. 2, P. 372.

82.   Chiche A., Christopher H., Stafford M., Cabral J.T., Soft Matter., 2008, V. 4, P. 2360-2364.

83.   Schweikart A., Fery A., Microchim Acta, 2009, V. 165, P. 249-263.

84.   Watanabe M., Mizukami K., Macromolecules, 2012, V. 45, P. 7128.

85.   Volynskii A.L., Grokhovskaya T.E., Kulebyakina A.I., Bolshakova A.V., Bakeev N.F., Vysokomolek. Soed. A, 2007, V. 49, No. 7, P. 1224-1238.

86.   Volynskii A.L. Grokhovskaya T.E., Kulebyakina A.I., Bolshakova A.V., Bakeev N.F., Vysokomolek. Soed. A, 2007, V. 49, No. 11, P. 1946-1958.

87.   Volynskii A.L. Grokhovskaya T.E., Kulebyakina A.I., Bolshakova A.V., Bakeev N.F., Vysokomolek. Soed. A, 2009, V. 51, No. 4, P. 598-609.

88.   Volynskii A.L. Yarysheva L.M., Bakeev N.F., Vysokomolek. Soed. A, 2011, V. 53, No. 10, P. 871-898.

89.   Volynskii A.L., Priroda, 2011, No. 7, P. 14-21.

90.   Stafford C.M., Harrison C., Beers K.L., Karim A., Amis E.J., Vanlandingham M.R., Kim H.-C., Volksen W., Miller R.D., Simonyi E.E., A buckling-based metrology for measuring the elastic moduli of polymeric thin films., Nat. Mater., 2004, August, 3(8), 545–50.

91.   Longley J.E., Chaudhury M.K., Macromolecules, 2010, V. 43, P. 6800-6810.

92.   Yang S., Khare K., Lin P.-C., Adv. Funct. Mater., 2010, V. 20, P. 2550-2564.

93.   Volynskii A.L., Bazhenov S.L., Lebedeva O.V., Ozerin A.N., Bakeev N.F., J. Appl. Polym. Sci., 1999, V. 72, No. 10, P. 1267.

94.   Volynskii A.L., Bazhenov S.L., Lebedeva O.V., Bakeev N.F., J. Mater. Sci., 2000, V. 35, No. 3, P. 547.

95.   Volynskii A.L., Bazhenov S.L., The European Physical Journal E - Soft Matter., 2007, V. 24, No. 4, P. 317–324.

96.   Volynskii A.L., Yarysheva L.M., Moses S.V., Bazhenov S.L., Bakeev N.F., Ros. Khim. Zh. (ZhVKhO im. D.I. Mendeleeva), 2006, Vol 50, No. 5, P. 126.

97.   Volynskii A.L., Priroda, 2007, V. 3, P. 10.

98.   Volynskii A.L. Bazhenov S.L., In: Advances in Materials Science Research, V. 7, Ed. M. Wythers, 2011, Nova Science Publishers Inc, Chapter 10.

Index

α-relaxation 99, 103, 104, 118, 121, 132, 297
α-transition 103, 117, 118, 121, 132, 155
β-relaxation 120, 121
β-transition 120, 121, 168

A

AAM (adsorption-active medium) 4, 5, 150, 151, 161, 181, 182, 183, 196,
 197, 198, 207, 208, 219, 221, 223, 225, 227, 228, 229, 230, 233,
 234, 236, 237, 239, 241, 243, 244, 245, 246, 247, 248, 249, 250,
 252, 253, 254, 255, 258, 259, 260, 261, 262, 265, 266, 268, 276,
 277, 278, 279, 280, 281, 282, 283, 284, 285, 286, 287, 288, 289,
 290, 291, 292, 293, 294, 295, 296, 297, 298, 300, 303, 305, 306,
 308, 309, 310, 311, 312, 313, 314, 315, 316, 320, 321, 322, 324,
 325, 328, 333, 335, 341, 342, 357, 358, 479, 480, 482, 483, 484,
 485, 486, 488, 493, 494, 495, 497, 498, 499, 500, 501, 502, 503,
 504, 505, 506, 507, 508, 515
ageing
 physical ageing 61, 110, 111, 114, 115, 116, 117, 118, 120, 121, 122, 123,
  124, 125, 134, 135, 136, 154
 thermal ageing 111, 120, 124, 137
anisodiametric solid 398, 399, 400
autohesion 41, 42, 44, 45, 208, 209

B

birefringence 82, 87, 127, 141, 142, 279

C

coefficient
 permeability coefficient 115
cold crystallisation 119, 301, 403
copolymer
 PS–PMMA copolymer 91
crazes 5, 33, 71, 72, 73, 74, 125, 140, 142, 145, 146, 147, 148, 150, 151,
 152, 160, 161, 162, 163, 164, 165, 166, 167, 168, 169, 180, 181,
 182, 183, 184, 185, 186, 187, 188, 189, 190, 196, 198, 207, 208,

211, 223, 229, 230, 232, 233, 234, 235, 236, 237, 238, 239, 240,
242, 243, 244, 245, 246, 247, 248, 249, 250, 251, 252, 253, 254,
255, 256, 257, 258, 259, 261, 262, 263, 264, 268, 269, 270, 271,
277, 278, 279, 280, 281, 283, 285, 286, 288, 289, 290, 291, 292,
293, 295, 296, 297, 298, 299, 300, 302, 303, 304, 305, 306, 308,
310, 313, 314, 315, 316, 320, 321, 322, 324, 325, 326, 327, 328,
329, 331, 333, 334, 335, 336, 341, 355, 357, 358, 401, 482, 483,
484, 485, 486, 492, 498, 500, 501, 502, 503, 504, 505, 506, 508
crazing 3, 4, 5, 7, 33, 67, 71, 72, 73, 74, 150, 160, 161, 163, 164, 166,
174, 180, 181, 185, 187, 188, 189, 190, 193, 196, 197, 198, 205,
206, 207, 208, 211, 212, 223, 229, 230, 231, 232, 233, 234, 237,
238, 239, 240, 241, 242, 243, 244, 245, 246, 249, 250, 251, 252,
253, 254, 256, 257, 258, 259, 260, 261, 262, 263, 264, 265, 266,
267, 268, 269, 270, 271, 272, 276, 281, 282, 284, 286, 287, 288,
289, 291, 293, 294, 295, 296, 297, 298, 299, 300, 301, 302, 303,
305, 310, 311, 312, 316, 317, 320, 321, 322, 332, 339, 341, 342,
343, 347, 348, 351, 355, 356, 357, 358, 359, 405, 406, 407, 479,
480, 482, 483, 484, 485, 487, 488, 489, 490, 492, 493, 494, 495,
496, 497, 498, 499, 500, 501, 502, 503, 505, 506, 507, 508, 509,
510, 515, 516
criterion
  Griffith criterion 254, 259, 270

D

deswelling 369
dewetting 50, 366, 367, 443
digitate aggregates 230, 366, 371
draw ratio 10, 11, 33, 69, 70, 73, 244, 245, 256, 261, 262, 277, 281, 284,
  287, 295, 298, 299, 312, 313, 314, 315, 316, 322, 324, 342, 347,
  417, 428, 434, 435, 484, 499, 505

E

effect
  Rehbinder effect 217, 218, 219, 220, 221, 223, 224, 225, 227, 228, 230,
    233, 234, 237, 238, 248, 249, 259, 260, 270, 271, 341
elastic aftereffect 126, 127
entropic elasticity 154, 187, 189, 200, 210, 291
equation
  Covey–Fergusson equation 113
  Eyring–Lazurkin equation 241
  Fox–Flory equation 84, 97
  Owens–Wendt equation 235
  Williams–Landel–Ferry equation 44
  Zhurkov equation 475, 476

Euler loss of stability 398
Eyring–Lazurkin approach 239

F

forced elasticity relaxation 153
forces
  surface tension forces 79, 80, 230
force softening 176, 179, 180, 181, 182, 187, 194, 200, 206, 210, 212, 232,
    276

G

Giant's Causeway 463
gutta-percha 177, 179

H

high-impact polystyrene 183, 184, 185, 211
Hookean section 125, 126, 127, 131, 134, 141

I

inhomogeneous swelling 28, 368, 370, 375
interpenetrating polymer networks 339, 345, 347
isometric annealing 3, 36, 37, 208, 513
isotactic polypropylene 197, 267, 268, 351

M

method
  DSC method 97, 113, 123, 149, 169, 299
  method of preliminary nucleation of crazes 502
  positron annihilation method 116, 117, 121, 122
model
  DeGennes reptation model 44
  Struik free volume model 117, 122

N

nanolayers 52, 419, 422

O

ortho-positron lifetime 82

P

peptisation 309, 310
phenomenon

Rehbinder phenomenon 4
plastic shear transformations 154
plasticizing 28, 65, 102, 218, 260, 267, 268, 423
PMMA 43, 44, 49, 50, 52, 54, 55, 57, 59, 60, 63, 65, 66, 67, 81, 84, 85,
      86, 87, 88, 91, 100, 113, 127, 128, 129, 130, 131, 141, 143, 144,
      145, 150, 167, 221, 252, 277, 294, 326, 328, 330, 344, 345, 346,
      348, 349, 351, 352, 353, 354, 372
polyaniline 355, 490, 491
polycarbonate 33, 34, 35, 61, 94, 97, 114, 117, 119, 121, 122, 123, 132,
      133, 160, 161, 182, 185, 186, 187, 188, 196, 221, 253, 277, 288,
      290, 294, 295, 296, 328
polydimethyl siloxane 69, 375
polyethylene 11, 14, 15, 17, 18, 30, 31, 32, 33, 36, 37, 53, 61, 62, 63, 92,
      102, 119, 122, 138, 139, 140, 146, 150, 151, 152, 156, 157, 158,
      160, 183, 190, 197, 204, 207, 208, 211, 219, 221, 225, 227, 228,
      234, 235, 236, 237, 239, 240, 241, 242, 244, 245, 246, 247, 248,
      250, 251, 253, 255, 260, 261, 262, 263, 267, 277, 278, 279, 282,
      283, 284, 288, 289, 290, 299, 300, 301, 302, 303, 304, 305, 306,
      307, 308, 309, 310, 311, 312, 313, 314, 315, 316, 321, 322, 323,
      324, 325, 328, 334, 335, 336, 337, 338, 343, 344, 348, 349, 350,
      355, 356, 357, 358, 377, 380, 381, 382, 383, 384, 385, 386, 387,
      388, 389, 393, 394, 395, 397, 403, 404, 405, 406, 420, 421, 422,
      423, 424, 425, 426, 427, 428, 429, 430, 431, 432, 433, 434, 435,
      439, 440, 443, 444, 447, 448, 450, 451, 452, 453, 454, 455, 456,
      462, 486, 488, 490, 491, 492, 494, 495, 496, 497, 499, 500, 502,
      503, 504, 505, 512, 513, 517
polyethylene terephthalate 11, 14, 15, 17, 18, 30, 31, 32, 33, 36, 37, 53,
      92, 119, 122, 138, 139, 140, 146, 150, 151, 152, 156, 157, 158, 160,
      183, 219, 221, 225, 227, 228, 234, 235, 236, 237, 239, 240, 241,
      242, 244, 245, 246, 247, 248, 250, 251, 253, 255, 260, 261, 262,
      263, 277, 278, 279, 282, 283, 284, 288, 289, 290, 299, 300, 301,
      302, 303, 304, 305, 306, 307, 308, 309, 310, 311, 312, 313, 314,
      315, 316, 321, 322, 323, 324, 325, 328, 334, 335, 336, 337, 338,
      344, 355, 356, 357, 358, 377, 380, 381, 382, 383, 384, 385, 386,
      387, 388, 389, 393, 394, 395, 397, 403, 404, 405, 406, 420, 421,
      422, 423, 424, 425, 426, 427, 428, 429, 430, 431, 432, 433, 434,
      435, 439, 440, 443, 444, 447, 448, 450, 451, 452, 453, 454, 455,
      456, 462, 486, 488, 490, 491, 492, 494, 495, 496, 497, 499, 500,
      502, 503, 504, 505, 512, 513, 517
polyimide 92, 116, 122
polymers
  crystallising polymer 267, 300
polystyrene 44, 48, 81, 82, 83, 84, 85, 86, 87, 88, 89, 90, 91, 92, 93, 94,
      95, 96, 98, 99, 100, 101, 102, 103, 115, 123, 124, 125, 126, 135,
      136, 141, 148, 149, 159, 160, 178, 181, 183, 184, 185, 186, 189,

211, 226, 277, 340, 344, 350, 365, 366, 370, 371, 373, 374, 376, 378, 379, 407, 441, 443
polytetrafluoroethylene 92, 221
polyvinyl acetate 141
polyvinyl chloride 17, 20, 21, 22, 23, 24, 25, 26, 27, 28, 29, 30, 31, 32, 33, 220, 260, 282, 299, 393, 397, 487, 513

R

rigid coating on a soft substratum 6, 393, 409, 410, 411, 416, 429, 448, 460, 461, 462, 463, 464, 466
rubber
  cis-polyisoprene rubber 28
  natural rubber 17, 22, 26, 27, 28, 129, 177, 465, 466
  synthetic isoprene rubber 20, 26
rubbery state 21, 30, 45, 91, 92, 108, 137, 157, 164, 168, 178, 180, 200, 208, 210, 212, 232, 277, 386, 388, 434, 455
rule
  Duclaux–Traube rule 217, 236, 253, 256

S

self-organisation 5, 6, 7, 333, 364, 368, 370, 374, 389
strain softening 3, 176, 177, 179, 180, 185, 186, 187, 190, 192, 193, 194, 197, 198, 199, 200, 210, 211
stress–strain curve 3, 109, 124, 125, 177, 435
structure
  fibrillar–porous structure 4, 206, 223, 228, 229, 262, 302, 326, 342, 359, 482
synaeresis 286

T

tacticity 87, 113
Taylor instability 231
Taylor meniscus instability 4
temperature
  glass transition temperature 2, 3, 10, 12, 13, 14, 15, 29, 34, 35, 41, 52, 54, 57, 63, 65, 68, 71, 72, 74, 81, 83, 84, 85, 87, 88, 89, 90, 91, 99, 100, 101, 102, 104, 108, 110, 111, 112, 114, 117, 120, 121, 124, 125, 128, 129, 130, 131, 136, 137, 141, 144, 146, 153, 154, 156, 157, 158, 160, 163, 164, 169, 170, 171, 177, 181, 183, 185, 186, 187, 188, 189, 190, 194, 195, 196, 202, 211, 212, 222, 223, 269, 277, 278, 290, 291, 292, 293, 294, 296, 297, 298, 299, 300, 301, 302, 303, 304, 306, 316, 317, 321, 333, 350, 351, 371, 423, 372, 388, 408, 421, 424, 427, 429, 430, 432, 453
  upper critical solution temperature 66

theory
  Fox–Flory theory 102

V

Vlodavets–Rehbinder cryptohererogeneous systems 310

X

x-ray photoelectronic spectroscopy 69, 84, 94, 96, 100

9 780367 575045